Iterative Receiver Design

Iterative processing is an important technique with applications in many fields, including digital communications, image processing, and machine learning. By exploiting the power of factor graphs, this detailed survey provides a general framework for systematically developing iterative algorithms for digital receivers, and highlights connections between important algorithms. Starting with basic concepts in digital communications, progressively more complex ideas are presented and integrated, resulting in the development of cutting-edge algorithms for iterative receivers. Real-world applications are covered in detail, including decoding for turbo and LDPC codes, and detection for multi-antenna and multi-user systems. This accessible framework will allow the reader to apply factor graphs to practical problems, leading to the design of new algorithms in applications beyond digital receivers.

With many examples and algorithms in pseudo-code, this book is an invaluable resource for graduate students and researchers in electrical engineering and computer science, and for practitioners in the communications industry.

Additional resources for this title are available online at www.cambridge.org/ 9780521873154.

HENK WYMEERSCH is a postdoctoral associate in the Laboratory for Information and Decision Systems at the Massachusetts Institute of Technology. He obtained his PhD in Electrical Engineering from Ghent University, Belgium, in 2005. In 2006 he received the Alcatel Bell Scientific Award for "an original study of Information & Communication Technology, Concepts and Multimedia applications."

Iterative Receiver Design

Henk Wymeersch
Massachusetts Institute of Technology

CAMBRIDGE
UNIVERSITY PRESS

University Printing House, Cambridge CB2 8BS, United Kingdom

Cambridge University Press is part of the University of Cambridge.

It furthers the University's mission by disseminating knowledge in the pursuit of education, learning and research at the highest international levels of excellence.

www.cambridge.org
Information on this title: www.cambridge.org/9780521873154

© Cambridge University Press 2007

First published 2007

A catalogue record for this publication is available from the British Library

ISBN 978-0-521-87315-4 Hardback

Contents

Preface   *page* ix
Abbreviations   xiii
Notations   xv
List of algorithms   xvii

1   Introduction   1
   1.1   Motivation   1
   1.2   The structure of this book   2

2   Digital communication   5
   2.1   Introduction   5
   2.2   Digital communication   6
   2.3   Single-user, single-antenna communication   9
   2.4   Multi-antenna communication   12
   2.5   Multi-user communication   15
   2.6   Goals and working assumptions   17
   2.7   Main points   17

3   Estimation theory and Monte Carlo techniques   19
   3.1   Introduction   19
   3.2   Bayesian estimation   19
   3.3   Monte Carlo techniques   25
   3.4   Main points   33

4   Factor graphs and the sum–product algorithm   35
   4.1   A brief history of factor graphs   35
   4.2   A ten-minute tour of factor graphs   37
   4.3   Graphs, factors, and factor graphs   42
   4.4   Marginals and the sum–product algorithm   50
   4.5   Normal factor graphs   62
   4.6   Remarks on factor graphs   66
   4.7   The *sum* and *product* operators   71
   4.8   Main points   74

5   **Statistical inference using factor graphs**   **77**
           5.1   Introduction   77
           5.2   General formulation   77
           5.3   Messages and their representations   86
           5.4   Loopy inference   100
           5.5   Main points   104

6   **State-space models**   **105**
           6.1   Introduction   105
           6.2   State-space models   106
           6.3   Hidden Markov models   113
           6.4   Linear Gaussian models   117
           6.5   Approximate inference for state-space models   127
           6.6   Main points   133

7   **Factor graphs in digital communication**   **135**
           7.1   Introduction   135
           7.2   The general principle   135
           7.3   Opening nodes   137
           7.4   Main points   141

8   **Decoding**   **143**
           8.1   Introduction   143
           8.2   Goals   144
           8.3   Block codes   145
           8.4   Repeat–accumulate codes   149
           8.5   Low-density parity-check codes   154
           8.6   Convolutional codes   163
           8.7   Turbo codes   170
           8.8   Performance illustration   174
           8.9   Main points   175

9   **Demapping**   **177**
           9.1   Introduction   177
           9.2   Goals   178
           9.3   Bit-interleaved coded modulation   178
           9.4   Trellis-coded modulation   182
           9.5   Performance illustration   184
           9.6   Main points   184

10   **Equalization–general formulation**   **187**
           10.1   Introduction   187
           10.2   Problem description   188
           10.3   Equalization methods   189
           10.4   Interaction with the demapping and the decoding node   203
           10.5   Performance illustration   203
           10.6   Main points   205

11   Equalization: single-user, single-antenna communication   207
11.1   Introduction   207
11.2   Single-carrier modulation   208
11.3   Multi-carrier modulation   213
11.4   Main points   216

12   Equalization: multi-antenna communication   217
12.1   Introduction   217
12.2   Single-carrier modulation   218
12.3   Multi-carrier modulation   223
12.4   Main points   225

13   Equalization: multi-user communication   227
13.1   Introduction   227
13.2   Direct-sequence code-division multiple access   228
13.3   Orthogonal frequency-division multiple access   234
13.4   Main points   236

14   Synchronization and channel estimation   237
14.1   Introduction   237
14.2   Channel estimation, synchronization, and factor graphs   237
14.3   An example   239
14.4   Main points   242

15   Appendices   243
15.1   Useful matrix types   243
15.2   Random variables and distributions   243
15.3   Signal representations   246

References   247
Index   253

Preface

In early 2002, I was absent-mindedly surfing the Internet, vaguely looking for a tutorial on turbo codes. My PhD advisor at Ghent University, Marc Moeneclaey, thought it wise for his new students to become familiar with these powerful error-correcting codes. Although I finally settled on W. E. Ryan's "A turbo code tutorial," my search led me (serendipitously?) to the PhD thesis of Niclas Wiberg. This thesis shows how to describe codes by means of a (factor) graph, and how to decode them by passing messages on this graph. Although interesting, the idea seemed a bit far-fetched, and I didn't fully appreciate or understand its significance. Nevertheless, Wiberg's thesis stayed in the back of my mind (or at least, I'd like to think so now).

During 2002 and 2003, I worked mainly on synchronization and estimation algorithms for turbo and LDPC codes. A colleague of mine, Justin Dauwels, who was at that time a PhD student of Andy Loeliger at the ETH in Zürich, had developed a remarkable synchronization algorithm for LDPC codes, based on Wiberg's factor graphs. Justin was interested in comparing his synchronization algorithm with ours, and became a visiting researcher in our lab for the first two months of 2004. I fondly remember many hours spent in the department lunchroom, with Justin (painstakingly) explaining the intricacies of factor graphs to me. His discussions motivated me to re-write my source code for decoding turbo and LDPC codes using the factor-graph framework. To my surprise, writing the code was an almost trivial process! Understanding factor graphs made understanding and programming algorithms so much easier. Why didn't more people use these graphs? As I became more engrossed in the topic, I realized that, despite the presence of some excellent tutorials on factor graphs, many researchers still felt intimidated by them. It would be very useful for someone to write an accessible tutorial on factor graphs, a "Factor graphs for dummies," as it were. By the time I felt I had a reasonable amount of experience on factor graphs, I had to write my PhD thesis and could not pursue this idea any further.

Late 2005, I was a postdoctoral fellow at MIT trying to survive my first winter in Boston. For some reason (perhaps tiredness, perhaps cerebral frostbite), I finally bit the bullet and decided to write a book on factor graphs in the context of digital communications. I was lucky enough to have a postdoc advisor, Moe Win, who graciously allowed me to work full-time on this project. The book you are holding now is pretty much the book I set out to write. Despite being somewhat longer than I originally intended, this book is limited in many ways, primarily due to time constraints and my limited understanding of the various topics. Many parts of this book reflect my own personal interests, as well as my inclination for completeness rather than succinctness.

There are many, many people I wish to thank, most importantly my PhD advisor Marc Moeneclaey at Ghent University and my postdoc advisor Moe Win at MIT. They have very different personalities, but both are truly excellent in their own fields, both are genuinely nice people, and I feel privileged to have had the opportunity to work with them. I am also greatly indebted to the Belgian American Educational Foundation, the organization that sponsored my US fellowship, and gave me the financial freedom to work on my own project at MIT. A great deal of my knowledge on factor graphs is a direct consequence of the technical discussions I have had with some very smart people. Obviously, Justin Dauwels springs to mind, my own *encyclopaedia factorgraphica*. Over the years he has always been more than willing to answer my unending list of questions. Also critical in shaping the contents of this book was Frederik Simoens, one of Marc's PhD students. My interactions with Frederik the past four years, from discussions on Kalman filters in the middle of the night in Stockholm, to getting wasted in dubious Boston nightclubs, have been nothing short of amazing. Another essential person in this list is Cédric Herzet, currently at UC Berkeley, a guy with an uncanny ability for lateral thinking. Some of the sweeter insights in this book are due to him. And then there is the long list of people who were at some point involved in my life this last crazy year, as a proof-reader, as a recipient of my many questions, or as a means to get away from it all. At MIT and surrounding areas, I thank Pedro Pinto, Faisal Kashif, Jaime Lien, Watcharapan Suwansantisuk, Damien Jourdan, Erik Sudderth, Wesley Gifford, Marco Chiani, Andrea Giorgetti, Gil Zussman, Alex Ihler, Hyundong Shin, Sejoon Lim, Sid Jaggi, Yuan Shen, Atilla Eryilmaz, Andrew Fletcher, Wee Peng Tay, Tony Quek, Eugene Baik, and the MIT Euroclub (in particular Bjorn-Mr. salsa-Maes). Back in Belgium, I would like to thank Xavier Jaspar, Valéry Ramon, Frederik Vanhaverbeke, Nele Noels, and Heidi Steendam. I am also grateful to have had the pleasure to work with Anna Littlewood and Phil Meyler at Cambridge University Press.

Finally, I wish to thank my family for their understanding, love, and patience, for helping me when times got tough (and boy, did they ever!), and for letting me pursue my dreams far from home. I miss you more than you know.

Henk Wymeersch

Cambridge, Massachusetts, October 2006

A Word on Notation

Vectors are written in bold, matrices in capital bold. The ith element in vector $\mathbf{x}$ is $[\mathbf{x}]_i$ or x_i. The element on the ith row, jth column of the matrix $\mathbf{A}$ is $[\mathbf{A}]_{ij}$ or $A_{i,j}$. To access elements l_1 through l_2 ($l_1 \leq l_2$) in $\mathbf{x}$, we write $[\mathbf{x}]_{l_1:l_2}$ or $\mathbf{x}_{l_1:l_2}$. Logarithms are to the base e unless specified otherwise.

Abbreviations

APD	a-posteriori distribution
BICM	bit-interleaved coded modulation
CDMA	code-division multiple access
DS	direct sequence
FDMA	frequency-division multiple access
GMD	Gaussian mixture density
HMM	hidden Markov model
iid	independent and identically distributed
LDPC	low-density parity check
MAP	maximum a-posteriori
MC	Monte Carlo
MCMC	Markov-chain Monte Carlo
MIMO	multi-input, multi-output (multi-antenna system)
MMSE	minimum mean-squared error
OFDM	orthogonal frequency-division multiplexing
OFDMA	orthogonal frequency-division multiple access
pdf	probability density function
pmf	probability mass function
P/S	parallel-to-serial conversion
S/P	serial-to-parallel conversion
SPA	sum–product algorithm
SSM	state-space model
TCM	trellis-coded modulation
TDMA	time-division multiple access

Notations

Common notations A

$\mathcal{X}$	domain of variable X, x
$\mathbf{x}$	a vector
$\|\mathbf{x}\|$	Frobenius norm of $\mathbf{x}$
$\mathbf{X}$	a vector random variable or a matrix
$\mathbf{X}^{\mathrm{H}}$	conjugate transpose of $\mathbf{X}$
$\mathbf{X}^{\mathrm{T}}$	transpose of $\mathbf{X}$
$\hat{\mathbf{x}}$	estimate of $\mathbf{x}$
$\mathbb{E}\{\cdot\}$	expectation operator
$|\mathcal{S}|$	cardinality of the set $\mathcal{S}$
$\mathbb{B}$	the set of binary numbers $\{0, 1\}$
$\mathbb{I}\{\cdot\}$	indicator function
δ_k	discrete Dirac distribution
$\delta(t)$	continuous Dirac distribution
$\boxed{=}(\cdot)$	equality function
$\mathcal{R}_L(p_X(\cdot))$	particle representation of distribution $p_X(\cdot)$ with L samples
$\mathrm{diag}\{\mathbf{x}\}$	a diagonal matrix with the elements of $\mathbf{x}$ on the diagonal
$f(x) \propto g(x)$	$f(x) = \gamma g(x), \forall x$, for some scalar γ
$\mathbf{I}_N$	an $N \times N$ identity matrix
$\mu_{X \to f}(x)$	message from edge X to node f, evaluated in $x \in \mathcal{X}$
$\mu_{f \to X}(x)$	message from node f to edge X, evaluated in $x \in \mathcal{X}$
$a \ll b$	a is much smaller than b
$\arg\max_x f(x)$	the value of x that achieves the maximum of $f(\cdot)$
$[a, b]$	the closed interval $\{x \in \mathbb{R} : a \le x \le b\}$
$[a, b)$	the half-open interval $\{x \in \mathbb{R} : a \le x < b\}$

Common notations B

$\mathbf{B}, \mathbf{b}$	information word
$\mathbf{C}, \mathbf{c}$	codeword
$\mathbf{A}, \mathbf{a}$	coded symbols
$\mathbf{Y}, \mathbf{y}$	observation
$\mathbf{n}$	noise vector
$\mathbf{h}(\cdot)$	channel function
$\mathbf{H}$	channel matrix
N_{b}	number of information bits per codeword
N_{c}	number of coded bits per codeword
N_{s}	number of coded symbols per codeword
N_{T}	number of transmit antennas
N_{R}	number of receive antennas
N_{u}	number of users
N_{FFT}	number of subcarriers per OFDM symbol
$\mathcal{M}$	the model
Ω	signaling constellation

List of algorithms

3.1	Importance sampling	30
3.2	Gibbs sampler	33
4.1	The sum–product algorithm	39
4.2	Variable partitioning: determine $S_{f_k}^{(X_n)}$	51
5.1	The sum–product rule for continuous variables – importance sampling	98
5.2	Sum–product rule for continuous variables – mixture sampling	99
6.1	Hidden Markov models: sum–product algorithm with message normalization	114
6.2	Hidden Markov models: sum–product algorithm with message normalization using vector and matrix representation	115
6.3	Hidden Markov models: max–sum algorithm	116
6.4	The forward phase of sum–product algorithm on a state-space model using particle representations	129
8.1	Decoding RA codes: forward–backward phase	157
8.2	Decoding RA codes: upward messages	157
8.3	Decoding RA codes: downward messages	157
8.4	Decoding RA codes: complete decoding algorithm	158
8.5	LDPC code: SPA in check nodes	161
8.6	Decoding LDPC codes	164
8.7	Convolutional code: building block	168
8.8	Convolutional code: building block in the log domain	168
8.9	Decoding of a convolutional code	169
8.10	Decoding of a PCCC/SCCC turbo code	174
9.1	Demapping for BICM	181
10.1	Gibbs sampler for $p(\mathbf{A}|\mathbf{Y} = \mathbf{y}, \mathcal{M})$	199

1 Introduction

1.1 Motivation

Claude E. Shannon was one of the great minds of the twentieth century. In the 1940s, he almost single-handedly created the field of information theory and gave the world a new way to look at information and communication. The channel-coding theorem, where he proved the *existence* of good error-correcting codes to transmit information at any rate below capacity with an arbitrarily small probability of error, was one of his fundamental contributions. Unfortunately, Shannon never described how to *construct* these codes. Ever since his 1948 landmark paper "A mathematical theory of communication" [1], the channel-coding theorem has tantalized researchers worldwide in their quest for the ultimate error-correcting code. After more than forty years, state-of-the-art error-correcting codes were still disappointingly far away from Shannon's theoretical capacity bound. No drastic improvement seemed to be forthcoming, and researchers were considering a more practical capacity benchmark, the cut-off rate [2], which could be achieved by practical codes.

In 1993, two until then little-known French researchers from the ENST in Bretagne, Claude Berrou and Alain Glavieux, claimed to have discovered a new type of code, which operated very close to Shannon capacity with reasonable decoding complexity. The decoding process consisted of two decoders passing information back and forth, giving rise to the name "turbo code." They first presented their results at the IEEE International Conference on Communications in Geneva, Switzerland [3]. Quite understandably, they were met with a certain amount of skepticism by the traditional coding community. Only when their findings were reproduced by other labs did the turbo idea really take off. It is fair to say that turbo codes have caused a paradigm shift in communications theory. The idea of passing information back and forth between different components in a receiver (so-called iterative processing or turbo processing) has become prevalent in state-of-the-art receiver design. Many books and international scientific conferences are currently devoted to turbo processing.

The original turbo decoder was developed in a somewhat ad-hoc way. An elegant mathematical framework was provided with the introduction of particular graphical models for statistical inference [4,5] at the end of the twentieth century, describing how iterative receivers can be designed in an almost automatic fashion. The goal of this book is to show how various tasks in the receiver can be cast in this graphical framework. Important practical problems such as decoding, equalization, and multi-user detection will be treated. Wherever possible, algorithms in pseudo-code will be provided to allow the reader to transfer knowledge from this book directly into a practical implementation. More importantly, it is my hope that the reader will discover the inherent beauty of these graphical models, and realize that they can be applied to a wide variety of problems far beyond Shannon's original coding problem.

1.2   The structure of this book

This book is organized as follows.

- In **Chapter 2**, we will give a brief overview of several important digital transmission schemes. We will cover single- and multi-carrier transmission, single- and multi-antenna transmission, and single- and multi-user transmission. For every one of these transmission schemes, a suitable receiver needs to be developed on the basis of some optimality criterion.
- **Chapter 3** deals with this optimality criterion. We will describe basic concepts from Bayesian estimation theory and Monte Carlo methods.
- In **Chapter 4** we will introduce the concept of factor graphs. Factor graphs are a way to represent graphically the factorization of a function. We will discuss factor graphs in detail in an abstract setting and show how marginals of functions can be computed by message-passing on the corresponding factor graph.
- **Chapter 5** ties together the knowledge from Chapter 3 and Chapter 4. We will show how to solve inference and estimation problems using factor graphs.
- Inference on state-space models appears in many engineering problems. **Chapter 6** is devoted to the application of factor graphs to such models. We will treat hidden Markov models, Kalman filters, and particle filters.
- After a detour into estimation theory and factor graphs, we return to receiver design in **Chapter 7.** We will show how, at least in principle, a digital receiver based on factor graphs can be built. Four critical functions will be revealed: decoding, demapping, equalization, and the conversion of the received waveform into a suitable observation.
- Decoding will be the topic of **Chapter 8**, where we will discuss four important types of error-correcting codes: repeat–accumulate codes, low-density parity-check codes, convolutional codes, and turbo codes.
- In **Chapter 9** we will cover demapping for bit-interleaved coded modulation and trellis-coded modulation in the factor-graph framework.
- Equalization techniques will be treated in **Chapter 10** in a general setting. A variety of general-purpose equalizers will be derived.
- **Chapter 11** deals with equalization for single-user, single-antenna transmission. For every transmission scheme, we show how to convert the received waveform into a suitable observation, and point out which equalizers from Chapter 10 can be applied.
- Equalization for multi-antenna transmission will be discussed in **Chapter 12**. Conversion to suitable observations and factor-graph equalization strategies will be covered.

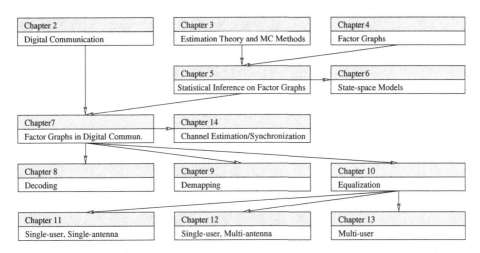

Figure 1.1. A graph representing the interdependencies between chapters.

- Multi-user transmission is the topic of **Chapter 13**. As in the previous two chapters, suitable observations will be determined and equalizers from Chapter 10 will be selected.
- **Chapter 14** will deal with channel estimation and synchronization. We will show how the unknown channel parameters can be incorporated into the factor-graph framework.
- The appendices make up **Chapter 15**. The reader is encouraged to browse through them at this point.

The logical relations between chapters are depicted in Fig. 1.1.

2 Digital communication

2.1 Introduction

As with any good story, it is best to start at the very beginning. Digital communication deals with the transmission of binary information (obtained as the output of a source encoder) from a *transmitter* to a *receiver*. The transmitter converts the binary information to an analog waveform and sends this waveform over a physical medium, such as a wire or open space, which we will call the *channel*. As we shall see, the channel modifies the waveform in several ways. At the receiver, this modified waveform is further corrupted due to thermal noise. Not only does the receiver have to recover the original binary information, but also it must deal with channel effects, thermal noise, and synchronization issues. All in all, the receiver has the bad end of the deal in digital communications. For this reason, this book deals mainly with receiver design, and only to a very small extent with the transmitter.

In this chapter we will describe several digital transmission schemes, detailing how binary information is converted into a waveform at the transmitter side, and how a corrupted version of this waveform arrives at the receiver side. Right now, our focus is not on how the corresponding receivers should be designed. Since there is a myriad of digital transmission schemes, we are obliged to limit ourselves to some of the most important ones. It is my hope that the receivers we will design for this limited set of transmission schemes will give the reader inspiration in developing novel receivers and understand existing receivers. We will assume that the reader has at least a passing familiarity with these transmission schemes. Textbooks and reference books on digital communications include [6–15].

This chapter is organized as follows.

- We will start with the basic principles of converting a sequence of bits into a baseband complex waveform in **Section 2.2**. This process consists of encoding the information, followed by mapping onto a signaling constellation and pulse-shaping.
- In **Section 2.3**, we then move on to the classical transmission scheme with a single transmitter using a single antenna. Both single-carrier and multi-carrier transmission will be described.
- These transmission schemes are then generalized to multi-antenna (**Section 2.4**) and multi-user (**Section 2.5**) scenarios.
- After this brief overview of transmission schemes, we will outline the main goals and working assumptions of this book in **Section 2.6**.

2.2   Digital communication

2.2.1   From bits to waveform

In digital communication, the goal is to convey a sequence of binary information digits (bits, belonging to the set $\mathbb{B} = \{0,1\}$) from the transmitter to the receiver. The binary information sequence (which may be infinitely long) is segmented into blocks of length N_b. A single block, say $\mathbf{b} \in \mathbb{B}^{N_b}$, is referred to as an information word. Every information word is protected against channel effects by encoding it with a channel encoder, which converts the N_b information bits into N_c coded bits, with $N_c > N_b$, using an encoding function $f_c: \mathbb{B}^{N_b} \to \mathbb{B}^{N_c}$. For instance, we can apply a convolutional code, a turbo code, or a low-density parity-check code. This results in a (longer) binary sequence $\mathbf{c} = f_c(\mathbf{b})$.

Example 2.1 (Repetition code). *Probably the easiest way of encoding an information stream is by using a repetition code, whereby we simply repeat every bit K times. For instance, when $N_b = 3$, $K = 2$, and $\mathbf{b} = [011]$, this yields*

$$\mathbf{c} = f_c([011])$$

$$= [001111]$$

so that $N_c = N_b \times K = 6$. In practice, N_b can be very large.

After encoding, the coded bits are converted to a sequence $\mathbf{a} = f_a(\mathbf{c})$ of N_s complex coded symbols using a mapping function f_a, where the kth symbol, a_k, belongs to a signaling constellation Ω_k. For instance, we can use bit-interleaved coded modulation (BICM) or trellis-coded modulation (TCM).

Example 2.2 (Mapping). *The most simple (and most common) way of mapping $\mathbf{c}$ to $\mathbf{a}$ is as follows. Suppose that we have a signaling constellation $\Omega = \{-1, +1, -j, +j\}$, where j is the imaginary unit (j $= \sqrt{-1}$). We select the same constellation for every symbol a_k. With every element in Ω, we associate a unique bit-string. Since there are $|\Omega| = 4$ elements in Ω, we can associate $\log_2 |\Omega| = 2$ bits with every element in Ω. This is known as quadrature phase-shift keying (QPSK). We can now define the following mapping $\phi: \mathbb{B} \times \mathbb{B} \to \Omega$*

$$\phi(0,0) = +1,$$

$$\phi(0,1) = -1,$$

$$\phi(1,0) = -j,$$

$$\phi(1,1) = +j.$$

We break up $\mathbf{c}$ into blocks of length $\log_2 |\Omega| = 2$ and map every block to a constellation point using the function $\phi(\cdot)$. In our case, using the sequence from the previous example

$\mathbf{c} = [001111]$, *we obtain*

$$\mathbf{a} = [\phi(0,0)\phi(1,1)\phi(1,1)]$$
$$= [+1 \ +j \ +j].$$

Observe that $N_s = N_c / \log_2 |\Omega|$.

Finally, once encoding and mapping are completed, the symbols in $\mathbf{a}$ are embedded in a (possibly infinitely long) data stream and pulse-shaped, giving rise to a complex baseband signal

$$s(t) = \sqrt{E_s} \sum_{k=-\infty}^{+\infty} a_k p_k(t), \qquad (2.1)$$

where E_s is the energy per transmitted coded symbol and $p_k(t)$ is a unit-energy transmit pulse corresponding to the kth symbol. This implies that

$$\int_{-\infty}^{+\infty} |p_k(t)|^2 dt = 1. \qquad (2.2)$$

The signal $s(t)$ is modulated onto a carrier waveform with carrier frequency f_C, yielding a real (as opposed to complex) radio-frequency (RF) signal

$$s_{RF}(t) = \Re\left\{\sqrt{2}s(t)e^{j2\pi f_C t}\right\}. \qquad (2.3)$$

The RF signal propagates through a physical medium (e.g., a wireless channel, an optical fiber), and is corrupted by thermal noise in the receiver, resulting in a received signal $r_{RF}(t)$. The received RF signal is down-converted to complex baseband, giving rise to an equivalent complex baseband received signal $r(t)$. We will follow the common practice of working only with equivalent baseband signals. This leads to the following observed signal at the receiver:

$$r(t) = \sqrt{E_s} \sum_{k=-\infty}^{+\infty} a_k h_k(t) + n(t), \qquad (2.4)$$

where $h_k(t)$ is the equivalent channel for the kth symbol, encompassing the transmit pulse and the equivalent baseband physical channel $h_{ch}(t)$, and $n(t)$ is a zero-mean complex white Gaussian process with power-spectral density $N_0/2$ both for the real and for the imaginary component:

$$\mathbb{E}\left\{N(t)N^*(u)\right\} = N_0\delta(t-u). \qquad (2.5)$$

The equivalent channel $h_k(t)$ can be written as the convolution of the transmit pulse and the physical channel:

$$h_k(t) = \int_{-\infty}^{+\infty} p_k(u) h_{ch}(t-u) du. \tag{2.6}$$

Example 2.3 (Pulse-shaping). *A simple type of complex baseband signal is formed by setting $p_k(t) = p(t-kT)$, where $p(t)$ is a unit-energy square pulse, defined as*

$$p(t) = \begin{cases} 1/\sqrt{T} & 0 \le t < T \\ 0 & else \end{cases}$$

so that we transmit one data symbol every T seconds. Going back to our previous example, where $N_s = 3$ and $\mathbf{a} = [+1 + j + j]$, this results in a signal shown in Fig. 2.1 (left-hand side), where we depict the real and imaginary parts of the signal $s(t)$. Let us assume that the equivalent baseband channel is given by $h_{ch}(t) = \exp(j\pi/4)\delta(t-\tau)$, for some propagation delay $\tau \in \mathbb{R}$, then

$$r(t) = \sqrt{E_s} \sum_{k=0}^{2} a_k e^{j\pi/4} p(t - kT - \tau) + n(t).$$

A noiseless version of $r(t)$ is also shown in Fig. 2.1 (right-hand side).

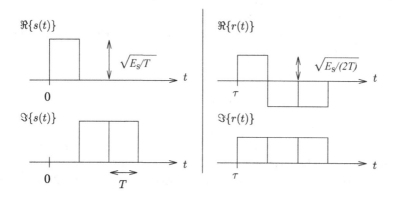

Figure 2.1. Pulse-shaping using square pulses. On the left is the transmitted signal with transmitted sequence $[+1 + j + j]$. On the right is the (noiseless) received signal, after a phase rotation of $\pi/4$ and a delay of τ.

2.2.2 Channel model

A common channel model for wireless communication is the multi-path model, whereby the channel impulse response consists of a number of distinct paths [16]:

$$h_{\mathrm{ch}}(t) = \sum_{l=0}^{L-1} \alpha_l \delta(t - \tau_l), \qquad (2.7)$$

where α_l and τ_l are the complex gain and the propagation delay of the lth path. We usually consider these paths to be resolvable (meaning that $\tau_{l+1} - \tau_l \gg 1/B$, where B represents the bandwidth of $p(t)$). This is without loss of generality, since unresolvable paths can be combined into a single path, and the complex gains added.

A channel is said to be *frequency-selective* when it has at least two resolvable paths. Otherwise the channel is frequency-non-selective (also known as a *frequency-flat channel*). The delay of the first path, τ_0, corresponds to the *propagation delay* of the signal through the channel.

2.2.3 Communication schemes

We can distinguish between single-antenna and multi-antenna transmission, between single-carrier and multi-carrier transmission, and between single-user and multi-user transmission. In the next few sections, we will describe these schemes in more detail. Our aim is not completeness, but rather to touch on a few selected schemes for which we will later design appropriate receivers.

2.3 Single-user, single-antenna communication

We first consider a system where there is only a single user transmitting. Both the transmitter and the receiver are equipped with a single antenna. The transmitter wishes to transmit a long data stream (consisting of many codewords).

2.3.1 Single-carrier modulation

In single-carrier modulation we transmit a symbol every T seconds over a single carrier. This gives rise to the following transmitted signal:

$$s(t) = \sqrt{E_{\mathrm{s}}} \sum_{k=-\infty}^{+\infty} a_k p(t - kT). \qquad (2.8)$$

The transmit pulse $p(t)$ is usually selected according to specific criteria. For instance, we like our pulses to have finite bandwidth and to be unit-energy square-root Nyquist pulses for a rate $1/T$. We remind the reader that a unit-energy square-root Nyquist pulse

$p(t)$ for a rate $1/T$ satisfies, for $k \in \mathbb{Z}$,

$$g(kT) = \begin{cases} 1 & k = 0, \\ 0 & \text{else,} \end{cases} \tag{2.9}$$

where

$$g(t) = \int_{-\infty}^{+\infty} p(u)p^*(t+u)\mathrm{d}u. \tag{2.10}$$

The pulse $g(t)$ is then a Nyquist pulse for a rate $1/T$. The received signal can be expressed as

$$r(t) = \sqrt{E_s} \sum_{k=-\infty}^{+\infty} a_k h(t-kT) + n(t), \tag{2.11}$$

where $h(t)$ is the equivalent channel (given by the convolution of the transmit pulse and the physical channel $h_{\mathrm{ch}}(t)$, see (2.6)) and $n(t)$ is a complex white Gaussian noise process.

2.3.2   Multi-carrier modulation – OFDM

Orthogonal frequency-division multiplexing (OFDM) is a popular multi-carrier transmission technique that avoids inter-symbol interference over frequency-selective channels [17,18]. Intuitively, the idea of OFDM is to break up the transmission bandwidth into narrow subbands (subcarriers) such that the channel is frequency-flat on every subcarrier. Inter-symbol interference is avoided by pre-appending a cyclic prefix to the transmitted symbols.

More formally, we break up the data stream into segments of length N_{FFT}, where N_{FFT} is usually a power of 2 (for instance, 256 or 1024). Let us call one such segment $\mathbf{a}$, an $N_{\mathrm{FFT}} \times 1$ vector. We multiply $\mathbf{a}$ by an N_{FFT}-point inverse discrete Fourier transform (IDFT) matrix $\mathbf{F}$, yielding a vector

$$\check{\mathbf{a}} = \mathbf{F}\mathbf{a}, \tag{2.12}$$

where

$$F_{m,n} = \frac{1}{\sqrt{N_{\mathrm{FFT}}}} \exp\left(\mathrm{j}2\pi \frac{m \times n}{N_{\mathrm{FFT}}}\right), \tag{2.13}$$

for $m, n \in \{0, \ldots, N_{\mathrm{FFT}} - 1\}$. Usually the operation (2.12) is performed by means of a computationally efficient inverse fast Fourier transform (FFT). We then pre-append the last N_{CP} of $\check{\mathbf{a}}$ symbols to $\check{\mathbf{a}}$, and obtain a vector of length $N_{\mathrm{FFT}} + N_{\mathrm{CP}}$, known as an *OFDM*

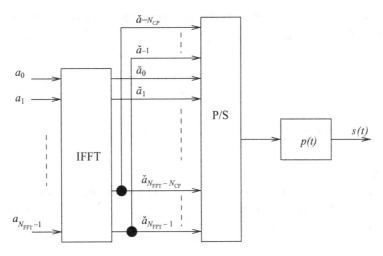

Figure 2.2. An OFDM transmitter. The data symbols are passed through an IFFT, a cyclic prefix is added and, after parallel-to-serial conversion, pulse-shaping is employed.

symbol:

$$
\left[\underbrace{\check{a}_{-N_{\text{CP}}} \ldots \check{a}_{-2}\,\check{a}_{-1}}_{\text{cyclic prefix}}\ \underbrace{\check{a}_0\,\check{a}_1\,\check{a}_2 \ldots \check{a}_{N_{\text{FFT}}-1}}_{\check{\mathbf{a}}^{\text{T}}} \right]^{\text{T}} \tag{2.14}
$$

where $\check{a}_{-l} = \check{a}_{N_{\text{FFT}}-l}$, for $l = 1, \ldots, N_{\text{CP}}$. The sequence $\left[\check{a}_{-N_{\text{CP}}} \ldots \check{a}_{-2}\check{a}_{-1}\right]^{\text{T}}$ of length N_{CP} is known as the *cyclic prefix*. Finally, we transmit one OFDM symbol using the following complex baseband signal (see Fig. 2.2):

$$
s(t) = \sqrt{\frac{E_{\text{s}}N_{\text{FFT}}}{N_{\text{FFT}} + N_{\text{CP}}}} \sum_{l=-N_{\text{CP}}}^{N_{\text{FFT}}-1} \check{a}_l p(t - lT), \tag{2.15}
$$

using a unit-energy transmit pulse $p(t)$. We can also transmit a sequence of OFDM symbols consecutively as

$$
s(t) = \sqrt{\frac{E_{\text{s}}N_{\text{FFT}}}{N_{\text{FFT}} + N_{\text{CP}}}} \sum_{k=-\infty}^{+\infty} \sum_{l=-N_{\text{CP}}}^{N_{\text{FFT}}-1} \check{a}_{l,k} p(t - lT - kT_{\text{OFDM}}), \tag{2.16}
$$

where $\check{a}_{l,k}$ is the lth component of the kth OFDM symbol and $T_{\text{OFDM}} = T(N_{\text{FFT}} + N_{\text{CP}})$, the symbol duration corresponding to a single OFDM symbol. The received signal is often written as

$$
r(t) = \sum_{k=-\infty}^{+\infty} \sum_{l=-N_{\text{CP}}}^{N_{\text{FFT}}-1} \check{a}_l h(t - lT - kT_{\text{OFDM}}) + n(t), \tag{2.17}
$$

where $h(t)$ again is the equivalent channel, including the transmit pulse, the physical channel, and the factor

$$\sqrt{\frac{E_s N_{FFT}}{N_{FFT} + N_{CP}}},$$

and $n(t)$ is a complex white Gaussian noise process.

Terminology

The quantities $\breve{a}_l$ are known as being in the time domain, whereas the symbols a_q are said to be in the frequency domain. We say that there are N_{FFT} subcarriers. The symbol a_q modulates the qth subcarrier.

2.4 Multi-antenna communication

Multi-antenna systems (or MIMO systems, for multiple input, multiple output; Fig. 2.3) were originally introduced by Winters [19] as a way to increase system capacity by the use of spatial diversity. Fundamental results are provided in [20, 21]; for pioneering work in practical coding schemes see [22–24]. Accessible works on MIMO include [25, 26].

2.4.1 Single-carrier modulation

In MIMO systems, we transmit simultaneously on different antennas. We denote by $a_k^{(n)}$ the symbol transmitted on the nth transmit antenna at the kth symbol duration. Suppose that we use N_T transmit antennas, we can then write the transmitted signal at a particular

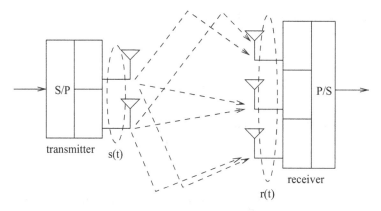

Figure 2.3. MIMO: a system with $N_T = 2$ transmit and $N_R = 3$ receive antennas.

time t as an $N_T \times 1$ vector $\mathbf{s}(t)$,

$$\mathbf{s}(t) = \sqrt{E_s} \sum_{k=-\infty}^{+\infty} \begin{bmatrix} a_k^{(1)} \\ \vdots \\ a_k^{(N_T)} \end{bmatrix} p(t - kT) \tag{2.18}$$

$$= \sqrt{E_s} \sum_{k=-\infty}^{+\infty} \mathbf{a}_k p(t - kT), \tag{2.19}$$

where $\mathbf{a}_k$ is the vector of transmitted symbols during the kth symbol period. At the receiver, equipped with N_R receive antennas, the received signal at the mth receive antenna is given by

$$r_m(t) = \sum_{k=-\infty}^{+\infty} \sum_{n=1}^{N_T} a_k^{(n)} h_m^{(n)}(t - kT) + n_m(t), \tag{2.20}$$

where $h_m^{(n)}(t)$ is the equivalent channel between transmit antenna n and receive antenna m. We can stack the received signals at time t in a vector of length N_R,

$$\mathbf{r}(t) = \sum_{k=-\infty}^{+\infty} \sum_{n=1}^{N_T} a_k^{(n)} \mathbf{h}^{(n)}(t - kT) + \mathbf{n}(t), \tag{2.21}$$

where $\mathbf{h}^{(n)}(t)$ represents the equivalent channel between the nth transmit antenna and the various receive antennas.

There are two common ways to transmit data in MIMO: space–time coding and spatial multiplexing. In space–time coding the goal is to achieve maximal exploitation of the diversity of the MIMO channel. In spatial multiplexing, the goal is to maximize the data rate [27].

Space–time coding
The most commonly used space–time codes are space–time block codes (STBC) [24]. Within the class of STBC, we focus on the Alamouti scheme [22]. Assume that we want to transmit two complex numbers (a and b) over an $N_T = 2$, $N_R = 2$ MIMO channel. The Alamouti scheme requires two symbol durations. During the first symbol duration, we transmit $\mathbf{a}_0 = [a \quad b]^T$. During the second symbol duration, we transmit $\mathbf{a}_1 = [-b^* \quad a^*]^T$. As we will see in Chapter 12, this way of transmitting the data leads to a very simple STBC decoder.

Spatial multiplexing
In spatial multiplexing we simply take our data stream $\mathbf{a}$, and demultiplex it (i.e., serial-to-parallel conversion) into N_T parallel data streams, one for each transmit antenna[1] [19,28].

---

[1] In some variations of spatial multiplexing there is an encoder for every transmit antenna (this is known as V-BLAST). This distinction is irrelevant for our purpose.

In other words, at time instant k we transmit

$$\mathbf{a}_k = \left[a_{kN_T} \ldots a_{(k+1)N_T-1}\right]^T \tag{2.22}$$

and then

$$\mathbf{a}_{k+1} = \left[a_{(k+1)N_T} \ldots a_{(k+2)N_T-1}\right]^T \tag{2.23}$$

and so forth. Observe that, for $N_T = 2$, we can transmit twice as much information per symbol duration as in the Alamouti scheme. The cost we pay is the need for a more computationally demanding receiver, and a potential loss in diversity.

2.4.2   Multi-carrier modulation – MIMO-OFDM

MIMO-OFDM combines the capacity gains of using multiple antennas and the simple receiver design of OFDM in a straightforward way [29, 30]. The data stream $\mathbf{a}$ is first demultiplexed onto the various transmit antennas.[2] This means that, as in spatial multiplexing, every antenna works on a unique part of the data stream: the nth transmit antenna operates on the stream $\mathbf{a}^{(n)}$. As shown in Fig. 2.4, at every transmit antenna, symbols are processed in blocks of size N_{FFT} using standard OFDM: they are passed through an IFFT, a cyclic prefix is attached and, after pulse-shaping, the signal is transmitted over the antenna. The transmitted signal at the nth transmit antenna can be written as

$$s^{(n)}(t) = \sqrt{\frac{E_s N_{FFT}}{N_{FFT} + N_{CP}}} \sum_{k=-\infty}^{+\infty} \sum_{l=-N_{CP}}^{N_{FFT}-1} \check{a}_{l,k}^{(n)} p^{(n)}(t - lT - kT_{OFDM}), \tag{2.24}$$

where $\check{a}_{l,k}^{(n)}$ is the lth time-domain component in the kth OFDM symbol, transmitted on the nth transmit antenna. At the receiver, equipped with N_R antennas, the received signal

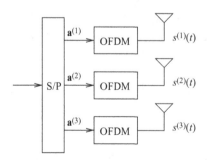

Figure 2.4. A MIMO-OFDM transmitter for $N_T = 3$.

[2] In some variations of MIMO-OFDM, there is an encoder for every transmit antenna (this is known as V-BLAST OFDM). This distinction is irrelevant for our purpose.

at time t can be expressed as an $N_R \times 1$ vector

$$\mathbf{r}(t) = \sum_{k=-\infty}^{+\infty} \sum_{n=1}^{N_T} \sum_{l=-N_{CP}}^{N_{FFT}-1} \check{a}_{l,k}^{(n)} \mathbf{h}^{(n)}(t - lT - kT_{OFDM}) + \mathbf{n}(t), \tag{2.25}$$

where $\mathbf{h}^{(n)}(t)$ is an $N_R \times 1$ vector representing the equivalent channel between the nth transmit antenna and the different receive antennas.

2.5   Multi-user communication

When multiple users transmit to the same receiver (see Fig. 2.5), we deal with multi-user communication and multi-user detection [31, 32]. The transmitted signal of the nth user can be written as

$$s^{(n)}(t) = \sqrt{E_s^{(n)}} \sum_{k=-\infty}^{+\infty} a_k^{(n)} p_k^{(n)}(t), \tag{2.26}$$

where $E_s^{(n)}$ is the nth user's energy per symbol, $a_k^{(n)}$ is the kth symbol of the nth user, and $p_k^{(n)}(t)$ is the corresponding transmit pulse. For a total of N_u users, the received signal can be written as

$$r(t) = \sum_{n=1}^{N_u} \sum_{k=-\infty}^{+\infty} a_k^{(n)} h_k^{(n)}(t) + n(t), \tag{2.27}$$

where $h_k^{(n)}(t)$ is the equivalent channel for the kth symbol of the nth user. Since multiple users are accessing the same medium, we require a way to share the available resources,

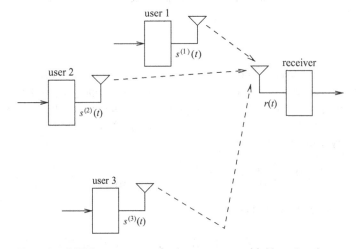

Figure 2.5. Multi-user communication: a system with $N_u = 3$ active users and one receiver.

and to ensure that the receiver can recover the information streams coming from each of the users. Over the past few decades many such schemes have been developed, including time-division multiple access (TDMA), frequency-division multiple access (FDMA), code-division multiple access (CDMA), and OFDM multiple access (OFDMA). In TDMA (FDMA), users are assigned different time-slots (frequency bands), such that they do not overlap in the time domain (frequency domain). The receiver simply listens to the different patches in the time (frequency) domain to recover the data streams. In other multiple-access schemes, the receiver has to perform multi-user detection to recover the superposition of the signals of the different users. Two important multiple-access schemes (CDMA and OFDMA) are described below. The corresponding receivers will be discussed in Chapter 13.

CDMA

We will describe direct-sequence CDMA (DS-CDMA) [33,34]. The kth user applies the following transmit pulse:

$$p_k^{(n)}(t) = \frac{1}{\sqrt{N_{SG}}} \sum_{i=0}^{N_{SG}-1} d_i^{(n)} p_S\left(t - i\frac{T}{N_{SG}} - kT\right). \tag{2.28}$$

Here $p_S(t)$ is a unit-energy square-root Nyquist pulse for a rate N_{SG}/T and

$$\mathbf{d}^{(n)} = \left[d_0^{(n)}, \ldots, d_{N_{SG}-1}^{(n)}\right]^{T}$$

is referred to as the spreading sequence of the nth user, with $d_i^{(n)}$ belonging to some discrete set (usually $\{-1,+1\}$ or $\{-1,+1,-j,+j\}$). The number N_{SG} is known as the spreading gain or the processing gain. Users differ in their spreading sequences. At the receiver, users will also differ in the equivalent channels. This diversity enables the receiver to recover the different data streams. The spreading sequences $\mathbf{d}^{(n)}$, $n = 1, \ldots, N_u$, are chosen in such a way that they have excellent auto- and cross-correlation properties [35].

OFDMA

OFDMA is a multi-carrier scheme whereby every user is assigned a set of subcarriers [36,37]. The subcarrier sets for different users are non-overlapping. Consider for instance a system with $N_u = 2$ users. During each OFDM symbol duration, both users transmit $N_{FFT}/2$ data symbols: $a_0^{(n)}, \ldots, a_{N_{FFT}/2-1}^{(n)}$, $n \in \{1,2\}$. User 1 then creates a vector of length N_{FFT} by using only the even carriers

$$\mathbf{a}^{(1)} = \left[a_0^{(1)} \quad 0 \quad a_1^{(1)} \quad 0 \quad \cdots \quad a_{N_{FFT}/2-1}^{(1)} \quad 0\right]^{T} \tag{2.29}$$

while user 2 uses only the odd carriers

$$\mathbf{a}^{(2)} = \left[0 \quad a_0^{(2)} \quad 0 \quad a_1^{(2)} \quad \cdots \quad 0 \quad a_{N_{FFT}/2-1}^{(2)}\right]^{T}. \tag{2.30}$$

Both users multiply their vectors by an N_{FFT}-point IDFT matrix, resulting in $\breve{\mathbf{a}}^{(1)}$ and $\breve{\mathbf{a}}^{(2)}$. They then transmit (after adding a cyclic prefix) the resulting time-domain sequence over the channel using a pulse $p(t)$. The received signal can be written as

$$r(t) = \sum_{n=1}^{N_u} \sum_{k=-\infty}^{+\infty} \sum_{l=-N_{\text{CP}}}^{N_{\text{FFT}}-1} \breve{a}_{l,k}^{(n)} h^{(n)}(t - lT - kT_{\text{OFDM}}) + n(t). \qquad (2.31)$$

As we will see in Chapter 13, the different signals can be separated efficiently as long the users are synchronized to a certain extent.

2.6 Goals and working assumptions

The reader may have noticed that we have described transmission schemes, but have not mentioned how the corresponding receivers should be designed. For each of these transmission schemes, there exist off-the-shelf receivers that are well understood. In this book, we will step away from these existing receivers and start from scratch (well, maybe not from scratch, but something close to it). The final goal of any receiver is to recover the information stream (the information word $\mathbf{b}$), in an optimal (or near-optimal) fashion. The final goal of this book is to design receivers that are able to perform this task, starting from basic principles of estimation theory, and exploiting the expressive power of factor graphs. To design these receivers, we will need make a number of assumptions.

- We consider only linear modulation. Extension to non-linear modulation is possible in some cases.
- The receiver knows the type of channel encoder and mapping used by the transmitter.
- The receiver knows the characteristics of the noise process. This is a reasonable assumption, since the noise characteristics usually do not change dramatically over time.
- The physical channel is linear. Extension to non-linear channels is possible in some cases.
- The receiver knows the equivalent channel $h_{\text{ch}}(t)$. This basically means that the receiver has estimated the channel prior to data detection. As long as the channel varies slowly over time, this is a reasonable assumption. We will remove this assumption (partially) in Chapter 14.
- The equivalent channel $h_{\text{ch}}(t)$ is static during the transmission of the N_s data symbols. Although this assumption is not necessary for most cases, it is helpful because it will keep the mathematical derivations simple.

2.7 Main points

In this chapter we have given a brief overview of digital communications, including channel coding, mapping, and pulse-shaping. Transmission schemes can be divided into various categories, depending on the number of active users, the number of antennas at receiver and/or transmitter, and whether or not multiple carriers are used. Before we can develop receivers for these different transmission schemes, we first need to understand what it means to recover the information bits in an optimal way and how exactly this recovery can be achieved. It is intuitively clear that we will need to formulate some type of estimation or optimization problem. To this end, we will now move away from our digital communication problem, and cover basic estimation theory using the language of factor graphs. We will return to digital communication from Chapter 7 onward.

3 Estimation theory and Monte Carlo techniques

3.1 Introduction

Before we can even consider designing iterative receivers, we have a lot of ground to cover. The ultimate goal of the receiver is to recover optimally the sequence of information bits that was sent by the transmitter. Since these bits were random in the first place, and they are affected by random noise at the receiver, we need to understand and quantify this randomness, and incorporate it into the design of our receiver. We need to specify what it means to recover *optimally* the information bits from the received signal. Optimal in what sense? To answer this problem, we call on estimation theory, which will allow us to formulate a suitable optimization problem, the solution of which will give our desired optimal data recovery. Unfortunately, as is common to most optimization problems of any practical relevance, the issue of finding a closed-form solution is intractable. There exists a set of tools that can solve these problems approximately by means of sampling. They go under the name of Monte Carlo (MC) techniques and can help in describing difficult distributions, and in obtaining their characteristics, by means of a list of samples. These MC methods will turn out to be very useful in the factor-graph framework.

This chapter is organized as follows.

- We will start with the basics of Bayesian estimation theory in **Section 3.2**, covering some important estimators both for discrete and for continuous parameters.
- In **Section 3.3**, we will provide a brief introduction to MC techniques, including particle representations of distributions, importance sampling, and Markov-chain Monte Carlo (MCMC) methods.

3.2 Bayesian estimation

Bayesian estimation is a branch of estimation theory whereby variables are considered to be random, with certain a-priori distributions. In this section, we will describe the problem of Bayesian estimation and derive several important estimators both for continuous and for discrete variables. Standard works on estimation theory can be found in [38–41].

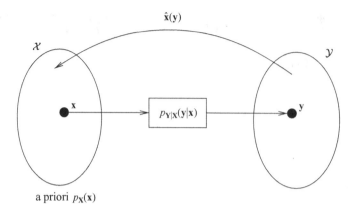

Figure 3.1. The general problem of estimation theory. The variable of interest **x** belongs to the variable space $\mathcal{X}$ and has the a-priori distribution $p_{\mathbf{X}}(\mathbf{x})$. The variable can be observed only through $\mathbf{y} \in \mathcal{Y}$. The probabilistic mapping $p_{\mathbf{Y}|\mathbf{X}}(\mathbf{y}|\mathbf{x})$ from $\mathcal{X}$ to $\mathcal{Y}$ is known.

3.2.1   Problem formulation

In Bayesian estimation theory, we are interested in estimating an instance (a realization) of a random variable, say **X**, belonging to a set $\mathcal{X}$, with an *a-priori distribution* $p_{\mathbf{X}}(\mathbf{x})$. The variable **X** can be discrete or continuous, and the a-priori distribution can be either given or assigned by us.[1] We do not observe the instance **x** directly, but only through an observation $\mathbf{y} \in \mathcal{Y}$. That is, we observe a realization **y** of a random variable **Y**, related to **x** by a known probabilistic mapping $p_{\mathbf{Y}|\mathbf{X}}(\mathbf{y}|\mathbf{x})$. All possible observations fall into an observation-space $\mathcal{Y}$. The variable **Y** can be discrete or continuous. Our goal is to design an estimator for **X**. The entire set-up is depicted in Fig. 3.1.

DEFINITION 3.1 (Estimator).   *An estimator is function from the observation-space $\mathcal{Y}$ to the variable-space $\mathcal{X}$, and is denoted $\hat{\mathbf{x}}(\mathbf{y})$. With every element $\mathbf{y} \in \mathcal{Y}$ we associate a unique element $\hat{\mathbf{x}}(\mathbf{y}) \in \mathcal{X}$. An estimator is designed so as to minimize an expected cost*[2]

$$\mathcal{C} = \mathbb{E}_{\mathbf{X},\mathbf{Y}}\{c(\mathbf{X}, \hat{\mathbf{x}}(\mathbf{Y}))\} \tag{3.1}$$

$$= \int_{\mathcal{X}} \int_{\mathcal{Y}} c(\mathbf{x}, \hat{\mathbf{x}}(\mathbf{y})) p_{\mathbf{X},\mathbf{Y}}(\mathbf{x}, \mathbf{y}) d\mathbf{x}\, d\mathbf{y} \tag{3.2}$$

$$= \int_{\mathcal{Y}} p_{\mathbf{Y}}(\mathbf{y}) \underbrace{\int_{\mathcal{X}} c(\mathbf{x}, \hat{\mathbf{x}}(\mathbf{y})) p_{\mathbf{X}|\mathbf{Y}}(\mathbf{x}|\mathbf{y}) d\mathbf{x}}_{\mathcal{C}(\mathbf{y})} d\mathbf{y}, \tag{3.3}$$

where $c(\cdot)$ is a cost function from $\mathcal{X} \times \mathcal{X} \to \mathbb{R}$, and depends on the specific problem.

---

[1] We will not go into the debate on Bayesian versus non-Bayesian estimation.
[2] Integrals should be replaced by summations, where appropriate.

We will consider only cost functions for which $c(\mathbf{x}, \hat{\mathbf{x}}(\mathbf{y})) \geq 0$. In that case, the expected cost is minimized by minimizing the integrand

$$C(\mathbf{y}) = \int_{\mathcal{X}} c(\mathbf{x}, \hat{\mathbf{x}}(\mathbf{y})) p_{\mathbf{X}|\mathbf{Y}}(\mathbf{x}|\mathbf{y}) d\mathbf{x}. \tag{3.4}$$

Terminology

For a given observation $\mathbf{y}$, $p_{\mathbf{Y}|\mathbf{X}}(\mathbf{y}|\mathbf{x})$ is a function of $\mathbf{x}$, and is known as the *likelihood function*. Note that the likelihood function is not a distribution since $\int p_{\mathbf{Y}|\mathbf{X}}(\mathbf{y}|\mathbf{x}) d\mathbf{x}$ is not necessarily equal to unity. On the other hand, for a fixed $\mathbf{y}$, $p_{\mathbf{X}|\mathbf{Y}}(\mathbf{x}|\mathbf{y})$ is a distribution and is known as the *a-posteriori distribution* (of $\mathbf{x}$).

3.2.2 Estimators for continuous variables

When $\mathbf{X}$ is a continuous random variable, the most important estimators are the minimum-mean-squared-error (MMSE) estimator, the linear-MMSE estimator, and the maximum a-posteriori (MAP) estimator.

3.2.2.1 The MMSE estimator

A natural cost function is one where we increase the cost proportionally to the squared distance between $\mathbf{x}$ and $\hat{\mathbf{x}}(\mathbf{y})$:

$$c(\mathbf{x}, \hat{\mathbf{x}}(\mathbf{y})) = \|\mathbf{x} - \hat{\mathbf{x}}(\mathbf{y})\|^2 \tag{3.5}$$

$$= (\mathbf{x} - \hat{\mathbf{x}}(\mathbf{y}))^{\mathsf{T}}(\mathbf{x} - \hat{\mathbf{x}}(\mathbf{y})). \tag{3.6}$$

Taking the derivative of $C(\mathbf{y})$ with respect to $\hat{\mathbf{x}}$ and equating to zero gives us (under some mild conditions)

$$\hat{\mathbf{x}}(\mathbf{y}) = \int_{\mathcal{X}} \mathbf{x} p_{\mathbf{X}|\mathbf{Y}}(\mathbf{x}|\mathbf{y}) d\mathbf{x}. \tag{3.7}$$

In other words, the MMSE estimate is given by the mean of the a-posteriori distribution $p_{\mathbf{X}|\mathbf{Y}}(\mathbf{x}|\mathbf{y})$.

Example 3.1 (Gaussian case).   *When $\mathbf{X}$ and $\mathbf{Y}$ are jointly Gaussian, we can write $\mathbf{Z} = [\mathbf{X}^{\mathsf{T}} \mathbf{Y}^{\mathsf{T}}]^{\mathsf{T}}$, where*

$$\mathbf{z} \sim \mathcal{N}_{\mathbf{z}}(\mathbf{m}, \Sigma)$$

with $\mathbf{m} = [\mathbf{m}_X^{\mathsf{T}} \mathbf{m}_Y^{\mathsf{T}}]^{\mathsf{T}}$ and

$$\Sigma = \left[\begin{array}{cc} \Sigma_{XX} & \Sigma_{XY} \\ \Sigma_{XY}^{\mathsf{T}} & \Sigma_{YY} \end{array} \right].$$

Here, $\Sigma_{XY} = \mathbb{E}\{(\mathbf{X} - \mathbf{m}_X)(\mathbf{Y} - \mathbf{m}_Y)^{\mathsf{T}}\}$, $\Sigma_{XX} = \mathbb{E}\{(\mathbf{X} - \mathbf{m}_X)(\mathbf{X} - \mathbf{m}_X)^{\mathsf{T}}\}$, and similarly for Σ_{YY}. The a-posteriori distribution is given by

$$p_{\mathbf{X}|\mathbf{Y}}(\mathbf{x}|\mathbf{y}) = \mathcal{N}_{\mathbf{x}}(\mathbf{m}_{X|Y}(\mathbf{y}), \Sigma_{X|Y}),$$

where (see appendices in Chapter 15)

$$\mathbf{m}_{X|Y}(\mathbf{y}) = \mathbf{m}_X + \Sigma_{XY}\Sigma_{YY}^{-1}(\mathbf{y} - \mathbf{m}_Y)$$

$$\Sigma_{X|Y} = \Sigma_{XX} - \Sigma_{XY}\Sigma_{YY}^{-1}\Sigma_{XY}^{\mathrm{T}}.$$

It then follows that the MMSE estimate of **x** *is*

$$\hat{\mathbf{x}}(\mathbf{y}) = \mathbf{m}_X + \Sigma_{XY}\Sigma_{YY}^{-1}(\mathbf{y} - \mathbf{m}_Y).$$

It is important to note that the estimator is a linear *function of the observation* **y**.

3.2.2.2   The linear MMSE estimator

As the example in the previous section indicates, the MMSE estimator is a linear function of the observation when **X** and **Y** are jointly Gaussian. When **X** and **Y** are not jointly Gaussian, we could still try to find the best linear estimator (best in an MMSE sense), that is, an estimator of the form

$$\hat{\mathbf{x}}(\mathbf{y}) = \mathbf{B}^{\mathrm{T}}\mathbf{y} + \mathbf{d}. \tag{3.8}$$

To fix **B** and **d**, this estimator should have the minimum mean-squared error among all linear estimators. It turns out that $\hat{\mathbf{x}}(\mathbf{y})$ is given by

$$\hat{\mathbf{x}}(\mathbf{y}) = \mathbf{m}_X + \Sigma_{XY}\Sigma_{YY}^{-1}(\mathbf{y} - \mathbf{m}_Y). \tag{3.9}$$

This estimator is the linear-MMSE (L-MMSE) estimator.

Example 3.2 (The linear model). *Consider the following model, with a linear relationship between* **x** *and* **y**:

$$\mathbf{y} = \mathbf{H}\mathbf{x} + \mathbf{n}.$$

H *is a known matrix,* $\mathbf{x} \sim p_{\mathbf{X}}(\cdot)$, *and* $\mathbf{n} \sim p_{\mathbf{N}}(\cdot)$, *zero-mean, with covariance matrix* Σ_{NN}. *Furthermore,* **N** *is independent from* **X**. *Let us determine the L-MMSE estimate of* **x** *from* **y**. *We know that*

$$\hat{\mathbf{x}}(\mathbf{y}) = \mathbf{m}_X + \Sigma_{XY}\Sigma_{YY}^{-1}(\mathbf{y} - \mathbf{m}_Y).$$

Now, $\mathbf{m}_Y = \mathbf{H}\mathbf{m}_X$ *and* $\Sigma_{XY} = \Sigma_{XX}\mathbf{H}^{\mathrm{T}}$, *while*

$$\Sigma_{YY} = \mathbb{E}\left\{(\mathbf{H}\mathbf{X} + \mathbf{N} - \mathbf{H}\mathbf{m}_X)(\mathbf{H}\mathbf{X} + \mathbf{N} - \mathbf{H}\mathbf{m}_X)^{\mathrm{T}}\right\}$$

$$= \mathbf{H}\Sigma_{XX}\mathbf{H}^{\mathrm{T}} + \Sigma_{NN}.$$

This all leads to the following L-MMSE estimate of **x**:

$$\hat{\mathbf{x}}(\mathbf{y}) = \mathbf{m}_X + \Sigma_{XX}\mathbf{H}^{\mathrm{T}}(\mathbf{H}\Sigma_{XX}\mathbf{H}^{\mathrm{T}} + \Sigma_{NN})^{-1}(\mathbf{y} - \mathbf{H}\mathbf{m}_X).$$

3.2.2.3　The MAP estimator

The maximum a-posteriori (or MAP) estimator is based on the following cost function, for an arbitrarily small $\delta > 0$:

$$c(\mathbf{x}, \hat{\mathbf{x}}(\mathbf{y})) = \begin{cases} 0 & \|\mathbf{x} - \hat{\mathbf{x}}(\mathbf{y})\| < \delta \\ 1 & \text{else} \end{cases} \tag{3.10}$$

so that

$$\mathcal{C}(\mathbf{y}) = \int_{\mathcal{X}} c(\mathbf{x}, \hat{\mathbf{x}}(\mathbf{y})) p_{\mathbf{X}|\mathbf{Y}}(\mathbf{x}|\mathbf{y}) d\mathbf{x} \tag{3.11}$$

$$= 1 - \int_{\mathcal{X}} (1 - c(\mathbf{x}, \hat{\mathbf{x}}(\mathbf{y}))) p_{\mathbf{X}|\mathbf{Y}}(\mathbf{x}|\mathbf{y}) d\mathbf{x} \tag{3.12}$$

$$= 1 - \int_{\hat{\mathbf{x}}(\mathbf{y})-\delta}^{\hat{\mathbf{x}}(\mathbf{y})+\delta} p_{\mathbf{X}|\mathbf{Y}}(\mathbf{x}|\mathbf{y}) d\mathbf{x}. \tag{3.13}$$

Hence, we need to *maximize* the second term with respect to the function $\hat{\mathbf{x}}(\mathbf{y})$ to minimize the expected cost. On letting $\delta \to 0$, we find that the cost is minimized when $\hat{\mathbf{x}}(\mathbf{y})$ is the maximum of $p_{\mathbf{X}|\mathbf{Y}}(\mathbf{x}|\mathbf{y})$. This explains the name maximum a-posteriori estimation:

$$\hat{\mathbf{x}}(\mathbf{y}) = \arg \max_{\mathbf{x} \in \mathcal{X}} p_{\mathbf{X}|\mathbf{Y}}(\mathbf{x}|\mathbf{y}). \tag{3.14}$$

Note that, when $\mathbf{X}$ and $\mathbf{Y}$ are jointly Gaussian, $p_{\mathbf{X}|\mathbf{Y}}(\mathbf{x}|\mathbf{y})$ is a Gaussian distribution, so its mode coincides with its mean. In other words, the MAP estimate coincides with the MMSE estimate.

Example 3.3 (Coin tossing).　*Suppose that we have a coin with bias x (the probability of heads is x and the probability of tails is $1 - x$). We toss the coins N times and observe the outcome* **y**, *a sequence of heads and tails. We wish to estimate x. We see that the likelihood is given by $p_{\mathbf{Y}|X}(\mathbf{y}|x) = (1-x)^{N_H} x^{N_T}$, where N_H (N_T) is the number of times heads (tails) appears in* **y**. *Owing to Bayes' rule, we find that*

$$p_{X|\mathbf{Y}}(x|\mathbf{y}) = C_{\mathbf{y}} p_X(x)(1-x)^{N_H} x^{N_T}$$

for some constant $C_{\mathbf{y}}$ (constant in the sense that it does not depend on x). The MAP estimate of x is given by

$$\hat{x}(\mathbf{y}) = \arg \max_{x \in [0,1]} p_X(x)(1-x)^{N_H} x^{N_T}.$$

Let us set $p_X(x)$ to a uniform distribution within some interval $\left[\frac{1}{2} - \varepsilon, \frac{1}{2} + \varepsilon\right]$. The MAP estimate is given by

$$\hat{x}(\mathbf{y}) = \begin{cases} \frac{1}{2} - \varepsilon & \frac{1}{2} - \varepsilon > N_H/N, \\ N_H/N & \frac{1}{2} - \varepsilon < N_H/N < \frac{1}{2} + \varepsilon, \\ \frac{1}{2} + \varepsilon & N_H/N > \frac{1}{2} + \varepsilon. \end{cases}$$

When $\varepsilon \to \frac{1}{2}$, we find the more conventional estimate

$$\hat{x}(\mathbf{y}) = \frac{N_H}{N}.$$

3.2.3   Estimators for discrete variables

When $\mathbf{X}$ is a discrete random variable, it is meaningful to say that an estimate is either right or wrong. Furthermore, the set $\mathcal{X}$ may be an abstract beast without much structure, such that the concept of distance (which we used for the MMSE estimator) might not exist (consider the set $\mathcal{X} = \{\text{banana}, \text{cow}, \text{Stata Center}\}$, for instance). In any case, the cost $\mathcal{C}(\mathbf{y})$ can be written as

$$\mathcal{C}(\mathbf{y}) = \sum_{\mathbf{x} \in \mathcal{X}} c(\mathbf{x}, \hat{\mathbf{x}}(\mathbf{y})) p_{\mathbf{X}|\mathbf{Y}}(\mathbf{x}|\mathbf{y}). \tag{3.15}$$

Suppose that we choose the following cost function: we assign a cost 1 when $\hat{\mathbf{x}}(\mathbf{y}) \neq \mathbf{x}$ and 0 otherwise. Such a cost function is always meaningful, no matter what $\mathcal{X}$ is. This leads to

$$\mathcal{C}(\mathbf{y}) = 1 - p_{\mathbf{X}|\mathbf{Y}}(\hat{\mathbf{x}}(\mathbf{y})|\mathbf{y}). \tag{3.16}$$

Hence, the cost is minimized when

$$\hat{\mathbf{x}}(\mathbf{y}) = \arg\max_{\mathbf{x} \in \mathcal{X}} p_{\mathbf{X}|\mathbf{Y}}(\mathbf{x}|\mathbf{y}). \tag{3.17}$$

We have found again the MAP estimator. In this case the expected cost $\mathcal{C}$ can be interpreted as an *error probability*. This can be seen as follows:

$$\mathcal{C} = \sum_{\mathbf{x} \in \mathcal{X}} \int_{\mathcal{Y}} c(\mathbf{x}, \hat{\mathbf{x}}(\mathbf{y})) p_{\mathbf{X}, \mathbf{Y}}(\mathbf{x}, \mathbf{y}) d\mathbf{y} \tag{3.18}$$

$$= \sum_{\mathbf{x} \in \mathcal{X}} p_{\mathbf{X}}(\mathbf{x}) \int_{\mathcal{Y}} c(\mathbf{x}, \hat{\mathbf{x}}(\mathbf{y})) p_{\mathbf{Y}|\mathbf{X}}(\mathbf{y}|\mathbf{x}) d\mathbf{y} \tag{3.19}$$

$$= \sum_{\mathbf{x} \in \mathcal{X}} p_{\mathbf{X}}(\mathbf{x}) \underbrace{\int_{\mathcal{Y}(\bar{\mathbf{x}})} p_{\mathbf{Y}|\mathbf{X}}(\mathbf{y}|\mathbf{x}) d\mathbf{y}}_{P_{\mathrm{e}}(\mathbf{x})} \tag{3.20}$$

where, for a given $\mathbf{x}$, $\mathcal{Y}(\bar{\mathbf{x}})$ is the part of the observation-space $\mathcal{Y}$ for which the estimator chooses an estimate $\hat{\mathbf{x}}$ different from $\mathbf{x}$. Now, $P_e(\mathbf{x})$ is clearly the probability of making an estimation error, given that $\mathbf{x}$ is the correct value: it is the probability of $\mathbf{y}$ falling within the region $\mathcal{Y}(\bar{\mathbf{x}})$, given that $\mathbf{x}$ is the correct value. Hence, the expected cost is the expected error probability: $\mathcal{C} = \mathbb{E}_{\mathbf{X}}\{P_e(\mathbf{X})\}$.

3.3 Monte Carlo techniques

While Bayesian estimation clearly depends on the possibility of determining the mode (for MAP) or the mean (for MMSE) of the a-posteriori distribution $p_{\mathbf{X}|\mathbf{Y}}(\mathbf{x}|\mathbf{y})$, this is in general not an easy task. It often requires integration of strange functions and solution of optimization problems well beyond the means of classical methods. Monte Carlo techniques form an attractive and powerful alternative to these classical methods, by representing distributions as a list of samples. Integration and optimization are based not on the entire distribution, but only on the samples.

In this section we will introduce the concept of a particle representation as a way to represent a distribution by a finite list of samples. Samples are obtained by sampling from an appropriate distribution. We will discuss several sampling techniques, including importance sampling and Gibbs sampling. Many other sampling techniques exist, but are beyond the scope of this book. The interested reader can consider [42–44] for further details.

3.3.1 Particle representations

3.3.1.1 Principles

A particle representation is a way to represent a distribution by a list of samples. The idea of particle representations is best understood through an example.

Example 3.4 (Monte Carlo integration). *Suppose that we have a continuous random variable $\mathbf{Z}$ with an associated distribution $p_{\mathbf{Z}}(\mathbf{z})$, and our goal is to determine*[3]

$$I = \mathbb{E}_{\mathbf{Z}}\{f(\mathbf{Z})\}$$

for some real-valued function $f(\mathbf{z})$. Let us draw L independent samples from $p_{\mathbf{Z}}(\mathbf{z})$, say $\mathbf{z}^{(1)}, \ldots, \mathbf{z}^{(L)}$. We can then approximate I by I_L, where

$$I_L = \frac{1}{L}\sum_{l=1}^{L} f\left(\mathbf{z}^{(l)}\right).$$

---

[3] For instance, $p_{\mathbf{Z}}(\mathbf{z})$ could be an a-posteriori distribution, and $f(\mathbf{z}) = \mathbf{z}$, in which case I is an MMSE estimate.

I_L is an approximation of I in the sense that (i) $\mathbb{E}_{\mathbf{Z}}\{I_L\} = I$ and (ii) $\mathbb{E}_{\mathbf{Z}}\{(I_L - I)^2\}$ decreases as L increases. This is seen as follows:

$$\mathbb{E}_{\mathbf{Z}}\{I_L\} = \mathbb{E}_{\mathbf{Z}}\left\{\frac{1}{L}\sum_{l=1}^{L}f(\mathbf{Z}^{(l)})\right\}$$

$$= \mathbb{E}_{\mathbf{Z}}\{f(\mathbf{Z})\}.$$

Furthermore, let us introduce $\sigma_f^2 = \mathbb{E}_{\mathbf{Z}}\{(f(\mathbf{Z}))^2\} - I^2$. Then, since the samples are independent,

$$\mathbb{E}_{\mathbf{Z}}\left\{(I_L - I)^2\right\} = \frac{\sigma_f^2}{L}.$$

In other words, the variance of the estimation error decreases with L. The samples $\mathbf{z}^{(1)}, \ldots, \mathbf{z}^{(L)}$ can be drawn without knowing the function $f(\cdot)$. The samples can thus be seen as a representation of $p_{\mathbf{Z}}(\mathbf{z})$. Note that, when the samples are not drawn independently, $\mathbb{E}_{\mathbf{Z}}\{I_L\} = I$, but the variance of the estimation error is usually much larger than with independent sampling (for a fixed L).

We see that a distribution can be represented by a finite number of independent samples in a meaningful way. This idea is generalized as follows.

DEFINITION 3.2 (Particle representation).   *We are given a random variable $\mathbf{Z}$, defined over a set $\mathcal{Z}$. A particle representation of a distribution $p_{\mathbf{Z}}(\mathbf{z})$ is a set of L couples $(w^{(l)}, \mathbf{z}^{(l)})$, with $\sum_l w^{(l)} = 1$ such that, for any integrable function $f(\mathbf{z})$ from $\mathcal{Z} \rightarrow \mathbb{C}$,*

$$I = \mathbb{E}_{\mathbf{Z}}\{f(\mathbf{Z})\} \tag{3.21}$$

can be approximated by

$$I_L = \sum_{l=1}^{L}w^{(l)}f\left(\mathbf{z}^{(l)}\right). \tag{3.22}$$

The approximation should be understood as $I_L \rightarrow I$ almost everywhere as $L \rightarrow +\infty$. We use the following notation: $\mathcal{R}_L(p_{\mathbf{Z}}(\cdot)) = \left\{(w^{(l)}, \mathbf{z}^{(l)})\right\}_{l=1}^{L}$ where $\mathbf{z}^{(l)}$ is named a properly weighted sample *with weight $w^{(l)}$.*

Note that we do not require the samples to be independent, and we make no claim as to unbiasedness, or variance of the estimation error. How samples and weights should be chosen will be the topic of Sections 3.3.2–3.3.3.

3.3.1.2 Alternative notations

Considering (3.21) and (3.22), it is meaningful to introduce an alternative notation for continuous $\mathbf{Z}$:

$$p_{\mathbf{Z}}(\mathbf{z}) \approx \sum_{l=1}^{L} w^{(l)} \delta\left(\mathbf{z} - \mathbf{z}^{(l)}\right), \tag{3.23}$$

where $\delta(\cdot)$ is the Dirac distribution. On the other hand, for discrete $\mathbf{Z}$, we can write

$$p_{\mathbf{Z}}(\mathbf{z}) \approx \sum_{l=1}^{L} w^{(l)} \mathbb{I}\left\{\mathbf{z} = \mathbf{z}^{(l)}\right\}, \tag{3.24}$$

where, for a proposition P, $\mathbb{I}\{P\}$ is the indicator function, defined as $\mathbb{I}\{P\} = 1$ when P is true and $\mathbb{I}\{P\} = 0$ when P is false. The notation "$\approx$" in (3.23)–(3.24) should be understood as follows: substitution of (3.23) or (3.24) into (3.21) yields the approximation (3.22). It will turn out to be convenient to use a more general notation by introducing the equality function:

$$\boxminus(\mathbf{x}_1, \ldots, \mathbf{x}_D) = \begin{cases} \prod_{k=1}^{D-1} \delta(\mathbf{x}_{k+1} - \mathbf{x}_k) & \mathbf{x}_k \text{ continuous} \\ \prod_{k=1}^{D-1} \mathbb{I}\{\mathbf{x}_{k+1} = \mathbf{x}_k\} & \mathbf{x}_k \text{ discrete} \end{cases} \tag{3.25}$$

so that we can write

$$p_{\mathbf{Z}}(\mathbf{z}) \approx \sum_{l=1}^{L} w^{(l)} \boxminus\left(\mathbf{z}, \mathbf{z}^{(l)}\right) \tag{3.26}$$

both for continuous and for discrete variables.

3.3.1.3 Fun with particle representations

Some important uses for particle representations are the following.

- **Determining marginals:** if we have a particle representation of a distribution of $\mathbf{Z} = [Z_1, Z_2, Z_3]$, say $\mathcal{R}_L(p_{\mathbf{Z}}(\cdot)) = \left\{\left(w^{(l)}, \left(z_1^{(l)}, z_2^{(l)}, z_3^{(l)}\right)\right)\right\}_{l=1}^{L}$, we immediately have particle distributions of the marginals, for instance $\mathcal{R}_L(p_{Z_1}(\cdot)) = \left\{\left(w^{(l)}, z_1^{(l)}\right)\right\}_{l=1}^{L}$ and $\mathcal{R}_L(p_{Z_1,Z_3}(\cdot, \cdot)) = \left\{\left(w^{(l)}, \left(z_1^{(l)}, z_3^{(l)}\right)\right)\right\}_{l=1}^{L}$.

- **Resampling:** given a particle representation $\mathcal{R}_L(p_{\mathbf{Z}}(\cdot)) = \left\{\left(w^{(l)}, \mathbf{z}^{(l)}\right)\right\}_{l=1}^{L}$, we can consider it as a probability mass function and draw samples from it. This leads to a new particle representation $\mathcal{R}_K(p_{\mathbf{Z}}(\cdot)) = \left\{\left(1/K, \mathbf{z}^{(l)}\right)\right\}_{l=1}^{K}$, where K can be greater or smaller than L.

- **Regularization:** a particle representation can be converted to a smooth density function by a process called regularization or density estimation. Given a particle representation

$$\mathcal{R}_L(p_{\mathbf{Z}}(\cdot)) = \left\{\left(w^{(l)}, \mathbf{z}^{(l)}\right)\right\}_{l=1}^{L},$$

we approximate $p_{\mathbf{Z}}(\cdot)$ by a Gaussian mixture[4]

$$p_{\mathbf{Z}}(\mathbf{z}) \approx \sum_{l=1}^{L} w^{(l)} \mathcal{N}_{\mathbf{z}}\left(\mathbf{z}^{(l)}, \Sigma\right). \tag{3.27}$$

The covariance matrix is a diagonal matrix $\Sigma = \sigma^2 \mathbf{I}_D$, where D is the dimensionality of $\mathbf{Z}$. The variance σ^2 is referred to as the bandwidth or size in density estimation. In our case [45], a good choice for the variance is $\sigma^2 = (4/(L(D+2)))^{1/(D+4)}$.

3.3.2   Sampling for small-dimensional systems

3.3.2.1   Introduction

While we now understand what a particle representation is, it is still unclear how the samples and the weights should be selected. One possible way is to draw L independent samples from $p_{\mathbf{Z}}(\mathbf{z})$: $\mathbf{z}^{(1)}, \dots, \mathbf{z}^{(L)}$. This leads to the particle representation $\mathcal{R}_L(p_{\mathbf{Z}}(\cdot)) = \{(1/L, \mathbf{z}^{(l)})\}_{l=1}^{L}$. Unfortunately, it is generally hard to sample from an arbitrary distribution. Secondly, when we are interested in computing an expectation $\mathbb{E}_{\mathbf{Z}}\{f(\mathbf{Z})\}$, it may be wise to select the samples only where $f(\mathbf{z})$ takes on significant values. Otherwise, many samples will not contribute to the summation in (3.22). There exists a method that deals with these problems, known as *importance sampling*.

3.3.2.2   Importance sampling

Importance sampling is useful in the following scenarios:

- sampling from $p_{\mathbf{Z}}(\mathbf{z})$ is difficult; and
- $p_{\mathbf{Z}}(\mathbf{z})$ is not well matched to $f(\mathbf{z})$. This occurs for instance when $f(\mathbf{z})$ takes on significant values only for those $\mathbf{z}$ where $p_{\mathbf{Z}}(\mathbf{z})$ is very small. Then, for many samples, $f(\mathbf{z}^{(l)})$ may be negligible, and a lot of computation is wasted.

Let us look at a short example.

Example 3.5 (Sampling from an a-posteriori distribution).   *In an estimation problem, we observe* $\mathbf{y}$*. We know the a-priori distribution* $p_{\mathbf{X}}(\mathbf{x})$ *the likelihood function* $p_{\mathbf{Y}|\mathbf{X}}(\mathbf{y}|\mathbf{x})$*. For a given observation* $\mathbf{y}$*, we wish to find a particle representation of the a-posteriori distribution* $p_{\mathbf{X}|\mathbf{Y}}(\mathbf{x}|\mathbf{y})$*. Bayes' rule tell us that*

$$p_{\mathbf{X}|\mathbf{Y}}(\mathbf{x}|\mathbf{y}) = \frac{p_{\mathbf{Y}|\mathbf{X}}(\mathbf{y}|\mathbf{x})p_{\mathbf{X}}(\mathbf{x})}{p_{\mathbf{Y}}(\mathbf{y})}.$$

Note that $\mathbf{y}$ *is fixed and that* $p_{\mathbf{Y}}(\mathbf{y})$ *does not depend on* $\mathbf{x}$*. When* $\mathbf{X}$ *is discrete and of a small dimensionality, we can find the pmf* $p_{\mathbf{X}|\mathbf{Y}}(\mathbf{x}|\mathbf{y})$ *by simply computing the numerator*

---

[4] A mixture of other distributions (kernels) is also possible.

for every $\mathbf{x} \in \mathcal{X}$, $\phi(\mathbf{x}_i) = p_{\mathbf{Y}|\mathbf{X}}(\mathbf{y}|\mathbf{x}_i)p_{\mathbf{X}}(\mathbf{x}_i)$, *so*

$$p_{\mathbf{X}|\mathbf{Y}}(\mathbf{x}|\mathbf{y}) = \frac{\phi(\mathbf{x})}{\sum_{\mathbf{x}_i \in \mathcal{X}} \phi(\mathbf{x}_i)}.$$

We can then sample from the pmf $p_{\mathbf{X}|\mathbf{Y}}(\mathbf{x}|\mathbf{y})$. When $\mathbf{X}$ is continuous, it is not clear how we can sample from $p_{\mathbf{X}|\mathbf{Y}}(\mathbf{x}|\mathbf{y})$ since $p_{\mathbf{Y}}(\mathbf{y})$ is in general very hard to determine.

Importance sampling avoids these problems as follows. Our goal is to find a particle representation of $p_{\mathbf{Z}}(\mathbf{z})$ (*the target distribution*). Take any distribution $q_{\mathbf{Z}}(\mathbf{z})$ from which it is easy to sample (*the importance sampling distribution*). We introduce an additional function $w(\mathbf{z}) = p_{\mathbf{Z}}(\mathbf{z})/q_{\mathbf{Z}}(\mathbf{z})$, and restrict $q_{\mathbf{Z}}(\mathbf{z})$ to being non-zero where $p_{\mathbf{Z}}(\mathbf{z})$ is non-zero. We can then write

$$I = \int_{-\infty}^{+\infty} f(\mathbf{z})p_{\mathbf{Z}}(\mathbf{z})\mathrm{d}\mathbf{z} \tag{3.28}$$

$$= \frac{\int f(\mathbf{z})w(\mathbf{z})q_{\mathbf{Z}}(\mathbf{z})\mathrm{d}\mathbf{z}}{\int w(\mathbf{z})q_{\mathbf{Z}}(\mathbf{z})\mathrm{d}\mathbf{z}}. \tag{3.29}$$

Let us now draw L iid samples from $q_{\mathbf{Z}}(\mathbf{z})$, $\mathbf{z}^{(1)}, \ldots, \mathbf{z}^{(L)}$ and approximate both the numerator and the denominator using Monte Carlo integration:

$$I \approx I_L \tag{3.30}$$

$$= \frac{\frac{1}{L}\sum_{l=1}^{L} f\left(\mathbf{z}^{(l)}\right) w\left(\mathbf{z}^{(l)}\right)}{\frac{1}{L}\sum_{k=1}^{L} w\left(\mathbf{z}^{(l)}\right)} \tag{3.31}$$

$$= \sum_{l=1}^{L} w^{(l)} f\left(\mathbf{z}^{(l)}\right), \tag{3.32}$$

where

$$w^{(l)} = \frac{w\left(\mathbf{z}^{(l)}\right)}{\sum_{k=1}^{L} w\left(\mathbf{z}^{(l)}\right)}. \tag{3.33}$$

It can then be shown that $\mathcal{R}_L(p_{\mathbf{Z}}(\cdot)) = \left\{\left(w^{(l)}, \mathbf{z}^{(l)}\right)\right\}_{l=1}^{L}$. We will often write

$$w^{(l)} \propto \frac{p_{\mathbf{Z}}\left(\mathbf{z}^{(l)}\right)}{q_{\mathbf{Z}}\left(\mathbf{z}^{(l)}\right)}, \tag{3.34}$$

which describes the weight up to a normalization constant (which can be determined once all the samples have been drawn). Now we can select a suitable $q_{\mathbf{Z}}(\mathbf{z})$ so that most of the samples have a non-zero weight. It is clear that we have solved our two problems: we don't need to sample from $p_{\mathbf{Z}}(\mathbf{z})$ directly, and we can tune $q_{\mathbf{Z}}(\mathbf{z})$ according to $f(\mathbf{z})$.

3.3.2.3   Importance sampling revisited

While importance sampling solves our two original problems, we get a new problem in return: we need to know $p_{\mathbf{Z}}(\mathbf{z})$ and $q_{\mathbf{Z}}(\mathbf{z})$ analytically in order to compute $w(\mathbf{z})$. In many cases we know $p_{\mathbf{Z}}(\mathbf{z})$ and $q_{\mathbf{Z}}(\mathbf{z})$ only up to a constant:

$$p_{\mathbf{Z}}(\mathbf{z}) = \frac{1}{C_p}\tilde{p}_{\mathbf{Z}}(\mathbf{z}), \tag{3.35}$$

$$q_{\mathbf{Z}}(\mathbf{z}) = \frac{1}{C_q}\tilde{q}_{\mathbf{Z}}(\mathbf{z}), \tag{3.36}$$

where $\tilde{p}_{\mathbf{Z}}(\mathbf{z})$ and $\tilde{q}_{\mathbf{Z}}(\mathbf{z})$ are known analytically, and C_p and C_q are unknown constants. We introduce $\tilde{w}(\mathbf{z}) = \tilde{p}_{\mathbf{Z}}(\mathbf{z})/\tilde{q}_{\mathbf{Z}}(\mathbf{z})$. Note that this function can be evaluated exactly for any $\mathbf{z}$. The weight of sample $\mathbf{z}^{(l)}$ is given by

$$w^{(l)} = \frac{w(\mathbf{z}^{(l)})}{\sum_{k=1}^{L} w(\mathbf{z}^{(l)})} \tag{3.37}$$

$$= \frac{\tilde{w}(\mathbf{z}^{(l)})}{\sum_{k=1}^{L} \tilde{w}(\mathbf{z}^{(l)})} \tag{3.38}$$

since the factor C_q/C_p cancels out in the numerator and the denominator. This implies that we can use importance sampling without knowledge of the constants C_p and C_q! Note that when $p_{\mathbf{Z}}(\mathbf{z}) = q_{\mathbf{Z}}(\mathbf{z})$ the weights are all equal to $1/L$. The final algorithm is described in Algorithm 3.1.

Algorithm 3.1 Importance sampling

1: **for** $l = 1$ to L **do**
2:     draw $\mathbf{z}^{(l)} \sim q_{\mathbf{Z}}(\mathbf{z})$
3:     compute $w^{(l)} = \tilde{p}_{\mathbf{Z}}(\mathbf{z}^{(l)})/\tilde{q}_{\mathbf{Z}}(\mathbf{z}^{(l)})$
4: **end for**
5: normalize weights
6: *output:* $\mathcal{R}_L(p_{\mathbf{Z}}(\cdot)) = \left\{(w^{(l)}, \mathbf{z}^{(l)})\right\}_{l=1}^{L}$

Example 3.6. *Let us return to our example where we wish to have a particle representation of*

$$p_{\mathbf{X}|\mathbf{Y}}(\mathbf{x}|\mathbf{y}) = \frac{p_{\mathbf{Y}|\mathbf{X}}(\mathbf{y}|\mathbf{x})p_{\mathbf{X}}(\mathbf{x})}{p_{\mathbf{Y}}(\mathbf{y})}$$

$$\propto p_{\mathbf{Y}|\mathbf{X}}(\mathbf{y}|\mathbf{x})p_{\mathbf{X}}(\mathbf{x}).$$

Let us draw L independent samples from the prior $p_\mathbf{X}(\mathbf{x})$: $\mathbf{x}^{(1)}, \ldots, \mathbf{x}^{(L)}$. The weight of $\mathbf{x}^{(l)}$ is then

$$w^{(l)} \propto \frac{p_{\mathbf{Y}|\mathbf{X}}(\mathbf{y}|\mathbf{x}^{(l)})\, p_\mathbf{X}(\mathbf{x}^{(l)})}{p_\mathbf{X}(\mathbf{x}^{(l)})}$$

$$= p_{\mathbf{Y}|\mathbf{X}}(\mathbf{y}|\mathbf{x}^{(l)}).$$

Hence, after normalization of the weights, $\mathcal{R}_L(p_{\mathbf{X}|\mathbf{Y}}(\cdot|\mathbf{y})) = \{(w^{(l)}, \mathbf{x}^{(l)})\}_{l=1}^{L}$, for a fixed observation $\mathbf{y}$.

3.3.3 Sampling for large-dimensional systems

Importance sampling is useful when the dimensionality of $\mathbf{Z}$ is small. In large-dimensional systems, importance sampling leads to unreliable representations because most of the weights will be close to zero [46]. One of the most important techniques for sampling from large-dimensional distributions is based on Markov chains. This leads to a set of algorithms that fall under the umbrella of Markov-chain Monte Carlo (MCMC) techniques. The MCMC approach dates back to the Second World War, and has been used in a wide variety of engineering applications ever since. The theory behind MCMC is too dense for our purpose, so we will only attempt to develop some intuition of MCMC, without any claim to rigorousness. The interested reader is referred to [42–44, 46–48] for additional information. While many MCMC sampling techniques exist, we will focus only on the well-known Gibbs sampler [49].

In this section, we will describe the concept of Markov chains and invariant distributions. We then show how to sample from a Markov chain. Finally, we will describe the Gibbs sampler. Our focus is on discrete random variables (both for convenience of the exposition and because we will not require MCMC techniques for continuous random variables in this book).

3.3.3.1 Markov chains

Consider a finite domain $\mathcal{Z} = \{\mathbf{a}_1, \ldots, \mathbf{a}_N\}$, containing N elements. A Markov chain is a sequence $\mathbf{Z}_0, \mathbf{Z}_1, \ldots$ of random variables over $\mathcal{Z}$, satisfying the Markov property

$$p_{\mathbf{Z}_n|\mathbf{Z}_0\ldots\mathbf{Z}_{n-1}}(\mathbf{z}_n|\mathbf{z}_0, \ldots, \mathbf{z}_{n-1}) = p_{\mathbf{Z}_n|\mathbf{Z}_{n-1}}(\mathbf{z}_n|\mathbf{z}_{n-1}). \tag{3.39}$$

The set $\mathcal{Z}$ is known as the *state space*, while $\mathbf{Z}_n$ is the *state* at time instant n, $n \geq 0$. A Markov chain is homogeneous when the transition probabilities $p_{\mathbf{Z}_n|\mathbf{Z}_{n-1}}(\mathbf{z}_n|\mathbf{z}_{n-1})$ do not depend on the time instant n. In a homogeneous Markov chain, we can define a so-called *transition kernel*

$$Q(\mathbf{z}'|\mathbf{z}) = p_{\mathbf{Z}_n|\mathbf{Z}_{n-1}}(\mathbf{z}'|\mathbf{z}). \tag{3.40}$$

We will write the distribution of the state of the system at time n as $p_n(\cdot)$. Suppose that we are given an initial distribution $p_0(\cdot)$, then the distribution of the state at time $n = 1$ is related to the distribution of the state at time 0 by

$$p_1(\mathbf{z}') = \sum_{\mathbf{z} \in \mathcal{Z}} Q(\mathbf{z}'|\mathbf{z}) p_0(\mathbf{z}). \tag{3.41}$$

Let us represent the distribution $p_n(\cdot)$ as an $N \times 1$ column-vector $\mathbf{p}_n$, with $[\mathbf{p}_n]_i = p_n(\mathbf{a}_i)$ and the transition kernel $Q(\mathbf{z}'|\mathbf{z})$ as an $N \times N$ matrix matrix $\mathbf{Q}$ with $[\mathbf{Q}]_{ij} = Q(\mathbf{a}_i|\mathbf{a}_j)$. Note that the elements in $\mathbf{p}_n$ always add up to unity, as do the columns of $\mathbf{Q}$. This implies that $\mathbf{Q}$ will have at least one eigenvalue equal to 1. We then have

$$\mathbf{p}_n = \mathbf{Q}\mathbf{p}_{n-1} \tag{3.42}$$

$$= \mathbf{Q}^n \mathbf{p}_0. \tag{3.43}$$

We say that a distribution $\mathbf{p}$ is *invariant* for a Markov chain when

$$\mathbf{p} = \mathbf{Q}\mathbf{p}. \tag{3.44}$$

In other words, it is a right eigenvector with eigenvalue $+1$. There exists at least one such eigenvector. A distribution $\mathbf{p}$ is *limiting* for a Markov chain when it is invariant and, for every initial distribution $\mathbf{p}_0$, $\mathbf{p} = \mathbf{Q}^n \mathbf{p}_0$, for some finite n.

3.3.3.2   Sampling from Markov chains

Suppose that we have a target distribution $p_{\mathbf{Z}}(\mathbf{z})$ we wish to draw samples from. We write $p_{\mathbf{Z}}(\mathbf{z})$ in vector notation $\mathbf{p}$. Assume that we can find a transition kernel $\mathbf{Q}$ for which $\mathbf{p}$ is a limiting distribution.

We draw $\mathbf{z}^{(0)}$ from some initial distribution $p_0(\cdot)$, then $\mathbf{z}^{(1)}$ from $Q(\cdot|\mathbf{z}^{(0)})$, then $\mathbf{z}^{(2)}$ from $Q(\cdot|\mathbf{z}^{(1)})$, and so forth. After some time (known as the burn-in time), we will eventually draw samples from the target distribution $\mathbf{p}$. The samples are usually not independent. Independent samples (or at least, approximately independent samples) can be obtained by decimating the MCMC samples (i.e., discarding $M - 1$ out of every M samples for some suitably large $M \in \mathbb{N}$).

3.3.3.3   Gibbs sampling

Gibbs sampling is a very attractive technique by which to sample from a distribution $p_{\mathbf{Z}}(\mathbf{z})$, where $\mathbf{Z}$ is a D-dimensional random variable: $\mathbf{Z} = [Z_1, Z_2, \ldots, Z_D]$. For simplicity, let us focus on $D = 3$. It is easily verified that the following relations hold:

$$\underbrace{p_{Z_1 Z_2 Z_3}(z_1', z_2', z_3')}_{p_{\mathbf{Z}}(\mathbf{z}')} = p_{Z_3|Z_1 Z_2}(z_3'|z_1', z_2') \, p_{Z_2|Z_1}(z_2'|z_1') \, p_{Z_1}(z_1')$$

$$= \sum_{z_1 z_2 z_3} \underbrace{p_{Z_1 Z_2 Z_3}(z_1, z_2, z_3)}_{p_{\mathbf{Z}}(\mathbf{z})}$$

$$\times \underbrace{p_{Z_3|Z_1 Z_2}(z_3'|z_1', z_2') p_{Z_2|Z_1 Z_3}(z_2'|z_1', z_3) p_{Z_1|Z_2 Z_3}(z_1'|z_2, z_3)}_{Q(\mathbf{z}'|\mathbf{z})}.$$

In other words, $p_\mathbf{Z}(\mathbf{z})$ is an invariant distribution for the transition kernel

$$Q(z_1', z_2', z_3' | z_1, z_2, z_3) = p_{Z_3 | Z_1 Z_2}(z_3' | z_1', z_2') p_{Z_2 | Z_1 Z_3}(z_2' | z_1', z_3) p_{Z_1 | Z_2 Z_3}(z_1' | z_2, z_3).$$

For a given z_1, z_2, z_3, sampling from the transition kernel is performed as follows:

- sample z_1' from $p_{Z_1 | Z_2 Z_3}(\cdot | z_2, z_3)$,
- sample z_2' from $p_{Z_2 | Z_1 Z_3}(\cdot | z_1', z_3)$,
- sample z_3' from $p_{Z_3 | Z_1 Z_2}(\cdot | z_1' z_2')$.

In other words, we need sample only from the conditional distributions. These one-dimensional conditional distributions are commonly much easier to sample from than the original D-dimensional distribution. Generalization to $D > 3$ is straightforward, and is described in Algorithm 3.2. It turns out that (under some general conditions), $p_\mathbf{Z}(\mathbf{z})$ is a limiting distribution for the Markov chain.

Algorithm 3.2 Gibbs sampler
___
1: initialization: choose an initial state $\mathbf{z}^{(-N_{\text{burn}}-1)}$
2: **for** $l = -N_{\text{burn}}$ to L **do**
3:     **for** $i = 1$ to D **do**
4:         draw $z_i^{(l)} \sim p\left(Z_i \middle| \mathbf{Z}_{1:i-1} = \mathbf{z}_{1:i-1}^{(l)}, \mathbf{Z}_{i+1:D} = \mathbf{z}_{i+1:D}^{(l-1)}\right)$
5:     **end for**
6: **end for**
7: output: $\mathcal{R}_L(p_\mathbf{Z}(\cdot)) = \left\{(1/L, \mathbf{z}^{(l)})\right\}_{l=1}^{L}$

3.4   Main points

In this chapter we have covered some basic ideas from Bayesian estimation theory. We have described the maximum a-posteriori (MAP) and minimum-mean-squared-error (MMSE) estimators. Estimation of both continuous and discrete random variables has been treated. We can now confidently define estimators and formulate the corresponding optimization problems for receivers in digital communications. This will be done in Chapter 7. Since solving such optimization problems is generally intractable, we have briefly covered some Monte Carlo techniques. Our focus was on particle representations, importance sampling, and Gibbs sampling.

Now we will leave communication theory and estimation theory for an even more abstract world, the world of factor graphs.

4 Factor graphs and the sum–product algorithm

4.1 A brief history of factor graphs

Factor graphs are a way to represent graphically the factorization of a function. The sum–product algorithm is an algorithm that computes marginals of that function by passing messages on its factor graph. The term and concept *factor graph* were originally introduced by Brendan Frey in the late 1990s, as a way to capture structure in statistical inference problems. They form an attractive alternative to Bayesian belief networks and Markov random fields, which have been around for many years. At the same time, factor graphs are strongly linked with coding theory, as a way to represent error-correcting codes graphically. They generalize concepts such Tanner graphs and trellises, which are the usual way to depict codes. The whole idea of seeing a code as a graph can be traced back to 1963, when Robert Gallager described low-density parity-check (LDPC) codes in his visionary PhD thesis at MIT. Although LDPC codes remained largely forgotten until fairly recently, the idea of representing codes on graphs was not, and led to the introduction of the concept *trellis* some ten years later, as well as Tanner graphs in 1981.

To get an idea of how factor graphs came about, let us take a look at the following timeline. It represents a selection of key contributions in the field. I have divided them into three categories: (S) for works on statistical inference, (C) for works on coding theory, and (G) for works linking these two fields.

- **1963**: (C) R. Gallager, "Low density parity check codes," MIT. In his PhD thesis, Gallager describes the relationships between coded bits using a graphical model [50].
- **1971**: (S) F. Spitzer, "Random fields and interacting particle systems," MAA Summer Seminar Notes. This is probably the first work on Markov random fields [51].
- **1973**: (C) G. D. Forney, "The Viterbi algorithm," *Proceedings of the IEEE*. Here, Forney introduces the concept "trellis" as a way to capture the temporal behavior of a finite-state machine [52].
- **1980**: (S) R. Kindermann and J. Snell, *Markov Random Fields and Their Applications*, American Mathematical Society. This book provides a readable tutorial on Markov random fields [53].
- **1981**: (C) M. Tanner, "A recursive approach to low complexity codes," *IEEE Transactions on Information Theory*. Tanner describes what are now known as Tanner graphs: a bipartite graph representation of codes using subcodes [54].

- **1988**: (S) S. L. Lauritzen and D. J. Spiegelhalter, "Local computations with probabilities on graphical structures and their application to expert systems," *Journal of the Royal Statistical Society*. In this paper, a type of Bayesian network is described, explicitly taking into account causality [55].
- **1988**: (S) J. Pearl, *Probabilistic Reasoning in Intelligent Systems*, Morgan-Kaufmann. This seminal work on probabilistic reasoning and inference describes Bayesian networks as acyclic directed graphs [56].
- **1996**: (G) N. Wiberg, "Codes and decoding on general graphs," PhD Thesis, Linköping University. Wiberg unites Forney's trellises and Tanner graphs with Bayesian belief networks. In this remarkable work, he describes the sum–product algorithm as a message-passing algorithm operating over a graph. His algorithms can be specialized to the Viterbi algorithm, to the LDPC decoding algorithm, and to the turbo-decoding algorithm [57].
- **1998**: (G) B. J. Frey, *Graphical Models for Machine Learning and Digital Communication*, MIT Press. Frey introduces the term factor graph. In this book, he gives an insightful overview of inference on graphical models [5].
- **2000**: (G) S. M. Aji and R.J. McEliece, "The Generalized Distributive Law," *IEEE Transactions on Information Theory*. A general framework for marginalizing a function on the basis of a graphical model is presented. This landmark paper employs a type of graphical model called junction trees [4].
- **2001**: (G) F. R. Kschischang, B. J. Frey, and H.-A. Loeliger, "Factor Graphs and the Sum–product Algorithm," *IEEE Transactions on Information Theory*. A very readable and insightful tutorial on factor graphs. Complementary to the work of Aji and McEliece. In many ways the inspiration for this book [58].
- **2001**: (C) G. D. Forney, "Codes on Graphs: Normal Realizations," *IEEE Transactions on Information Theory*. Forney introduces normal factor graphs [59].
- **2004**: (G) H.-A. Loeliger, "An Introduction to factor graphs," *IEEE Signal Processing Magazine*. Factor graphs are becoming mainstream [60].
- **2005**: (G), J. S. Yedidia, W. T. Freeman, and Y. Weiss, "Constructing free energy approximations and generalized belief propagation algorithms," *IEEE Transactions on Information Theory*. This remarkable paper reveals an important link between free-energy minimization in statistical physics and the sum–product algorithm.

This brief timeline is woefully incomplete. Important contributions in this field were made (and are being made) by M. I. Jordan and M. Wainwright, K. Murphy, R. Kötter, D. MacKay, and many, many others.

At this point, we are ready to embark on our journey into the world of factor graphs. I have chosen to give a bottom-up view of factor graphs. While this is somewhat less elegant than the conventional top-down derivation, I hope that it will provide more insight. This chapter is structured as follows.

- We will start with a basic introduction of factor graphs and the sum–product algorithm in **Section 4.2**. This section is intended to give the reader the possibility to skip the remainder of this chapter and move on to Chapter 5.

- In **Section 4.3** we explain the basics of graphs, factorizations, and factor graphs in detail.
- With this knowledge, we then introduce the sum–product algorithm in **Section 4.4** as an efficient way to compute the marginals of a function. We also show how the sum–product algorithm can be seen as a message-passing algorithm on the corresponding factor graph.
- In **Section 4.5** Forney's normal factor graphs will be introduced. They are equivalent to the more conventional-style factor graph, but have some notational advantages.
- Some of the finer points of factor graphs will be discussed in **Section 4.6**.
- Finally, in **Section 4.7**, the sum–product algorithm is extended to a more general setting, and the max-sum algorithm is derived.

Comment

The reader should be warned that parts of this chapter are somewhat abstract. Sections 4.4.2, 4.4.3, and 4.4.4 are marked with asterisks, and can be skipped without losing any essential information. The primary goal of this chapter is to familiarize the reader with factor graphs and the sum–product algorithm.

4.2 A ten-minute tour of factor graphs

In this section, we will cover the essentials of factor graphs and the sum–product algorithm (SPA). The remainder of this chapter delves deeper into various aspects, but is not strictly necessary to understand the rest of the book.

4.2.1 Factor graphs

Let us dive right in and introduce the concept of (normal) factor graphs. Suppose that we have a function $f\colon \mathcal{X}_1 \times \mathcal{X}_2 \times \cdots \times \mathcal{X}_N \to \mathbb{R}$, which is factorized into K factors,

$$f(x_1, x_2, \ldots, x_N) = \prod_{k=1}^{K} f_k(s_k), \tag{4.1}$$

where $s_k \subseteq \{x_1, \ldots, x_N\}$ is the kth variable subset, and $f_k(\cdot)$ is a real-valued function. We create a factor graph of this factorization as follows. We create a node (drawn as a circle or a square) for every factor, and an edge (drawn as a line) for every variable. We attach a certain edge to a certain node when the corresponding variable appears in the corresponding factor. Since an edge can be attached only to two nodes, we need to take special measures for variables that appear in more than two factors. For a variable (say X_l) that appears in $D > 2$ factors, we introduce a special equality node and D dummy variables $X_l^{(1)}, \ldots, X_l^{(D)}$. For each of the dummy variables we create an edge and attach it to the equality node. The equality node corresponds to the following function (see also

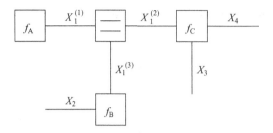

Figure 4.1. A normal factor-graph representation of the factorization $f_A(x_1)f_B(x_1,x_2)f_C(x_1,x_3,x_4)$.

Section 3.3.1 in Chapter 3):

$$\boxminus(x_l^{(1)},\ldots,x_l^{(D)}) = \begin{cases} \prod_{k=1}^{D-1} \delta(x_l^{(k+1)} - x_l^{(k)}) & x_l \text{ continuous,} \\ \prod_{k=1}^{D-1} \mathbb{I}\{x_l^{(k+1)} = x_l^{(k)}\} & x_l \text{ discrete.} \end{cases} \tag{4.2}$$

Example 4.1. *Consider the function $f(x_1,x_2,x_3,x_4,x_5)$ with the following factorization:*

$$f(x_1,x_2,x_3,x_4) = f_A(x_1)f_B(x_1,x_2)f_C(x_1,x_3,x_4).$$

The factor graph is depicted in Fig. 4.1. There is a node for every factor (f_A, f_B, and f_C) as well as an equality node for variable X_1.

4.2.2   Marginals and the sum–product algorithm

Our goal is to compute marginals of the function $f(\cdot)$ with respect to the N variables. The marginal of X_n is given by

$$g_{X_n}(x_n) = \sum_{\sim\{x_n\}} f(x_1,x_2,\ldots,x_N), \tag{4.3}$$

where the notation $\sim\{x_n\}$ refers to all the variables except x_n. In other words, this marginalization requires the summation over all possible values of all variables except x_n. To avoid this cumbersome computation, we make use of the factorization (4.1). The N marginals can be determined in a computationally attractive way by passing messages over the edges of the factor graph, according to the SPA. These messages are functions of the corresponding variables. The details of the SPA can be found in Algorithm 4.1.

In Algorithm 4.1, we treat the equality nodes just like any other factor. It should be noted that, to compute the marginal of a variable that appears in $D > 2$ factors, we can simply compute the marginal of any of the D corresponding dummy variables.

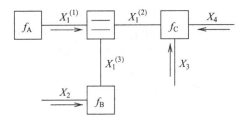

Figure 4.2. The sum–product algorithm (SPA): initialization.

Algorithm 4.1 The sum–product algorithm
1: **Initialization** ∀ nodes f_k connected to a single edge X_m, transmit message
   $\mu_{f_k \to X_m}(x_m) = f_k(x_m), \forall x_m \in X_m$
2: **Initialization** ∀ edges X_n connected to a single node f_l, transmit message
   $\mu_{X_n \to f_l}(x_n) = 1, \forall x_n \in X_n$
3: **repeat**
4:   select a node f_k (say $f_k(x_1, x_2, \ldots, x_D)$, connected to D edges) which has received
     incoming messages $\{\mu_{X_l \to f_k}(x_l)\}$ on at least $D - 1$ edges
5:   compute the outgoing message on the remaining edge on the basis of f_k and the
     incoming messages on the other edges:

$$\mu_{f_k \to X_m}(x_m) = \sum_{\sim\{x_m\}} f_k(x_1, x_2, \ldots, x_D) \prod_{n \neq m} \mu_{X_n \to f_k}(x_n)$$

   $\forall x_m \in X_m$
6: **until** all messages are computed
7: **for** $n = 1$ to N **do**
8:   select any node f_k attached to edge X_n
9:   the marginal of X_n is given by

$$g_{X_n}(x_n) = \mu_{f_k \to X_n}(x_n)\mu_{X_n \to f_k}(x_n)$$

   $\forall x_n \in X_n$
10: **end for**

Example 4.2. *Consider again the function $f(x_1, x_2, x_3, x_4, x_5)$ with the following factorization:*

$$f(x_1, x_2, x_3, x_4) = f_A(x_1)f_B(x_1, x_2)f_C(x_1, x_3, x_4).$$

The factor graph is depicted in Fig. 4.1. Assume that the variables can take on only the values $+1$ and -1, so that $X_k = \{-1, +1\}$, $k = 1, 2, 3, 4$. This implies that all messages can be represented by vectors with two elements: the value of the message in -1 and the value of the message in $+1$. The SPA proceeds as follows:

(1) *We initialize with the message* $\mu_{f_A \to X_1^{(1)}}(x_1) = f_A(x_1)$, *the message* $\mu_{X_2 \to f_B}(x_2) = 1$, *the message* $\mu_{X_3 \to f_C}(x_3) = 1$, *and the message* $\mu_{X_4 \to f_C}(x_4) = 1$. *These four messages are shown in Fig. 4.2. Observe that the messages can be written in two ways. For instance,* $\mu_{f_A \to X_1^{(1)}}(x_1)$ *can be written as* $\mu_{X_1^{(1)} \to \boxminus}(x_1)$.

(2) *We can now compute that the message* $\mu_{f_B \to X_1^{(3)}}(x_1)$ *is*

$$\mu_{f_B \to X_1^{(3)}}(x_1) = \sum_{x_2} f_B(x_1, x_2) \mu_{X_2 \to f_B}(x_2)$$

$$= f_B(x_1, -1)\mu_{X_2 \to f_B}(-1) + f_B(x_1, 1)\mu_{X_2 \to f_B}(1)$$

for $x_1 \in \{-1, +1\}$. *Similarly,* $\mu_{f_C \to X_1^{(2)}}(x_1)$ *is given by*

$$\mu_{f_C \to X_1^{(2)}}(x_1) = \sum_{x_3, x_4} f_C(x_1, x_3, x_4) \mu_{X_3 \to f_C}(x_3) \mu_{X_4 \to f_C}(x_4)$$

for $x_1 \in \{-1, +1\}$. *These two messages are depicted in Fig. 4.3.*

(3) *In the third phase, we can compute all outgoing messages of the equality node:*

$$\mu_{\boxminus \to X_1^{(3)}}(x_1) = \sum_{x_1^{(1)}, x_1^{(2)}} \boxminus (x_1^{(1)}, x_1^{(2)}, x_1) \mu_{X_1^{(1)} \to \boxminus}\left(x_1^{(1)}\right) \mu_{X_1^{(2)} \to \boxminus}\left(x_1^{(2)}\right)$$

$$= \mu_{X_1^{(1)} \to \boxminus}(x_1) \mu_{X_1^{(2)} \to \boxminus}(x_1)$$

and, similarly,

$$\mu_{\boxminus \to X_1^{(2)}}(x_1) = \mu_{X_1^{(1)} \to \boxminus}(x_1) \mu_{X_1^{(3)} \to \boxminus}(x_1)$$

and

$$\mu_{\boxminus \to X_1^{(1)}}(x_1) = \mu_{X_1^{(2)} \to \boxminus}(x_1) \mu_{X_1^{(3)} \to \boxminus}(x_1).$$

We see that, for equality nodes, an outgoing message over a certain edge is simply the pointwise multiplication of the incoming messages on all other edges. The three messages are depicted in Fig. 4.4.

(4) *In the fourth phase, we compute the messages* $\mu_{f_C \to X_3}(x_3)$, $\mu_{f_C \to X_4}(x_4)$, *and* $\mu_{f_B \to X_2}(x_2)$, *as shown in Fig. 4.5.*

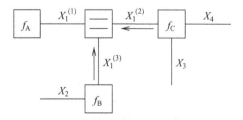

Figure 4.3. The sum–product algorithm (SPA): message computation.

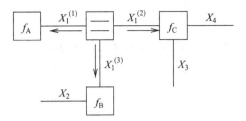

Figure 4.4. The sum–product algorithm (SPA): message computation.

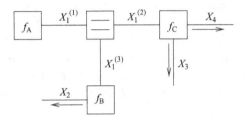

Figure 4.5. The sum–product algorithm (SPA): message computation.

(5) *At this point all the messages have been computed and the SPA terminates. We can now compute the marginals of the four variables. For instance*

$$g_{X_3}(x_3) = \mu_{f_C \to X_3}(x_3)\mu_{X_3 \to f_C}(x_3)$$

for $x_3 \in \{-1,+1\}$ and

$$g_{X_1}(x_1) = \mu_{X_1^{(k)} \to \boxminus}(x_1)\mu_{\boxminus \to X_1^{(k)}}(x_1)$$

for $x_1 \in \{-1,+1\}$, where we can select any $k \in \{1,2,3\}$.
Let us focus on X_3 and verify that the correct marginals are computed. Multiplying the two messages on edge X_3 gives us

$$\mu_{f_C \to X_3}(x_3)\underbrace{\mu_{X_3 \to f_C}(x_3)}_{=1} = \mu_{f_C \to X_3}(x_3)$$

$$= \sum_{x_1,x_4} f_C(x_1,x_3,x_4)\mu_{X_1^{(2)} \to f_C}(x_1)\underbrace{\mu_{X_4 \to f_C}(x_4)}_{=1},$$

where

$$\mu_{X_1^{(2)} \to f_C}(x_1) = \mu_{X_1^{(1)} \to \boxminus}(x_1)\mu_{X_1^{(3)} \to \boxminus}(x_1)$$

$$= f_A(x_1)\sum_{x_2} f_B(x_1,x_2)\underbrace{\mu_{X_2 \to f_B}(x_2)}_{=1},$$

so that

$$\mu_{f_C \to X_3}(x_3)\mu_{X_3 \to f_C}(x_3) = \sum_{x_1, x_4} f_C(x_1, x_3, x_4) f_A(x_1) \sum_{x_2} f_B(x_1, x_2)$$

$$= g_{X_3}(x_3).$$

Comments

- It can be shown that the SPA computes the correct marginals when the factor graph has no cycles (in other words, when the factor graph is a tree). When cycles are present, the SPA runs into problems: due to the cyclic dependencies, the SPA has no natural initialization or termination. Initialization can be forced by introducing artificial messages, while termination can be achieved by simply aborting the SPA after some time. Unfortunately, the SPA is no longer guaranteed to give the correct marginals (see Section 4.6.6).
- The notions of "sum" and "product" can be generalized as long as certain conditions are satisfied. This leads to algorithms such as the max-sum algorithm (see Section 4.7).

At this point the reader can move on to the next chapter. The remainder of this chapter is devoted to the intricacies of factor graphs and the sum–product algorithm and is intended as background information.

4.3   Graphs, factors, and factor graphs

As the title may suggest, this section consists of three parts: we begin with a brief revision of some basic graph theory. This is followed by the introduction of functions and their factorizations. We end by combining these, resulting in factor-graph representations of factorizations.

4.3.1   Some basic graph theory

DEFINITION 4.1 (Graph). *A graph is a pair $G = (V, E)$ of sets such that $E \subseteq V \times V$. The elements in E are (unordered) two-element subsets of V; V represents the set of* vertices[1] *of the graph, while E is the set of* edges. *Graphs are usually depicted by drawing a point, circle or square for every vertex $v \in V$, and a line for every edge $e \in E$ connecting the two corresponding circles. We will restrict our attention to graphs for which there is at most one edge between any two vertices.*

DEFINITION 4.2 (Adjacency). *Given an edge $e \in E$, there are two vertices v_1, v_2 such that $e = (v_1, v_2)$. We say that v_1 and v_2 are adjacent to one another. The set of vertices adjacent to vertex v is denoted $\mathcal{A}(v)$.*

[1] Most of the time, I will use the term "node" instead of "vertex." My apologies to graph theorists.

DEFINITION 4.3 (Degree).   *The degree of a vertex is the number of vertices adjacent to it.*

DEFINITION 4.4 (Incidence).   *Given an edge $e \in E$, there are two vertices v_1, v_2 such that $e = (v_1, v_2)$. We say that v_1 and v_2 are incident with e. At the same time, e is said to be incident with v_1 and v_2. The set of vertices incident with an edge e is denoted $\mathcal{I}(e)$. Similarly, the set of edges incident with a vertex v is denoted $\mathcal{I}(v)$. Clearly, $v \in \mathcal{I}(e) \iff e \in \mathcal{I}(v)$.*

DEFINITION 4.5 (Bipartite graph).   *A bipartite graph is a graph in which the set of vertices V can be partitioned into two subsets (classes), V_X and V_Y, such that $E \subseteq V_X \times V_Y$. In other words, there are no edges between vertices of the same class.*

Example 4.3.   *In Fig. 4.6 we show a graph $G = (V, E)$ with vertex set $V = \{v_1, v_2, \ldots, v_7\}$ and edge set $E = \{(v_1, v_5), (v_1, v_7), (v_6, v_5), (v_6, v_7), (v_4, v_3), (v_4, v_2)\}$. There are two classes of vertices, $V_X = \{v_1, v_4, v_6\}$ (drawn as circles) and $V_X = \{v_2, v_3, v_5, v_7\}$ (drawn as squares). Since there are only edges between vertices of different classes, the graph is bipartite. Clearly, v_1 and v_5 are adjacent to the edge (v_1, v_5), the degree of vertex v_1 is 2, and $\mathcal{I}(v_1) = \{(v_1, v_5), (v_1, v_7)\}$. In Fig. 4.7 an equivalent graph, obtained by moving the vertices is shown. Observe that the figure represents the same graph G as in Fig. 4.6.*

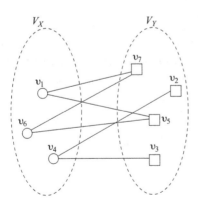

Figure 4.6. A bipartite graph with seven vertices and six edges.

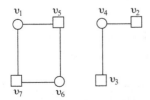

Figure 4.7. A different depiction of the same bipartite graph as in Fig. 4.6.

DEFINITION 4.6 (Path).   *A path is a sequence of vertices, with each couple of successive vertices belonging to the set of edges E. A path of length N consists of N + 1 vertices and is written as $p = v_{k_1}, v_{k_2}, \ldots, v_{k_{N+1}}$, with $\left(v_{k_l}, v_{k_{l+1}}\right) \in E$.*

DEFINITION 4.7 (Cycle).   *Given a path p of length N (N > 1) in a graph G, a cycle in G can be formed when $\left(v_{k_{N+1}}, v_{k_1}\right) \in E$, by adding v_1 to the path. The length of the cycle is N + 1. A graph without cycles is called acyclic.*

DEFINITION 4.8 (Connected graph).   *A graph G is said to be connected if there exists a path between any two vertices.*

DEFINITION 4.9 (Forests, trees, and leaves).   *A graph without cycles is called a forest. A connected graph without cycles is a called a tree. Vertices of degree 1 are called leaves.*

Example 4.4.   *The graph in Fig. 4.7 is clearly disconnected: there exists no path between vertex v_4 and v_1. There is a cycle between vertices v_1, v_5, v_6, and v_7. The graph is not a tree, nor is it a forest. In Fig. 4.8 we show a connected graph without cycles (a tree). The graph is connected since there is a path between any two vertices. A path ($p = v_1, v_3, v_2$) between v_1 and v_2 is shown in bold. In Fig. 4.9, a graph with a cycle is shown. This graph is again connected, but it is not a tree. However, we can still consider v_1 as a leaf-vertex, since it has degree 1.*

4.3.2   Functions, factorizations, and marginals

In this section we will introduce the concept of acyclic factorizations of a function. Consider a function $f : \mathcal{X}_1 \times \mathcal{X}_2 \times \cdots \times \mathcal{X}_N \to \mathbb{R}$. We will write the names of the N variables in capitals and any instantiation of these variables by small letters. The function (the abstract entity) is written as

$$f(X_1, X_2, \ldots, X_N) \tag{4.4}$$

and the function evaluated in $X_1 = x_1, \ldots, X_N = x_N$ as

$$f(X_1 = x_1, X_2 = x_2, \ldots, X_N = x_N). \tag{4.5}$$

For notational convenience we will usually write (4.5) as $f(x_1, x_2, \ldots, x_N)$. Now consider a factorization of the function (4.4) into K factors:

$$f(X_1, X_2, \ldots, X_N) = \prod_{k=1}^{K} f_k(S_k), \tag{4.6}$$

where $S_k \subseteq \{X_1, \ldots, X_N\}$ is the kth variable subset, and $f_k(\cdot)$ is a real-valued function. A given function $f(\cdot)$ may have many such factorizations. There exists always a trivial factorization with one factor ($f(\cdot)$ itself) and $S_1 = \{X_1, \ldots, X_N\}$.

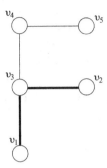

Figure 4.8. A graph that is a tree. The path $p = v_1, v_3, v_2$ of length 2 is shown in bold.

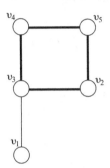

Figure 4.9. A graph that is not a tree. The cycle of length 4 is shown in bold.

DEFINITION 4.10 (Marginal). *The marginal of $f(\cdot)$ with respect to variable X_n (often simply called the marginal of X_n) is a function from $\mathcal{X}_n$ to $\mathbb{R}$ and is denoted by $g_{X_n}(X_n)$. It is obtained by summing out all other variables. More specifically, the marginal evaluated in $x_n \in \mathcal{X}_n$ is given by*

$$g_{X_n}(x_n) = \sum_{x_1 \in \mathcal{X}_1} \cdots \sum_{x_{n-1} \in \mathcal{X}_{n-1}} \sum_{x_{n+1} \in \mathcal{X}_{n+1}} \cdots \sum_{x_N \in \mathcal{X}_N} f(X_1 = x_1, \ldots, X_n = x_n, \ldots, X_N = x_N).$$

$$(4.7)$$

For notational convenience, we will often use the following shorthand

$$g_{X_n}(x_n) = \sum_{\sim\{x_n\}} f(x_1, x_2, \ldots, x_N) \tag{4.8}$$

$$= \sum_{\mathbf{x}:x_n} f(\mathbf{x}), \tag{4.9}$$

where $\sim \{x_n\}$ denotes the set of all possible values of all variables, except X_n which is fixed to $X_n = x_n$. Similarly, $\mathbf{x} : x_n$ denotes all possible sequences $\mathbf{x} = [x_1, \ldots, x_N]$ with nth component equal to x_n.

Example 4.5. *Consider the following function of $N = 5$ variables:*

$$f(X_1, X_2, X_3, X_4, X_5) = f_1(X_1, X_3)f_2(X_2, X_3)f_3(X_1, X_2)f_4(X_2, X_4) \tag{4.10}$$

with $S_1 = \{X_1, X_3\}$, $S_2 = \{X_2, X_3\}$, $S_3 = \{X_1, X_2\}$, and $S_4 = \{X_2, X_4\}$. The marginal with respect to X_1 is given by

$$g_{X_1}(x_1) = \sum_{\sim\{x_1\}} f(X_1, X_2, X_3, X_4, X_5)$$

$$= \sum_{x_2, x_3, x_4, x_5} f_1(X_1, X_3)f_2(X_2, X_3)f_3(X_1, X_2)f_4(X_2, X_4).$$

Factorizations are not unique: by simply lumping together the first three factors, we can also have another factorization:

$$f(X_1, X_2, X_3, X_4, X_5) = \tilde{f}_1(X_1, X_2, X_3)f_4(X_2, X_4) \tag{4.11}$$

with $\tilde{S}_1 = \{X_1, X_2, X_3\}$ and $\tilde{S}_2 = \{X_2, X_4\}$.

DEFINITION 4.11 (Acyclic factorization). *A factorization $\prod_{k=1}^{K} f_k(S_k)$ of a function $f(X_1, X_2, \ldots, X_N)$ contains a cycle of length $L \geq 2$ when there exists a list of L distinct couples $\{X_{i_1}, X_{i_2}\}, \{X_{i_2}, X_{i_3}\}, \ldots, \{X_{i_L}, X_{i_1}\}$, and L distinct variable subsets $S_{k_1}, \ldots, S_{k_L}$, such that $\{X_{i_l}, X_{i_{l+1}}\} \subseteq S_{k_l}$, for all $l \in \{1, \ldots, L\}$. A factorization of a function is acyclic when it contains no cycles of any length $L \geq 2$.*

Example 4.6. *The factorization (4.10) of $f(X_1, X_2, X_3, X_4, X_5)$ has a cycle of length $L = 3$: $\{X_1, X_3\} \subseteq S_1$, $\{X_3, X_2\} \subseteq S_2$, $\{X_2, X_1\} \subseteq S_3$. On the other hand, the factorization (4.11) of the same function has no cycles; this factorization is acyclic.*

DEFINITION 4.12 (Connected factorization). *A factorization $\prod_{k=1}^{K} f_k(S_k)$ of a function $f(X_1, X_2, \ldots, X_N)$ is said to be disconnected when we can group factors $\prod_{k=1}^{K} f_k(S_k) = f_A(S_A)f_B(S_B)$ such that $S_A \cap S_B = \phi$. When no such grouping is possible, the factorization is said to be connected.*

Example 4.7. *Consider a function $f(X_1, X_2, X_3, X_4, X_5)$. Suppose that this function can be factored into three factors as follows:*

$$f(X_1, X_2, X_3, X_4, X_5) = f_1(X_1, X_2)f_2(X_3, X_4)f_3(X_4, X_5). \tag{4.12}$$

This factorization is disconnected since we can group factors into $f_A(X_1, X_2) = f_1(X_1, X_2)$ and $f_B(X_3, X_4, X_5) = f_2(X_3, X_4)f_3(X_4, X_5)$ with $\{X_1, X_2\} \cap \{X_3, X_4, X_5\} = \phi$.

We can obtain a connected factorization as follows: we introduce $\tilde{f}(X_1, X_2, X_3, X_4) = f_1(X_1, X_2)f_2(X_3, X_4)$, then we have a factorization of f into two factors:

$$f(X_1, X_2, X_3, X_4, X_5) = \tilde{f}(X_1, X_2, X_3, X_4)f_3(X_4, X_5).$$

This factorization is connected.

All these definitions are rather tedious. In the next section, we will consider factorizations from a graphical point of view. Concepts such as acyclic and connected factorizations will then have obvious graphical counterparts.

4.3.3 Factor graphs

Factor graphs are a type of graph that represents the factorization of a function. Before we formally describe factor graphs, it is useful to introduce the concept of neighbors and neighborhoods.

DEFINITION 4.13 (Neighborhood). *For a given factorization, we associate a neighborhood with each factor and with each variable. For a factor f_k, the neighborhood of f_k, denoted by $\mathcal{N}(f_k)$, is the set of variables that appear in f_k. Similarly, for a variable X_n, the neighborhood of X_n, denoted by $\mathcal{N}(X_n)$, is the set of factors that have X_n as a variable. Clearly, $X_n \in \mathcal{N}(f_k) \iff f_k \in \mathcal{N}(X_n)$.*

Example 4.8. *Revisiting the factorization (4.10) of $f(X_1, X_2, X_3, X_4, X_5)$, we find that $\mathcal{N}(X_1) = \{f_1, f_3\}$ and that $\mathcal{N}(f_2) = \{X_2, X_3\}$.*

DEFINITION 4.15 (Factor graph). *Given a factorization of a function*

$$f(X_1, X_2, \ldots, X_N) = \prod_{k=1}^{K} f_k(S_k), \tag{4.13}$$

the corresponding factor graph $G = (V, E)$ is a bipartite graph, created as follows:

- *for every variable X_n, we create a vertex (a variable node): $X_n \in V$;*
- *for every factor f_k, we create a vertex (a function node): $f_k \in V$; and*
- *for factor f_k, and every variable $X_n \in \mathcal{N}(f_k)$, we create an edge $e = (X_n, f_k) \in E$.*

The definition of neighborhood is equivalent to the graph-theoretic notion of adjacency: when a variable X_n appears in a factor f_k, then $X_n \in \mathcal{N}(f_k)$ and $f_k \in \mathcal{N}(X_n)$. In terms of the factor graph, the X_n-node will be adjacent to the f_k-node, so we may write (with a slight abuse of notation) $X_n \in \mathcal{A}(f_k)$ and $f_k \in \mathcal{A}(X_n)$. In the remainder of this chapter, we will not distinguish between a factor f_k in a factorization and the corresponding vertex/node in the factor graph of that factorization. The same goes for variables X_k and variable nodes.

There is a one-to-one mapping between factorizations and factor graphs. For a given factorization, the factor graph is generally very easy to draw. Given a factor graph, various properties of the factorization can be obtained by simply looking at the graph. The concepts of connectedness and cyclic factorizations translate in a natural way to connected factor graphs and cycles. When we see that a factor graph is a tree, we know that it corresponds to a connected acyclic factorization.

This brings us to a most important point, the key property of factor graphs: they are graphs. This may seem a trivial remark, but it is not. Because they are graphs, it is easy for us humans to reason about them and understand their structure, much easier than understanding a bunch of equations.

Example 4.9.   *In the previous section, we have seen several factorizations of functions. Let us recall some of them. First, we return to Eqs. (4.10) and (4.11), which were two factorizations of the same function,*

$$f(X_1, X_2, X_3, X_4, X_5) = f_1(X_1, X_3)f_2(X_2, X_3)f_3(X_1, X_2)f_4(X_2, X_4) \tag{4.14}$$

and

$$f(X_1, X_2, X_3, X_4, X_5) = \tilde{f}_1(X_1, X_2, X_3)f_4(X_2, X_4). \tag{4.15}$$

We have shown that (4.14) is a cyclic factorization, whereas (4.15) is acyclic. The corresponding factor graphs are shown in Figs. 4.10 and Fig. 4.11. Observe the following: the factor graph of the cyclic factorization (4.14) is cyclic, whereas the factor graph of the acyclic factorization (4.15) is acyclic.

Example 4.10.   *In Eq. (4.12), we considered the following factorization of another function:*

$$f(X_1, X_2, X_3, X_4, X_5) = f_1(X_1, X_2)f_2(X_3, X_4)f_3(X_4, X_5). \tag{4.16}$$

We have shown that this factorization was disconnected. The factor graph of (4.16) is shown in Fig. 4.12. Observe that the factor graph is disconnected.

Example 4.11 (Running example).   *We will use the following example throughout this chapter: consider a function of N = 7 variables that can be factored into K = 6 factors as follows:*

$$f(X_1, X_2, X_3, X_4, X_5, X_6, X_7) = f_1(X_1, X_2, X_3)f_2(X_1, X_4)f_3(X_1, X_6, X_7)$$

$$\times f_4(X_4, X_5)f_5(X_4)f_6(X_2). \tag{4.17}$$

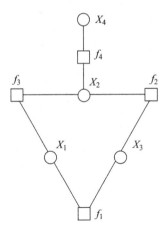

Figure 4.10. A factor graph of (4.14). The graph contains a cycle.

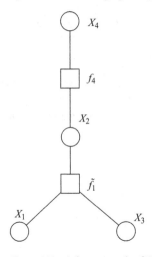

Figure 4.11. A factor graph of (4.15). The graph is a tree.

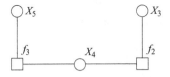

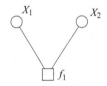

Figure 4.12. A factor graph of (4.16). The graph is disconnected, but acyclic. It is a forest.

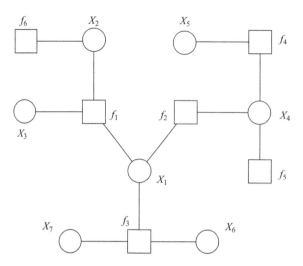

Figure 4.13. A factor graph of the factorization $f_1(X_1, X_2, X_3)f_2(X_1, X_4)f_3(X_1, X_6, X_7) \times f_4(X_4, X_5)f_5(X_4)f_6(X_2)$.

It is readily verified that this factorization is connected and acyclic. The corresponding factor graph is shown in Fig. 4.13. From the graph, it is immediately obvious that $\{X_1, X_2, X_3\} = \mathcal{N}(f_1)$, that $\{f_1, f_2, f_3\} = \mathcal{N}(X_1)$, and that the graph is connected and acyclic.

4.4   Marginals and the sum–product algorithm

4.4.1   Marginals of connected acyclic factorizations

As we will see in the next chapter, an important problem is the computation of marginals of a function, on the basis of a connected acyclic factorization. Let us start from such a connected acyclic factorization

$$f(X_1, X_2, \ldots, X_N) = \prod_{k=1}^{K} f_k(S_k). \tag{4.18}$$

Suppose that we are interested in computing $g_{X_1}(X_1)$. Computing $g_{X_1}(x_1)$ directly requires an $(N-1)$-fold summation over $\mathcal{X}_2 \times \cdots \times \mathcal{X}_N$ for any $x_1 \in \mathcal{X}_1$. Our goal is to find a way to compute the marginals in a computationally attractive way by exploiting the connected acyclic factorization. The algorithm we will obtain is called the *sum–product algorithm* (SPA). This algorithm describes how marginals can be computed in an automated and efficient way. We will focus on a specific variable (say, X_n), and proceed as follows.

(1) We first re-write the factorization (4.18) explicitly as a function of X_n.

(2) We compute the marginal $g_{X_n}(X_n)$ using a recursive technique.

At every step we will first describe the tedious mathematical intricacies, and then relate these to the fun factor-graph interpretation. We remind the reader that sections marked with an asterisk are slightly abstract and can be skipped at first reading.

4.4.2 Step 1: variable partitioning

4.4.2.1 The math way*

Consider a factor f_k and a variable $X_n \in \mathcal{N}(f_k)$. With this couple (X_n, f_k), we associate a set of variables $S_{f_k}^{(X_n)}$, defined recursively as shown in Algorithm 4.2.

Algorithm 4.2 Variable partitioning: determine $S_{f_k}^{(X_n)}$

1: *input:* X_n and f_k

2: initialization: $S_{f_k}^{(X_n)} = \phi$

3: **for** $X_m \in \mathcal{N}(f_k) \setminus \{X_n\}$ **do**

4:     add X_m to $S_{f_k}^{(X_n)}$

5:     **for** $f_l \in \mathcal{N}(X_m) \setminus \{f_k\}$ **do**

6:         determine $S_{f_l}^{(X_m)}$ and add to set $S_{f_k}^{(X_n)}$

7:     **end for**

8: **end for**

9: *output:* $S_{f_l}^{(X_m)}$

Intuitively, this algorithm corresponds to adding all variables that appear in f_k (except X_n) to the set $S_{f_k}^{(X_n)}$. Then we look at all the functions (except f_k) where these variables appear as an argument. We then add all the variables of those functions to $S_{f_k}^{(X_n)}$, and so forth.

Example 4.12 (Running example).   *We return to our example, with*

$$f(X_1, X_2, X_3, X_4, X_5, X_6, X_7) = f_1(X_1, X_2, X_3) f_2(X_1, X_4) f_3(X_1, X_6, X_7)$$
$$\times f_4(X_4, X_5) f_5(X_4) f_6(X_2).$$

Let us determine $S_{f_1}^{(X_1)}$. We need to find the variables (except X_1) that appear in f_1, then find all functions of those variables (except f_1) and add all the variables of those functions, etc.

- *Clearly, $\mathcal{N}(f_1) \setminus \{X_1\} = \{X_2, X_3\}$, so X_2 and X_3 belong in $S_{f_1}^{(X_1)}$.*
- *$\mathcal{N}(X_3) \setminus \{f_1\} = \phi$, so this branch of the recursion ends.*
- *$\mathcal{N}(X_2) \setminus \{f_1\} = \{f_6\}$, so we need to add $S_{f_6}^{(X_2)}$ to $S_{f_1}^{(X_1)}$. Since $\mathcal{N}(f_6) \setminus \{X_2\} = \phi$, $S_{f_6}^{(X_2)} = \phi$, so this branch of the recursion ends.*

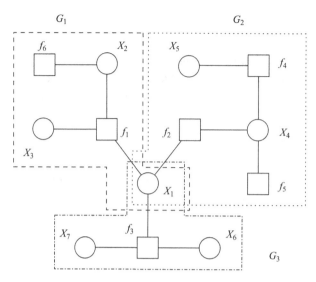

Figure 4.14. Factor graph of the factorization
$f_1(X_1,X_2,X_3)f_2(X_1,X_4)f_3(X_1,X_6,X_7)f_4(X_4,X_5)f_5(X_4)f_6(X_2)$, with variable partitioning.

In conclusion, we find that $S_{f_1}^{(X_1)} = \{X_2,X_3\}$. *Similarly,* $S_{f_2}^{(X_1)} = \{X_4,X_5\}$ *and* $S_{f_3}^{(X_1)} = \{X_6,X_7\}$. *We can easily verify that* $S_{f_1}^{(X_1)} \cup S_{f_2}^{(X_1)} \cup S_{f_3}^{(X_1)} \cup \{X_1\} = \{X_1,X_2,\ldots,X_7\}$ *and* $S_{f_{k'}}^{(X_1)} \cap S_{f_k}^{(X_1)} = \phi$ *for* $k' \neq k$.

4.4.2.2   The factor-graph way

Variable partitioning can be achieved in a very straightforward way using factor graphs. Take a variable node X_n. This variable node is adjacent to $|\mathcal{N}(X_n)|$ function nodes. Removing the node X_n results in $|\mathcal{N}(X_n)|$ factor graphs. Suppose $f_k \in \mathcal{N}(X_n)$, then one of these factor graphs will have node f_k as leaf node. The set of variables in that factor graph is exactly $S_{f_k}^{(X_n)}$.

Example 4.13 (Running example).   *The factor graph of our running example is shown again in Fig. 4.14. We focus on variable node X_1, with $\mathcal{N}(X_1) = \{f_1,f_2,f_3\}$. Removing the node X_1 results in three factor graphs. The factor graph containing f_1 has as variable nodes $S_{f_1}^{(X_1)} = \{X_2,X_3\}$. Let us add the node X_1 again to each of the three factor graphs. The three resulting factor graphs are denoted G_1, G_2, and G_3 in Fig. 4.14. The variables in G_1 are (apart from X_1) $S_{f_1}^{(X_1)} = \{X_2,X_3\}$. The variables in G_2 are (apart from X_1) $S_{f_2}^{(X_1)} = \{X_4,X_5\}$. The variables in G_3 are (apart from X_1) $S_{f_3}^{(X_1)} = \{X_6,X_7\}$. Each of these three factor graphs represents the factorization of a function. Let us denote the function represented by G_1 as $h_{f_1}^{(X_1)}(X_1,S_{f_1}^{(X_1)})$, the function represented by G_2 as $h_{f_2}^{(X_1)}(X_1,S_{f_2}^{(X_1)})$,* and the function represented by G_3 as $h_{f_3}^{(X_1)}(X_1, S_{f_3}^{(X_1)})$. Obviously,

$$f(X_1, X_2, X_3, X_4, X_5, X_6, X_7) = \prod_{f_k \in \mathcal{N}(X_1)} h_{f_k}^{(X_1)}(X_1, S_{f_k}^{(X_1)}).$$

Note that we can apply the same technique to any other variable. For instance, considering X_2,

$$f(X_1, X_2, X_3, X_4, X_5, X_6, X_7) = h_{f_6}^{(X_2)}\left(X_2, S_{f_6}^{(X_2)}\right) h_{f_1}^{(X_2)}\left(X_2, S_{f_1}^{(X_2)}\right).$$

Observe that

$$h_{f_1}^{(X_1)}\left(X_1, S_{f_1}^{(X_1)}\right) = f_1(X_1, X_2, X_3) h_{f_6}^{(X_2)}\left(X_6, S_{f_6}^{(X_2)}\right).$$

In the next section we will formalize this last observation.

4.4.3 Step 2: grouping factors

4.4.3.1 The math way*

When the factorization is acyclic, it follows that $S_{f_k}^{(X_n)} \cap S_{f_{k'}}^{(X_n)} = \phi$, $k' \neq k$.

When the factorization is connected, it also follows that $\left\{\bigcup_{f_k \in \mathcal{N}(X_n)} S_{f_k}^{(X_n)}\right\} \cup \{X_n\} = \{X_1, X_2, \ldots, X_N\}$. These properties allow us to express $f(X_1, X_2, \ldots, X_N)$ as follows:

$$f(X_1, X_2, \ldots, X_N) = \prod_{f_k \in \mathcal{N}(X_n)} h_{f_k}^{(X_n)}(X_n, S_{f_k}^{(X_n)}), \tag{4.19}$$

where $h_{f_k}^{(X_n)}(X_n, S_{f_k}^{(X_n)})$ is defined recursively as

$$h_{f_k}^{(X_n)}\left(X_n, S_{f_k}^{(X_n)}\right) = f_k(\{X_m\}_{X_m \in \mathcal{N}(f_k)}) \prod_{X_m \in \mathcal{N}(f_k) \setminus \{X_n\}} \left\{ \prod_{f_l \in \mathcal{N}(X_m) \setminus \{f_k\}} h_{f_l}^{(X_m)}\left(X_m, S_{f_l}^{(X_m)}\right) \right\}. \tag{4.20}$$

In (4.20), $\{X_m\}_{X_m \in \mathcal{N}(f_k)}$ represents the set of variables appearing in f_k. The recursion ends when either $\mathcal{N}(f_k) \setminus \{X_n\}$ or $\mathcal{N}(X_m) \setminus \{f_k\}$ are empty.

Example 4.14 (Running example). *Returning to our example from Section 4.4.2, we can now express the function $f(\cdot)$ as the product of three factors:*

$$f(X_1, X_2, X_3, X_4, X_5, X_6, X_7) = h_{f_1}^{(X_1)}\left(X_1, S_{f_1}^{(X_1)}\right) h_{f_2}^{(X_1)}\left(X_1, S_{f_2}^{(X_1)}\right) h_{f_3}^{(X_1)}\left(X_1, S_{f_3}^{(X_1)}\right)$$

with

$$h_{f_1}^{(X_1)}(X_1, S_{f_1}^{(X_1)}) = f_1(X_1, X_2, X_3) \underbrace{h_{f_6}^{(X_2)}(X_2, S_{f_6}^{(X_2)})}_{=f_6(X_2)},$$

$$h_{f_2}^{(X_1)}(X_1, S_{f_2}^{(X_1)}) = f_2(X_1, X_4) \underbrace{h_{f_4}^{(X_4)}(X_4, S_{f_4}^{(X_4)})}_{=f_4(X_4, X_5)} \underbrace{h_{f_5}^{(X_4)}(X_4, S_{f_5}^{(X_4)})}_{=f_5(X_4)},$$

$$h_{f_3}^{(X_1)}(X_1, S_{f_3}^{(X_1)}) = f_3(X_1, X_6, X_7).$$

Now, while the above equations are undoubtedly true, they are not very insightful. Let us see whether factor graphs can help.

4.4.3.2   The factor-graph way

We start again from a factor graph of a factorization of $f(\cdot)$. When we focus on a particular variable node (say X_n), we can break up the factor graph into $|\mathcal{N}(X_n)|$ trees that have the node X_n as leaf node. Each of these trees is of course a factor graph of a function. Take one of those trees. Then X_n will be adjacent to exactly one function node, say f_k. The tree is a factor graph of the function $h_{f_k}^{(X_n)}(X_n, S_{f_k}^{(X_n)})$. Let us consider a generic example (see Fig. 4.15). Clearly, $\mathcal{N}(X_n) = \{f_1, f_2, f_3\}$. The factor graphs of the functions $h_{f_3}^{(X_n)}\left(X_n, S_{f_3}^{(X_n)}\right)$ and $h_{f_2}^{(X_n)}\left(X_n, S_{f_2}^{(X_n)}\right)$ are marked in dashed boxes. They are easily found

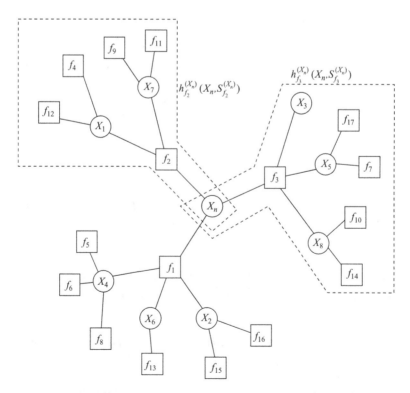

Figure 4.15. Variable partitioning and factor grouping on a factor graph.

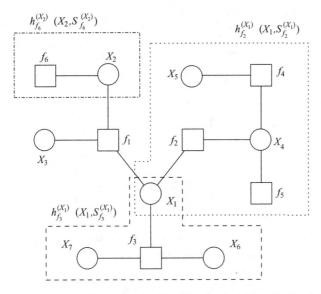

Figure 4.16. The factor graph of the factorization $f_1(X_1, X_2, X_3)f_2(X_1, X_4)f_3(X_1, X_6, X_7) \times f_4(X_4, X_5)f_5(X_4)f_6(X_2)$, with variable partitioning and factor grouping. The factor graph of the function $h_{f_k}^{(X_n)}(X_n, S_{f_k}^{(X_n)})$ is the part of the graph behind node f_k, from the viewpoint of node X_n.

on sight by considering the tree behind the functions f_2 and f_3, from the viewpoint of node X_n. The recursive relation (4.20) can be obtained immediately from the factor graph.

Example 4.15 (Running example). *The factor graph of our running example is shown again in Fig. 4.16. The graphs G_2 and G_3 from Fig. 4.14 correspond to the the functions $h_{f_2}^{(X_1)}(X_1, S_{f_2}^{(X_1)})$ and $h_{f_3}^{(X_1)}(X_1, S_{f_3}^{(X_1)})$ in Fig. 4.16 (marked in dashed boxes). Also depicted in Fig. 4.16 is the function $h_{f_6}^{(X_2)}(X_2, S_{f_6}^{(X_2)})$ (note that $S_{f_6}^{(X_2)}$ is the empty set). We see that $h_{f_1}^{(X_1)}(X_1, S_{f_1}^{(X_1)}) = f_1(X_1, X_2, X_3)h_{f_6}^{(X_2)}(X_2, S_{f_6}^{(X_2)})$.*

4.4.4 Step 3: computing marginals

4.4.4.1 A key observation

The marginal $g_{X_n}(X_n)$, evaluated in $x_n \in \mathcal{X}_n$, is given by

$$g_{X_n}(x_n) = \sum_{\sim\{x_n\}} f(x_1, x_2, \ldots, x_N). \tag{4.21}$$

Substituting (4.19) into this gives us

$$g_{X_n}(x_n) = \sum_{\sim\{x_n\}} \left\{ \prod_{f_k \in \mathcal{N}(X_n)} h_{f_k}^{(X_n)}(x_n, s_{f_k}^{(X_n)}) \right\} \tag{4.22}$$

$$= \prod_{f_k \in \mathcal{N}(X_n)} \left\{ \sum_{s_{f_k}^{(X_n)}} h_{f_k}^{(X_n)}(x_n, s_{f_k}^{(X_n)}) \right\} \tag{4.23}$$

$$= \prod_{f_k \in \mathcal{N}(X_n)} \left\{ \sum_{\sim\{x_n\}} h_{f_k}^{(X_n)}(x_n, s_{f_k}^{(X_n)}) \right\}. \tag{4.24}$$

The first transition is due to the acyclic factorization: the sets $\{S_{f_k}^{(X_n)}\}_{f_k \in \mathcal{N}(X_n)}$ are non-overlapping. Ponder this for a few moments. This is a non-trivial result.

Example 4.16 (Running example).   *From Section 4.4.3, we know that the function in our running example can be written as*

$$f(X_1, X_2, X_3, X_4, X_5, X_6, X_7) = h_{f_1}^{(X_1)}(S_{f_1}^{(X_1)}, X_1) h_{f_2}^{(X_1)}(S_{f_2}^{(X_1)}, X_1) h_{f_3}^{(X_1)}(S_{f_3}^{(X_1)}, X_1).$$

The marginal of X_1 is given by

$$g_{X_1}(x_1) = \sum_{\sim\{x_1\}} \left\{ h_{f_1}^{(X_1)}\left(s_{f_1}^{(X_1)}, x_1\right) h_{f_2}^{(X_1)}\left(s_{f_2}^{(X_1)}, x_1\right) h_{f_3}^{(X_1)}\left(s_{f_3}^{(X_1)}, x_1\right) \right\}$$

$$= \sum_{s_{f_1}^{(X_1)}, s_{f_2}^{(X_1)}, s_{f_3}^{(X_1)}} \left\{ h_{f_1}^{(X_1)}\left(s_{f_1}^{(X_1)}, x_1\right) h_{f_2}^{(X_1)}\left(s_{f_2}^{(X_1)}, x_1\right) h_{f_3}^{(X_1)}\left(s_{f_3}^{(X_1)}, x_1\right) \right\}$$

$$= \left\{ \sum_{s_{f_1}^{(X_1)}} h_{f_1}^{(X_1)}\left(s_{f_1}^{(X_1)}, x_1\right) \right\} \left\{ \sum_{s_{f_2}^{(X_1)}} h_{f_2}^{(X_1)}\left(s_{f_2}^{(X_1)}, x_1\right) \right\} \left\{ \sum_{s_{f_3}^{(X_1)}} h_{f_3}^{(X_1)}\left(s_{f_3}^{(X_1)}, x_1\right) \right\}$$

$$= \prod_{f_k \in \mathcal{N}(X_1)} \left\{ \sum_{\sim\{x_1\}} h_{f_k}^{(X_1)}\left(x_1, s_{f_k}^{(X_1)}\right) \right\}.$$

4.4.4.2   Two special functions: the math way*

Let us introduce two special functions of variable X_n: for any $f_k \in \mathcal{N}(X_n)$, we define $\mu_{X_n \to f_k} : \mathcal{X}_n \to \mathbb{R}$ and $\mu_{f_k \to X_n} : \mathcal{X}_n \to \mathbb{R}$ as follows:

$$\mu_{f_k \to X_n}(x_n) = \sum_{\sim\{x_n\}} h_{f_k}^{(X_n)}(x_n, s_{f_k}^{(X_n)}) \tag{4.25}$$

and

$$\mu_{X_n \to f_k}(x_n) = \prod_{f_l \in \mathcal{N}(X_n) \backslash \{f_k\}} \left\{ \sum_{\sim \{x_n\}} h_{f_l}^{(X_n)}\left(x_n, s_{f_l}^{(X_n)}\right) \right\}. \tag{4.26}$$

Note that $\mu_{f_k \to X_n}(x_n) = f_k(x_n)$ when $\mathcal{N}(f_k) = \{X_n\}$, and that $\mu_{X_n \to f_k}(x_n) = 1$ when $\mathcal{N}(X_n) = \{f_k\}$. The functions $\mu_{X_n \to f_k}(X_n)$ and $\mu_{f_k \to X_n}(X_n)$ can be related as follows. First of all, we can substitute (4.25) into (4.26), yielding

$$\mu_{X_n \to f_k}(x_n) = \prod_{f_l \in \mathcal{N}(X_n) \backslash \{f_k\}} \mu_{f_l \to X_n}(x_n). \tag{4.27}$$

Secondly, substituting (4.20) into (4.25) gives us, due to the acyclic factorization,

$\mu_{f_k \to X_n}(x_n)$

$$= \sum_{\sim \{x_n\}} f_k(\{X_m = x_m\}_{X_m \in \mathcal{N}(f_k)}) \prod_{X_m \in \mathcal{N}(f_k) \backslash \{X_n\}} \left\{ \prod_{f_l \in \mathcal{N}(X_m) \backslash \{f_k\}} h_{f_l}^{(X_m)}\left(x_m, s_{f_l}^{(X_m)}\right) \right\}$$

$$= \sum_{\sim \{x_n\}} f_k(\{X_m = x_m\}_{X_m \in \mathcal{N}(f_k)}) \prod_{X_m \in \mathcal{N}(f_k) \backslash \{X_n\}} \left\{ \prod_{f_l \in \mathcal{N}(X_m) \backslash \{f_k\}} \sum_{\sim \{x_m\}} h_{f_l}^{(X_m)}\left(x_m, s_{f_l}^{(X_m)}\right) \right\}$$

$$= \sum_{\sim \{x_n\}} f_k(\{X_m = x_m\}_{X_m \in \mathcal{N}(f_k)}) \prod_{X_m \in \mathcal{N}(f_k) \backslash \{X_n\}} \mu_{X_m \to f_k}(x_m). \tag{4.28}$$

These special functions allow us to express the marginal $g_{X_n}(X_n)$, evaluated in x_n, in the following two ways, thanks to (4.24):

$$g_{X_n}(x_n) = \prod_{f_k \in \mathcal{N}(X_n)} \mu_{f_k \to X_n}(x_n) \tag{4.29}$$

$$= \mu_{f_l \to X_n}(x_n) \mu_{X_n \to f_l}(x_n), \tag{4.30}$$

where the last line is valid for any $f_l \in \mathcal{N}(X_n)$.

4.4.4.3 Two special functions: the factor-graph way

In the context of factor graphs, the functions $\mu_{f_k \to X_n}(X_n)$ and $\mu_{X_n \to f_k}(X_n)$ can be interpreted as *messages*, passed over the edges of the factor graph. Imagine each of the nodes in the factor graph to be a small computer, and each of the edges to be a communication link. There are two types of computers: computers associated with variables (X_n), and computers associated with functions (f_k). A computer can transmit messages over each of its links. We denote the message from variable node X_n to function node $f_k \in \mathcal{N}(X_n)$ by $\mu_{X_n \to f_k}(X_n)$. Similarly, the message from function node f_k to variable node X_n is denoted by $\mu_{f_k \to X_n}(X_n)$.

The *messages are functions of the associated variable*. An outgoing message from X_n to f_k is computed on the basis of incoming messages $\mu_{f_l \to X_n}(X_n), f_l \in \mathcal{N}(X_n) \backslash \{f_k\}$, as described in (4.27). This computation is visualized in Fig. 4.17. Similarly, an outgoing

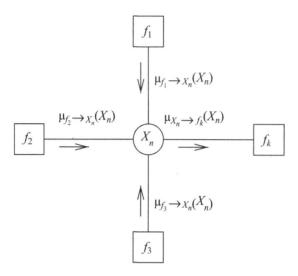

Figure 4.17. The sum–product algorithm (SPA): the message-computation rule from variable node to function node.

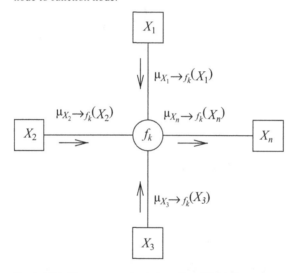

Figure 4.18. The sum–product algorithm (SPA): the message-computation rule from function node to variable node.

message from f_k to X_n is computed on the basis of the function f_k, as well as the incoming messages $\mu_{X_m \to f_k}(X_m)$, with $X_m \in \mathcal{N}(f_k)\backslash\{X_n\}$, as described in (4.28). This computation is visualized in Fig. 4.18.

4.4.5   The sum–product algorithm

Computing marginals based on the functions $\mu_{f_k \to X_n}(X_n)$ and $\mu_{X_n \to f_k}(X_n)$ is known as the *sum–product algorithm* (SPA). The name is due to presence of the summation and product in relation (4.28). The functions $\mu_{f_k \to X_n}(X_n)$ can be computed *recursively* using (4.27) and (4.28). Once all the functions $\mu_{f_k \to X_n}(X_n)$ and $\mu_{X_n \to f_k}(X_n)$ have been computed (i.e., for every X_n and every $f_k \in \mathcal{N}(X_n)$), the marginals of the variables can be found using (4.30).

For convenience, the algorithm will be described starting from the leaves of the factor graph. There are three phases: an initialization phase, during which messages at the leaf nodes are computed; a computation phase, during which all other messages are computed; and a termination phase, during which the marginals for all the variables are determined. The details are as follows.

Initialization
- For every leaf function node f_k, with $\{X_m\} = \mathcal{N}(f_k)$, transmit message to variable node X_m, with $\mu_{f_k \to X_m}(x_m) = f_k(x_m)$, $x_m \in \mathcal{X}_m$.
- For every leaf variable node X_n, with $\{f_l\} = \mathcal{N}(X_n)$, transmit message to function node f_l, with $\mu_{X_n \to f_l}(x_n) = 1$, $\forall x_n \in \mathcal{X}_n$.

Message-computation rules
Perform until all $\mu_{X_n \to f_k}(X_n)$ and all $\mu_{f_k \to X_n}(X_n)$ have been computed. Every message is computed only once.

- For any function node f_k of degree D: when f_k has received incoming messages from $D - 1$ distinct variable nodes $X_n \in \mathcal{N}(f_k)$, then node f_k can transmit an outgoing message $\mu_{f_k \to X_m}(X_m)$ to the remaining variable node X_m, according to (4.28),

$$\mu_{f_k \to X_m}(x_m) = \sum_{\sim\{x_m\}} f_k\left(\{X_n = x_n\}_{X_n \in \mathcal{N}(f_k)}\right) \prod_{X_n \in \mathcal{N}(f_k) \setminus \{X_m\}} \mu_{X_n \to f_k}(x_n). \quad (4.31)$$

The message computation is visualized in Fig. 4.18.
- For any variable node X_n of degree D: when X_n has received incoming messages from $D - 1$ distinct function nodes $f_k \in \mathcal{N}(X_n)$, then node X_n can transmit an outgoing message $\mu_{X_n \to f_l}(X_n)$ to the remaining function node f_l, according to (4.27),

$$\mu_{X_n \to f_l}(x_n) = \prod_{f_k \in \mathcal{N}(X_n) \setminus \{f_l\}} \mu_{f_k \to X_n}(x_n). \quad (4.32)$$

The message computation is visualized in Fig. 4.17.

Termination
- To compute the marginal of X_n in $x_n \in \mathcal{X}_n$, take any $f_k \in \mathcal{N}(X_n)$, then multiply the two messages on the edge (X_n, f_k), so that, according to (4.30),

$$g_{X_n}(x_n) = \mu_{f_k \to X_n}(x_n) \mu_{X_n \to f_k}(x_n). \quad (4.33)$$

Note that the sets $S_{f_l}^{(X_m)}$ and the functions $h_{f_l}^{(X_m)}\left(X_m, S_{f_l}^{(X_m)}\right)$ no longer need to be determined explicitly.

Example 4.17. *Let us return to the running example from Section 4.4.1. We have a function of $N = 7$ variables that can be factored into $K = 6$ factors as follows:*

$$f(X_1, X_2, X_3, X_4, X_5, X_6, X_7) = f_1(X_1, X_2, X_3) f_2(X_1, X_4) f_3(X_1, X_6, X_7)$$
$$\times f_4(X_4, X_5) f_5(X_4) f_6(X_2) \qquad (4.34)$$

with a factor graph shown in Fig. 4.13. Let us now apply the SPA.

- **Step 1** *We first compute messages from leaf nodes. The leaf nodes are X_3, X_5, X_6, X_7 and f_5, f_6. The messages $\mu_{X_3 \to f_1}(X_3)$, $\mu_{X_5 \to f_4}(X_5)$, $\mu_{X_6 \to f_3}(X_6)$, and $\mu_{X_7 \to f_3}(X_7)$ are all equal to 1 over their respective domains, whereas*

$$\mu_{f_6 \to X_2}(x_2) = f_6(x_2),$$
$$\mu_{f_5 \to X_4}(x_4) = f_5(x_4).$$

These messages are depicted in Fig. 4.19.

- **Step 2** *Now we can compute three new messages. For instance, function node f_3 has degree 3 and has received two messages (one from X_6 and one from X_7). It can now compute a message $\mu_{f_3 \to X_1}(X_1)$ to the remaining variable node (X_1) using (4.31):*

$$\mu_{f_3 \to X_1}(x_1) = \sum_{\sim\{x_1\}} f_3(x_1, x_6, x_7) \mu_{X_6 \to f_3}(x_6) \mu_{X_7 \to f_3}(x_7).$$

Similarly, node f_4 can transmit a message $\mu_{f_4 \to X_4}(X_4)$ to X_4, with

$$\mu_{f_4 \to X_4}(x_4) = \sum_{\sim\{x_4\}} f_4(x_4, x_5) \mu_{X_5 \to f_4}(x_5).$$

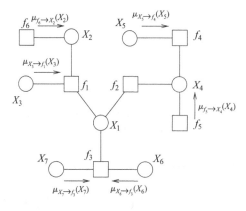

Figure 4.19. Step 1. Messages are sent from the leaf nodes.

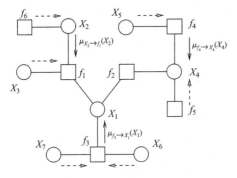

Figure 4.20. Step 2. Message computations based on available messages.

At the same time, variable node X_2 has received an incoming message from f_6, so it can transmit a message $\mu_{X_2 \to f_1}(X_2)$ to f_1 using (4.32):

$$\mu_{X_2 \to f_1}(x_2) = \mu_{f_6 \to X_2}(x_2).$$

These messages are depicted in Fig. 4.20. Messages computed during the previous step are marked using dashed arrows.

- **Step 3** *The third step is depicted in Fig. 4.21. Using (4.31), function node f_1 sends a message $\mu_{f_1 \to X_1}(X_1)$ to variable node X_1, with*

$$\mu_{f_1 \to X_1}(x_1) = \sum_{\sim\{x_1\}} f_1(x_1, x_2, x_3) \mu_{X_2 \to f_1}(x_2) \mu_{X_3 \to f_1}(x_3).$$

At the same time, variable node X_4 is now in a position to send a message $\mu_{X_4 \to f_2}(X_4)$ to function node f_2 using (4.32):

$$\mu_{X_4 \to f_2}(x_4) = \mu_{f_4 \to X_4}(x_4) \mu_{f_5 \to X_4}(x_4).$$

- **Step 4** *The next step is shown in Fig. 4.22, where f_2 sends a message $\mu_{f_2 \to X_1}(X_1)$ to X_1 using (4.31),*

$$\mu_{f_2 \to X_1}(x_1) = \sum_{\sim\{x_1\}} f_2(x_1, x_4) \mu_{X_4 \to f_2}(x_4),$$

and X_1 sends a message $\mu_{X_1 \to f_2}(X_1)$ to f_2 using (4.32),

$$\mu_{X_1 \to f_2}(x_1) = \mu_{f_1 \to X_1}(x_1) \mu_{f_3 \to X_1}(x_1).$$

Note that at this point we are able to compute the marginal $g_{X_1}(X_1)$ by simple pointwise multiplication of the messages over the edge (X_1, f_2), for $x_1 \in \mathcal{X}_1$:

$$g_{X_1}(x_1) = \mu_{X_1 \to f_2}(x_1) \mu_{f_2 \to X_1}(x_1).$$

The subsequent steps are not depicted, but the reader can easily verify that at step 7 all the messages have been computed and all the marginals can be obtained.

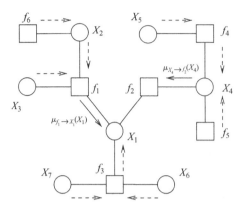

Figure 4.21. Step 3. Message computations based on available messages.

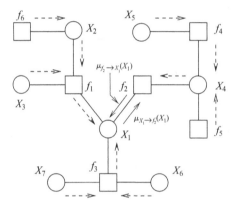

Figure 4.22. Step 4. Message computations based on available messages.

4.5   Normal factor graphs

4.5.1   Motivation

Two different types of factor graphs can be found in the technical literature. The type we have just covered is the more conventional type. Another type, introduced by Forney, is known as *normal factor graphs*. The latter type is equivalent to the conventional factor graphs in the sense that exactly the same messages are computed. Normal factor graphs require (at first sight) some strange modifications to the graph structure. Strictly speaking, normal factor graphs are not even graphs! Personally, I always use normal factor graphs for two simple reasons. First of all, normal factor graphs require fewer vertices/nodes than conventional factor graphs do. This makes normal factor graphs easier to understand. Secondly, there is only one type of vertex/node in normal factor graphs: function nodes. As a result, there is only a single message computation rule. Before we describe normal factor graphs explicitly, let us make the following observations regarding conventional factor graphs.

- Every variable node X_n of degree 1 sends the message "1" over its domain $\mathcal{X}_n$.
- Every variable node X_n of degree 2 simply forwards any incoming message unmodified. To see this, consider (4.32) for a variable node of degree $D = 2$.
- Every variable node X_n of degree greater than 2 performs pointwise multiplication of incoming messages to obtain an outgoing message. On the other hand, consider a function node f_k of degree D, from $\mathcal{X}^D \to \mathbb{R}$, defined as

$$f_k(x_1, \ldots, x_D) = \boxminus(x_1, \ldots, x_D), \tag{4.35}$$

where $\boxminus(\cdot)$ was defined in Section 3.3.1:

$$\boxminus(x_1, \ldots, x_D) = \begin{cases} \prod_{k=1}^{D-1} \delta(x_{k+1} - x_k) & x_k \text{ continuous,} \\ \prod_{k=1}^{D-1} \mathbb{I}\{x_{k+1} = x_k\} & x_k \text{ discrete,} \end{cases} \tag{4.36}$$

where, for a proposition P, $\mathbb{I}\{P\}$ is the indicator function, which is defined as $\mathbb{I}\{P\} = 1$ when P is true and $\mathbb{I}\{P\} = 0$ when P is false. For such a function node, what is the message-computation rule? Suppose that incoming messages $\mu_{X_n \to f_k}(X_n)$ for $n = 2, \ldots, D$ are available, and we wish to compute an outgoing message $\mu_{f_k \to X_1}(X_1)$. Using (4.31), we find

$$\mu_{f_k \to X_1}(x_1) = \sum_{\sim\{x_1\}} f_k(x_1, x_2, \ldots, x_D) \prod_{n=2}^{D-1} \mu_{X_n \to f_k}(x_n) \tag{4.37}$$

$$= \prod_{n=2}^{D-1} \mu_{X_n \to f_k}(x_1). \tag{4.38}$$

We see that *such a function node performs exactly the same pointwise multiplication of incoming messages as a variable node.* For obvious reasons, we call such a node an *equality node*.

4.5.2 Definition

Let us now formally introduce normal factor graphs. We will associate a vertex with every factor and an edge with every variable.

DEFINITION 4.14 (Normal factor graph). *Given a factorization of a function,*

$$f(X_1, X_2, \ldots, X_N) = \prod_{k=1}^{K} f_k(S_k), \tag{4.39}$$

and a corresponding conventional factor graph $G = (V, E)$, the normal factor graph is created as follows.

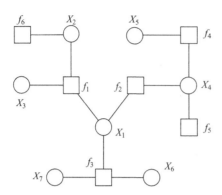

Figure 4.23. A conventional factor graph.

- *Every variable node X_n of degree 1 is removed. This creates a so-called half-edge (connected to only a single node). This half-edge is labeled X_n.*
- *Every variable node X_n of degree 2, with $\{f_k, f_l\} = \mathcal{N}(X_n)$ is removed, together with the two adjacent edges. We connect the nodes f_k and f_l directly with an edge and label this edge X_n.*
- *Every variable node X_n of degree $D > 2$ with $\{f_{k_1}, \dots, f_{k_D}\} = \mathcal{N}(X_n)$ is replaced by an equality node $\boxminus \left(X_n^{(1)}, \dots, X_n^{(D)}\right)$, where $X_n^{(1)}, \dots, X_n^{(D)}$ are dummy variables, defined over the domain $\mathcal{X}_n$ of X_n. The edge between the equality node and the node f_{k_m} is labeled $X_n^{(m)}$.*

Example 4.18 (Running example). *Let us return to the running example. We have a function of $N = 7$ variables that can be factored into $K = 6$ factors as follows:*

$$f(X_1, X_2, X_3, X_4, X_5, X_6, X_7) = f_1(X_1, X_2, X_3) f_2(X_1, X_4) f_3(X_1, X_6, X_7)$$
$$\times f_4(X_4, X_5) f_5(X_4) f_6(X_2) \qquad (4.40)$$

with a factor graph shown in Fig. 4.23. The corresponding normal factor graph is shown in Fig. 4.24. We see that it has fewer nodes, and that the variable nodes for X_1 and X_4 have each been replaced by an equality node and three dummy variables.

4.5.3 The sum–product algorithm on normal factor graphs

We will now describe the SPA for normal factor graphs. For the sake of clarity, we make no distinction between dummy variables and real variables. This allows us to treat the equality nodes just like any other function node. Also, we need to take some extra care when naming the messages: given an edge (X_n) incident with two function nodes (f_k and f_l), the message from function node f_k over the edge X_n can have the following names: $\mu_{f_k \to X_n}(X_n)$, $\mu_{X_n \to f_l}(X_n)$, or even $\mu_{f_k \to f_l}(X_n)$.

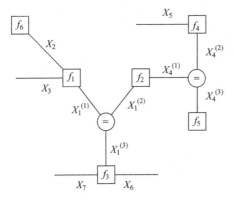

Figure 4.24. The normal factor graph corresponding to the conventional factor graph from Fig. 4.23.

Initialization

- For every leaf node f_k with incident edge X_m, transmit message $\mu_{f_k \to X_m}(X_m)$, with $\mu_{f_k \to X_m}(x_m) = f_k(x_m)$, $\forall x_m \in \mathcal{X}_m$.
- For every half-edge X_n, with incident node f_l, transmit message $\mu_{X_n \to f_l}(x_n) = 1$, $\forall x_n \in \mathcal{X}_n$.

Message-computation rule

Perform until all $\mu_{X_n \to f_k}(X_n)$ and all $\mu_{f_k \to X_n}(X_n)$ have been computed. Every message is computed only once.

- For any node f_k of degree D: when f_k has received incoming messages from $D - 1$ distinct incident edges $X_n \in \mathcal{N}(f_k)$, then node f_k can transmit an outgoing message $\mu_{f_k \to X_m}(X_m)$ on the remaining edge X_m, with (see Fig. 4.25)

$$\mu_{f_k \to X_m}(x_m) = \sum_{\sim\{x_m\}} f_k\big(\{X_n = x_n\}_{X_n \in \mathcal{N}(f_k)}\big) \prod_{X_n \in \mathcal{N}(f_k)\setminus\{X_m\}} \mu_{X_n \to f_k}(x_n). \qquad (4.41)$$

Termination

- To compute the marginal of X_n, take any $f_k \in \mathcal{N}(X_n)$. Then multiply the two messages on the edge X_n to obtain the marginal $g_{X_n}(X_n)$,

$$g_{X_n}(x_n) = \mu_{f_k \to X_n}(x_n)\mu_{X_n \to f_k}(x_n). \qquad (4.42)$$

Example 4.19. *The reader can easily verify that executing the SPA on the normal factor graph in Fig. 4.24 yields exactly the same results as executing the SPA on the conventional factor graph in Fig. 4.23.*

From this point onward, we will use only normal factor graphs.

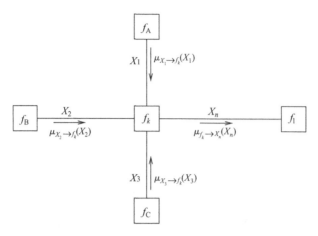

Figure 4.25. The message-computation rule for normal factor graphs. Observe that $\mu_{X_1 \to f_k}(X_1)$ is just another name for $\mu_{f_A \to X_1}(X_1)$.

4.6   Remarks on factor graphs

4.6.1   Why factor graphs?

What's the big deal? Why do we need these graphs? And who's interested in marginals anyway? Factor graphs have some very attractive properties that will become apparent as we proceed in this book.

- They are simple: constructing a factor graph from a factorization is easy.
- They are easy to read: for a human, interpreting a picture (a graph) is much more natural than reading a bunch of equations.
- Using the SPA, they allow us to compute marginals in an efficient, automated way.
- They allow hierarchical modeling and functional decomposition: by grouping together parts of a factor graph into a single super-vertex, we abstract away a certain part of the function.
- They are compatible with conventional block diagrams and flow charts.

The last two points are important, so let's look into them in some more detail.

4.6.2   Opening and closing nodes

Given a function node representing the factor $f_k(X_1, \ldots, X_D)$, we can replace this node by a factor graph representation of the acyclic factorization of another function, $g(X_1, \ldots, X_D, U_1, \ldots, U_K)$, as long as the following conditions are met:

- the variables $U_1, \ldots, U_K$ appear nowhere else in the graph; and
- the functions f_k and g satisfy

$$\sum_{u_1, \ldots, u_K} g(x_1, \ldots, x_D, u_1, u_2, \ldots, u_K) = f_k(x_1, \ldots, x_D).$$

It is easy to see that this replacement will not affect the messages computed on the edges $X_1, \ldots, X_D$. This process is known as *opening* the node f_k. The inverse operation is known as *closing* nodes.

Opening of nodes will be performed frequently in this book since it allows hierarchical modeling, and allows one to reveal or hide the structure of nodes.

Example 4.20 (Storing intermediate results). *We are given a function* $f_k(x_1, x_2, x_3)$, *structured as*

$$f_k(x_1, x_2, x_3) = h(\varphi(x_1, x_2), x_3),$$

with a factor-graph representation of $f_k(x_1, x_2, x_3)$ *shown on the left in Fig. 4.26. We can introduce an additional variable* u, *with* $u = \varphi(x_1, x_2)$. *This enables us to replace the node* $f_k(x_1, x_2, x_3)$ *by a node representing the factorization:*

$$g(x_1, x_2, x_3, u) = \boxed{=}(u, \varphi(x_1, x_2)) \times h(u, x_3).$$

We see that

$$\sum_u g(x_1, x_2, x_3, u) = \sum_u \boxed{=}(u, \varphi(x_1, x_2)) \times h(u, x_3)$$

$$= h(\varphi(x_1, x_2), x_3)$$

$$= f_k(x_1, x_2, x_3).$$

We open the node f_k *to reveal its structure (see Fig. 4.26, right): we see an additional variable,* U, *and two nodes,* $h(u, x_3)$ *and* $\tilde{\varphi}(u, x_1, x_2)$, *with*

$$\tilde{\varphi}(u, x_1, x_2) = \boxed{=}(u, \varphi(x_1, x_2)).$$

The messages over the edges X_1, X_2, *and* X_3 *will be the same in both graphs.*

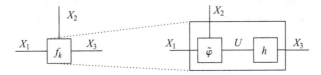

Figure 4.26. Opening a node: $f_k(x_1, x_2, x_3) = h(\varphi(x_1, x_2), x_3)$ and $\tilde{\varphi}(u, x_1, x_2) = \boxed{=}(u, \varphi(x_1, x_2))$. Messages on the X_k edges remain the same.

4.6.3   Computing joint marginals

When we focus on a factor $f_k(S_k)$ in the factorization of $f(\cdot)$, we can always write

$$f(X_1, X_2, \ldots, X_N) = \prod_{k=1}^{K} f_k(S_k) \tag{4.43}$$

$$= f_k(S_k) \prod_{X_n \in \mathcal{N}(f_k)} \prod_{f_l \in \mathcal{N}(X_n) \setminus \{f_k\}} h_{f_l}^{X_n}\left(X_n, S_{f_l}^{X_n}\right). \tag{4.44}$$

The marginal of $S_k = \{X_n\}_{X_n \in \mathcal{N}(f_k)}$ is given by

$$g_{S_k}(s_k) = \sum_{\sim \{s_k\}} f(x_1, x_2, \ldots, x_N) \tag{4.45}$$

$$= f_k(s_k) \prod_{X_n \in \mathcal{N}(f_k)} \left\{ \prod_{f_l \in \mathcal{N}(X_n) \setminus \{f_k\}} \sum_{\sim \{x_n\}} h_{f_l}^{X_n}\left(x_n, s_{f_l}^{X_n}\right) \right\} \tag{4.46}$$

$$= f_k(s_k) \prod_{X_n \in \mathcal{N}(f_k)} \mu_{X_n \to f_k}(x_n), \tag{4.47}$$

where the last transition is due to (4.26). In other words, the SPA can help us in computing *joint* marginals of multiple variables [61].

4.6.4   Complexity considerations

Since a function can have many factorizations, it can have many factor graphs. As long as the factorizations are acyclic (or, equivalently, the factor graphs have no cycles), the same marginals will be computed. However, some factorizations may lead to a computationally more complex SPA. To gain some insight into the computational complexity of the SPA, consider the following scenario.

We have a function of N variables, all defined over the same domain $\mathcal{X}$, with an acyclic factorization in K factors, resulting in a (conventional) factor graph with K function nodes. Suppose that, of the function nodes, d_1 have degree 1, d_2 have degree $2, \ldots, d_D$ have degree D, where D is the maximal degree of the function nodes in the graph. For simplicity, let us assume that the message-computation time for variable nodes can be neglected and that it takes 1 CPU cycle to evaluate any factor for any value of its arguments. In that case the total computational complexity (expressed in cycles) can be approximated by

$$C_1 = \sum_{k=1}^{D} k d_k |\mathcal{X}|^k, \tag{4.48}$$

whereas computing the marginals by simply summing out all other variables has a complexity

$$C_2 = N|\mathcal{X}|^N . \tag{4.49}$$

Generally $D \ll N$, so $C_1 \ll C_2$.

4.6.5   Dealing with continuous variables

Although we have not considered this explicitly, the variables X_n may be defined over a continuous domain $\mathcal{X}_n$. In this case, all derivations are still valid, provided that we replace summations by integrals, so that

$$\sum_{\sim\{x_k\}} f(x_1, \ldots, x_D) \tag{4.50}$$

becomes an integration over all variables except x_k, written concisely as

$$\int f(x_1, \ldots, x_D) \sim \{dx_k\}. \tag{4.51}$$

Observe that for equality nodes (say, of degree D)

$$\mu_{\boxminus \to X_k}(x_k) = \int \boxminus (x_1, \ldots, x_D) \prod_{X_l \neq X_k} \mu_{X_l \to \boxminus}(x_l) \sim \{dx_k\} \tag{4.52}$$

$$= \prod_{X_l \neq X_k} \mu_{X_l \to \boxminus}(x_k). \tag{4.53}$$

4.6.6   Disconnected and cyclic factorizations

In some cases we have a factorization that is not acyclic or connected. When the factor graph is a forest (i.e., disconnected and acyclic), marginals can be found by performing the SPA on the various trees, followed by a simple scaling. Details are left to the industrious reader. If the factor graph is connected, but has cycles, the SPA runs into problems. This is best seen using a simple example.

Example 4.21. *We have two variables, X and Y, defined over finite domains $\mathcal{X}$ and $\mathcal{Y}$, respectively. Let us assume that these domains both contain L elements: $\mathcal{X} = \{x_1, x_2, \ldots, x_L\}$ and $\mathcal{Y} = \{y_1, y_2, \ldots, y_L\}$. The function we consider can be factorized as follows:*

$$f(X, Y) = f_A(X, Y) f_B(X, Y).$$

In this case, the normal factor graph has a cycle, as shown in Fig. 4.27. Because $\mathcal{X}$ and $\mathcal{Y}$ are finite, the messages can be represented by a list of L numbers; the lth entry in the list corresponds to the message evaluated in the lth element of the domain $\mathcal{X}$ or

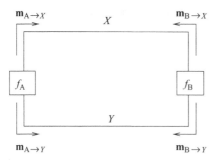

Figure 4.27. A normal factor graph of the function $f_A(X,Y)f_B(X,Y)$. This graph has a cycle.

$\mathcal{Y}$. *Hence, we represent* $\mu_{f_A \to X}(X) = \mu_{X \to f_B}(X)$ *by the* $L \times 1$ *column-vector* $\mathbf{m}_{A \to X}$. *Similarly,* $\mu_{f_B \to X}(X)$ *is represented by* $\mathbf{m}_{B \to X}$, $\mu_{f_B \to Y}(Y)$ *by* $\mathbf{m}_{B \to Y}$, *and* $\mu_{f_A \to Y}(Y)$ *by* $\mathbf{m}_{A \to Y}$. *We evaluate* $f_A(x,y)$, $\forall (x,y) \in \mathcal{X} \times \mathcal{Y}$ *and place the result in an* $L \times L$ *matrix* $\mathbf{A}$ *with*

$$[\mathbf{A}]_{ij} = f_A(x_i, y_j).$$

Similarly, $f_B(x,y)$, $\forall (x,y) \in \mathcal{X} \times \mathcal{Y}$ *is placed into an* $L \times L$ *matrix* $\mathbf{B}$. *Using these notations, we see that the sum-product rules now correspond to two disjunct sets of message computations:* $\mathbf{m}_{A \to X}$ *and* $\mathbf{m}_{B \to Y}$ *are updated clockwise in Fig. 4.27,*

$$\mathbf{m}_{B \to Y} = \mathbf{B}\mathbf{m}_{A \to X},$$

$$\mathbf{m}_{A \to X} = \mathbf{A}^\mathsf{T}\mathbf{m}_{B \to Y},$$

whereas the messages $\mathbf{m}_{A \to Y}$ *and* $\mathbf{m}_{B \to X}$ *are updated counterclockwise in Fig. 4.27,*

$$\mathbf{m}_{A \to Y} = \mathbf{A}^\mathsf{T}\mathbf{m}_{B \to X},$$

$$\mathbf{m}_{B \to X} = \mathbf{B}\mathbf{m}_{A \to Y}.$$

Because of the cyclic dependencies, the SPA is unable to start: for instance, the message $\mathbf{m}_{B \to Y}$ *requires the message* $\mathbf{m}_{A \to X}$, *and vice versa.*

Owing to the cycles, the SPA is unable to compute all the messages. A natural way to circumvent this problem is by artificially inserting the missing messages (for instance as a message that is equal to "1" over the corresponding domain). This allows the SPA to continue. However, this creates a new problem: the SPA will now run indefinitely, so it can never terminate. To solve this problem, we artificially terminate the SPA after a certain time (after which we hope the messages will have converged in some sense), and compute (what we hope to be) the marginals. Let us return to our example.

Example 4.22. *If we choose an initial message* $\mathbf{m}_{A \to X}^{(0)}$, *then the clockwise messages can be updated iteratively. On choosing an initial message* $\mathbf{m}_{B \to X}^{(0)}$, *the same can be done* *for the counterclockwise messages. After n such iterations, $\mathbf{m}_{A \to X}$ becomes*

$$\mathbf{m}_{A \to X}^{(n)} = \mathbf{C}^n \mathbf{m}_{A \to X}^{(0)},$$

where $\mathbf{C} = \mathbf{A}\mathbf{B}^T$. Suppose that we can find a set of eigenvectors and corresponding eigenvalues[2] $\{(\lambda_k, \mathbf{v}_k)\}_{k=1}^L$ that span $\mathbb{R}^L$. Possibly some eigenvalues are counted multiple times, and some eigenvalues are zero. We can then express $\mathbf{m}_{A \to X}^{(0)}$ as a linear combination of the eigenvectors,

$$\mathbf{m}_{A \to X}^{(0)} = \sum_{k=1}^{L} \alpha_k \mathbf{v}_k,$$

so that

$$\mathbf{m}_{A \to X}^{(n)} = \sum_{k=1}^{L} \alpha_k \lambda_k^n \mathbf{v}_k.$$

We see that, as the iterations progress, the terms for which $|\lambda_k| < 1$ tend to zero, whereas terms for which $|\lambda_k| > 1$ tend to be dominated by the eigenvalue with the largest absolute value. Except for some rare cases, messages will generally converge to zero or diverge.

The conclusion is that applying the SPA on a cyclic factor graph is unlikely to yield the correct marginals. So, this probably means that the factor graphs with cycles have no practical importance... Without any doubt the most ironic aspect of factor graphs is that the most exciting applications are precisely those in which the factor graph has cycles! More details will be given in the next chapter.

4.7   The *sum* and *product* operators

4.7.1   Generalizations

In this chapter we have considered functions from a domain $\mathcal{X}_1 \times \cdots \times \mathcal{X}_N \to \mathbb{R}$. The sum (denoted by $\sum$, $\int$, or $+$) and product ($\prod$ or $\times$) are the sum and product we are all familiar with. However, the SPA can be extended to more general scenarios. In this chapter, we required only that $(\mathbb{R}, +, \times)$ forms a commutative semi-ring. The SPA can be applied to general abstract sets F endowed with suitable "sum" ($\oplus$) and "product" ($\otimes$) operations, such that $(F, \oplus, \otimes)$ forms a commutative semi-ring. This means that

- $\oplus$ is associative and commutative (there exists an identity element for $\oplus$: $e_\oplus$);
- $\otimes$ is associative and commutative (there exists an identity element for $\otimes$: $e_\otimes$); and
- $\otimes$ is distributive over $\oplus$: $a \otimes (b \oplus c) = (a \otimes b) \oplus (a \otimes c)$ for any $a, b, c \in F$.

The SPA now uses the new sum and product operations, but remains essentially unmodified. The only subtle point is that, in the initialization step of the SPA, half-edges transmit the message $e_\otimes$ over the corresponding domain.

---

[2] We remind the reader that, for a matrix $\mathbf{C}$ with eigenvalue λ and (column) eigenvector $\mathbf{v}$, we have $\mathbf{C}\mathbf{v} = \lambda \mathbf{v}$.

Example 4.23. *The triple* $(\mathbb{R}, \max, +)$ *forms a commutative semi-ring. Let us verify this.*
- max *is associative,*

$$\max(a, \max(b, c)) = \max(\max(a, b), c).$$

- max *is commutative,*

$$\max(a, b) = \max(b, a).$$

- *There exists an identity element for* max, $e_\oplus$, *such that* $\max(a, e_\oplus) = a$, $\forall a \in \mathbb{R}$. *We see that* $e_\oplus = -\infty$ *satisfies this condition.*
- $+$ *is associative and commutative with* $e_\otimes = 0$.
- $+$ *is distributive over* max:

$$a + \max(b, c) = \max(a + b, a + c).$$

In this case, the sum–product algorithm is usually given a more appropriate name: the max–sum algorithm. It is easily verified that, similarly to $(\mathbb{R}, \max, +)$, $(\mathbb{R}^+, \max, \times)$ *also forms a commutative semi-ring with* $e_\oplus = -\infty$ *and* $e_\otimes = 1$, *leading to the max–product algorithm. More examples can be found in [4].*

4.7.2 The max–sum algorithm

Let us consider the max-sum algorithm. We have a function from $\mathcal{X}_1 \times \cdots \times \mathcal{X}_N \to \mathbb{R}$, with the following acyclic "factorization":

$$f(X_1, X_2, \ldots, X_N) = \sum_{k=1}^{K} f_k(S_k). \tag{4.54}$$

We can then construct a factor graph of this factorization, and apply the max–sum algorithm as follows.

Initialization
- For every leaf node f_k with incident edge X_m, transmit message $\mu_{f_k \to X_m}(X_m)$, with $\mu_{f_k \to X_m}(x_m) = f_k(x_m)$, $x_m \in \mathcal{X}_m$.
- For every half-edge X_n, with incident node f_l, transmit message $\mu_{X_n \to f_l}(x_n) = 0$, $\forall x_n \in \mathcal{X}_n$.

Message-computation rule
- For any node f_k of degree D: when f_k has received incoming messages from $D - 1$ distinct incident edges $X_n \in \mathcal{N}(f_k)$, then node f_k can transmit an outgoing message

$\mu_{f_k \to X_m}(X_m)$ on the remaining edge X_m, with

$$\mu_{f_k \to X_m}(x_m) = \bigoplus_{\sim \{x_m\}} \left\{ f_k\left(\{X_n = x_n\}_{X_n \in \mathcal{N}(f_k)}\right) \otimes \bigotimes_{X_n \in \mathcal{N}(f_k) \backslash \{X_m\}} \mu_{X_n \to f_k}(x_n) \right\} \tag{4.55}$$

$$= \max_{\sim \{x_m\}} \left\{ f_k\left(\{X_n = x_n\}_{X_n \in \mathcal{N}(f_k)}\right) + \sum_{X_n \in \mathcal{N}(f_k) \backslash \{X_m\}} \mu_{X_n \to f_k}(x_n) \right\}. \tag{4.56}$$

Termination

- To compute the marginal of X_n, take any $f_k \in \mathcal{N}(X_n)$. Then add the messages on the edge X_n to obtain the marginal $g_{X_n}(X_n)$, so that for any $x_n \in \mathcal{X}_n$

$$g_{X_n}(x_n) = \mu_{f_k \to X_n}(x_n) \otimes \mu_{X_n \to f_k}(x_n) \tag{4.57}$$

$$= \mu_{f_k \to X_n}(x_n) + \mu_{X_n \to f_k}(x_n). \tag{4.58}$$

The marginals $g_{X_n}(X_n)$ are to be interpreted as

$$g_{X_n}(x_n) = \bigoplus_{\sim \{x_n\}} f(x_1, x_2, \ldots, x_N) \tag{4.59}$$

$$= \max_{\sim \{x_n\}} f(x_1, x_2, \ldots, x_N) \tag{4.60}$$

so that the overall maximum is given by, for any n,

$$\max_{x_1, \ldots, x_N} f(x_1, x_2, \ldots, x_N) = \max_{x_n} g_{X_n}(x_n). \tag{4.61}$$

This is an important result: the max–sum algorithm allows us to determine the maximum of a function. Note that this is not the same as determining the values of $\mathbf{x} = [x_1, \ldots, x_N]$ that achieve this maximum! In optimization problems we are usually interested in finding not the maximum of a function, but rather the values $\mathbf{x}$ that achieve this maximum. Can the max–sum algorithm help? Let us denote by $\hat{x}_n$ a value that maximizes $g_{X_n}(X_n)$:

$$\hat{x}_n = \arg\max_{x_n} g_{X_n}(x_n). \tag{4.62}$$

It then follows that

$$g_{X_n}(\hat{x}_n) = \max_{x_1, \ldots, x_N} f(x_1, x_2, \ldots, x_N). \tag{4.63}$$

When we construct a vector $\hat{\mathbf{x}} = [\hat{x}_1, \hat{x}_2, \ldots, \hat{x}_N]$, then we are guaranteed that, when $\hat{\mathbf{x}}$ is unique,

$$\hat{\mathbf{x}} = \arg\max_{x_1, \ldots, x_N} f(x_1, x_2, \ldots, x_N). \tag{4.64}$$

When $f(\cdot)$ has multiple maxima, the arguments that achieve these maxima can (at least in principle) be found by considering all combinations of the arguments that maximize the marginals, and then evaluating $f(x_1, x_2, \ldots, x_N)$ for all those combinations. We retain only those combinations that maximize $f(\cdot)$. Hence, the max–sum algorithm allows us to locate efficiently the maximum of a function, as well as the values of the variables that achieve this maximum.

4.8   Main points

In this chapter we have taken a leisurely tour in the land of factor graphs and the sum–product algorithm (SPA). Factor graphs are a way of graphically representing the factorization of a function. A (normal) factor graph is created as follows: we create an edge for every variable and a vertex (a node) for every factor in the factorization. We connect an edge to a node if and only if the corresponding variable appears as an argument in the corresponding factor. Variables that appear in more than two factors require a special equality node.

The SPA is an algorithm that efficiently computes the marginals of a function by passing messages over the edges of the corresponding factor graph. We denote by $\mu_{f_k \to X_m}(X_m)$ the message from node f_k over edge X_m. Messages are functions of the corresponding variables. The SPA starts from the leaf nodes (with $\mu_{f_l \to X_m}(x_m) = f_l(x_m)$) and half-edges (with $\mu_{X_n \to f_k}(x_n) = 1$ for all $x_n \in \mathcal{X}_n$) in the graph. Nodes accept incoming messages and compute outgoing messages. The message-computation rule relating incoming messages $\mu_{X_n \to f_k}(X_n)$ to an outgoing message $\mu_{f_k \to X_m}(X_m)$ is given

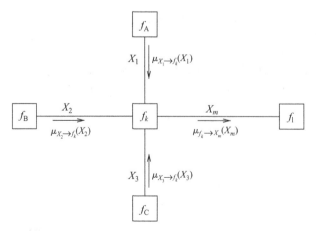

Figure 4.28. The update rule for normal factor graphs. Observe that $\mu_{X_1 \to f_k}(X_n)$ is just another name for $\mu_{f_A \to X_1}(X_1)$.

by (see also Fig. 4.28)

$$\mu_{f_k \to X_m}(x_m) = \sum_{\sim \{x_m\}} f_k\left(\{X_n = x_n\}_{X_n \in \mathcal{N}(f_k)}\right) \prod_{X_n \in \mathcal{N}(f_k) \backslash \{X_m\}} \mu_{X_n \to f_k}(x_n). \qquad (4.65)$$

This computation rule is undoubtedly the most important relation in this book. Read it a few times, and remember it by heart. Once all the messages have been computed, the marginal of any variable (say, X_n) can be found by point wise multiplication of the two messages over the corresponding edge. When the factor graph contains cycles, the SPA runs into difficulties.

We have also introduced the concept of opening a node, and have illustrated how the SPA can be generalized to arbitrary commutative semi-rings. One particular extension is the max–sum algorithm, which can be interpreted as an optimization technique/algorithm.

5 Statistical inference using factor graphs

5.1 Introduction

In the previous two chapters we have introduced estimation theory and factor graphs. Although these two topics may seem disparate, they are closely linked. In this chapter we will use factor graphs to solve estimation problems and, more generally, inference problems. In the context of statistical inference, factor graphs are important for two reasons. First of all, they allow us to reformulate several important inference algorithms in a very elegant way with an all-encompassing, well-defined notation and terminology. As we will see in the future chapters, well-known algorithms such as the forward–backward algorithm, the Viterbi algorithm, the Kalman filter, and the particle filter can all be cast in the factor-graph framework in a very natural way. Secondly, deriving new, optimal (or near-optimal) inference algorithms is fairly straightforward in the factor-graph framework. Applying the SPA on a factor graph relies solely on local computations in basic building blocks. Once we understand the basic building blocks, we need remember only one rule: the sum–product rule. In this chapter, we will go into considerable detail on how to perform inference using factor graphs. Certain aspects of this chapter were inspired by [62].

This chapter is organized as follows.

- We start by explaining the various problems of statistical inference in **Section 5.2**, and then provide a general factor-graph-based framework for solving these problems.
- We will deal with how messages should be represented (a topic we glossed over in Chapter 4) in **Section 5.3**. The representation has many important implications, as will become apparent throughout this book.
- We end with **Section 5.4**, covering the fun topic of loopy inference, that is inference on factor graphs with cycles.

5.2 General formulation

5.2.1 Five problems of statistical inference

Let us introduce five important problems of statistical inference. We have a number of random variables that we would like to infer from a number of observations. Let us group the variables of interest in a vector $\mathbf{X} = [X_1, \ldots, X_N]$ defined over $\mathcal{X}_1 \times \cdots \times \mathcal{X}_N$. The observation is a vector $\mathbf{y} = [y_1, \ldots, y_M]$, which is a realization of a random variable $\mathbf{Y}$. Additionally, there is usually some underlying model describing additional assumptions related to $\mathbf{X}$ and $\mathbf{Y}$. This model will be captured in a parameter $\mathcal{M}$, where $p(\mathbf{Y}|\mathbf{X}, \mathcal{M})$ is a known distribution. In statistical inference, we are usually interested in answering one or more of the following questions [63].

1. What is $p(\mathbf{Y} = \mathbf{y}|\mathcal{M})$, the likelihood of the model $\mathcal{M}$? Note that, for a given $\mathbf{y}$, the answer is a positive real number (i.e., in $\mathbb{R}^+$).
2. What is the marginal a-posteriori distribution $p(X_k|\mathbf{Y} = \mathbf{y}, \mathcal{M})$, given a certain model $\mathcal{M}$?
3. What are the characteristics (mode, moments, etc.) of the distribution $p(X_k|\mathbf{Y} = \mathbf{y}, \mathcal{M})$?
4. What is the joint a-posteriori distribution $p(\mathbf{X}|\mathbf{Y} = \mathbf{y}, \mathcal{M})$, given a certain model $\mathcal{M}$?
5. What are the characteristics (mode, moments, etc.) of the distribution $p(\mathbf{X}|\mathbf{Y} = \mathbf{y}, \mathcal{M})$?

5.2.2 Opening nodes

Before we start applying factor graphs to solve these inference problems, it is instructive to re-cap the concept of opening nodes in the context of inference. Opening nodes will happen many times in the course of this book, so let us devote a little time to it.

In inference problems, nodes will usually represent distributions (say, $p(X_1, \ldots, X_D)$) or likelihood functions (say, $p(\mathbf{Y} = \mathbf{y}|X_1, \ldots, X_D)$). In either case, from Section 4.6.2 we know that we can replace a node by a factorization of a distribution with additional variables, as long as these variables do not appear elsewhere in the factor graph. In other words, the node representing $p(X_1, \ldots, X_D)$ can be replaced (opened up) by a (cycle-free) factor graph of a factorization of

$$p(X_1, \ldots, X_D, U_1, \ldots, U_{D'}) \tag{5.1}$$

as long as the U_k are new variables. Similarly, the node representing $p(\mathbf{Y} = \mathbf{y}|X_1, \ldots, X_D)$ can be replaced by a factor graph of a factorization of

$$p(\mathbf{Y} = \mathbf{y}, U_1, \ldots, U_{D'} |X_1, \ldots, X_D). \tag{5.2}$$

Opening of the node does not change the messages (either incoming or outgoing) computed on the edges $X_1, \ldots, X_D$.

5.2.3 Inference on factor graphs

Let us attempt to solve the five inference problems in a general, abstract way, and see how factor graphs may help. After that, in Section 5.2.4, we will cover some basic examples to get a feel of the expressive power of factor graphs in the context of inference.

5.2.3.1 Problem 1 – likelihood of the model

- **Step 1.** Consider the joint distribution of the observations and the unknown parameters, evaluated in $\mathbf{Y} = \mathbf{y}$:

$$p(\mathbf{X}, \mathbf{Y} = \mathbf{y}|\mathcal{M}) = p(\mathbf{Y} = \mathbf{y}|\mathbf{X}, \mathcal{M}) p(\mathbf{X}|\mathcal{M}). \tag{5.3}$$

This is a function from $\mathcal{X}_1 \times \cdots \times \mathcal{X}_N \to \mathbb{R}^+$. The first factor $p(\mathbf{Y} = \mathbf{y}|\mathbf{X}, \mathcal{M})$ is the *likelihood function* (note that the likelihood function is a function not of $\mathbf{Y}$, but of $\mathbf{X}$). The second factor $p(\mathbf{X}|\mathcal{M})$ is the *a-priori distribution* of $\mathbf{X}$, given a certain model.
- **Step 2.** Factorize both the likelihood function and the a-priori distribution:

$$p(\mathbf{Y} = \mathbf{y}|\mathbf{X}, \mathcal{M}) = \prod_{k=1}^{K} f_k(S_k), \tag{5.4}$$

where $S_k \subseteq \{X_1, \ldots, X_N\}$, and

$$p(\mathbf{X}|\mathcal{M}) = \prod_{l=1}^{L} g_l(R_l), \tag{5.5}$$

where $R_l \subseteq \{X_1, \ldots, X_N\}$. This factorization may require the introduction of additional variables and opening of nodes.
- **Step 3.** Create a cycle-free, connected factor graph of the factorization of $p(\mathbf{X}, \mathbf{Y} = \mathbf{y}|\mathcal{M})$. Remember that $\mathbf{y}$ is fixed, and does not appear as a variable (i.e., an edge) in the graph.
- **Step 4.** Perform the SPA on this graph. This gives us the marginals $g_{X_k}(X_k) = p(X_k, \mathbf{Y} = \mathbf{y}|\mathcal{M})$, for $k = 1, \ldots, N$.
- **Step 5 – solution.** Take any k between 1 and N, then

$$p(\mathbf{Y} = \mathbf{y}|\mathcal{M}) = \sum_{x_k \in \mathcal{X}_k} p(X_k = x_k, \mathbf{Y} = \mathbf{y}|\mathcal{M}). \tag{5.6}$$

5.2.3.2 Problem 2 – a-posteriori distribution of X_k

We can repeat the first four steps from Section 5.2.3.1, so that we end up with the marginals $p(X_k, \mathbf{Y} = \mathbf{y}|\mathcal{M})$, for $k = 1, \ldots, N$.

- **Step 5 – solution.** Clearly,

$$p(X_k|\mathbf{Y} = \mathbf{y}, \mathcal{M}) = \frac{p(X_k, \mathbf{Y} = \mathbf{y}|\mathcal{M})}{p(\mathbf{Y} = \mathbf{y}|\mathcal{M})}. \tag{5.7}$$

Hence, $p(X_k|\mathbf{Y} = \mathbf{y}, \mathcal{M})$ evaluated in $x_k \in \mathcal{X}_k$ is given by

$$p(X_k = x_k|\mathbf{Y} = \mathbf{y}, \mathcal{M}) = \frac{p(X_k = x_k, \mathbf{Y} = \mathbf{y}|\mathcal{M})}{\sum_{x \in \mathcal{X}_k} p(X_k = x, \mathbf{Y} = \mathbf{y}|\mathcal{M})}. \tag{5.8}$$

In other words, the marginal a-posteriori distribution $p(X_k | \mathbf{Y} = \mathbf{y}, \mathcal{M})$ is obtained by *normalizing* the marginal $g_{X_k}(X_k)$. Note that, to determine the marginal a-posteriori distribution $p(X_k | \mathbf{Y} = \mathbf{y}, \mathcal{M})$, the marginals $g_{X_k}(X_k)$ need be known only up to a multiplicative constant. This may seem a trivial remark, but will turn out to be important.

5.2.3.3   Problem 3–characteristics of the a-posteriori distribution of X_k

Once we have determined the marginal a-posteriori distribution $p(X_k | \mathbf{Y} = \mathbf{y}, \mathcal{M})$, its characteristics can usually easily be found. For instance, the mode

$$\hat{x}_k = \arg \max_{x \in \mathcal{X}_k} p(X_k = x | \mathbf{Y} = \mathbf{y}, \mathcal{M}) \tag{5.9}$$

can be found through complete enumeration (for discrete variables), or using standard optimization techniques.

5.2.3.4   Problem 4–a-posteriori distribution of $\mathbf{X}$

This problem can be solved only when we consider the trivial factorization

$$p(\mathbf{X}, \mathbf{Y} = \mathbf{y} | \mathcal{M}) = p(\mathbf{Y} = \mathbf{y} | \mathbf{X}, \mathcal{M}) \, p(\mathbf{X} | \mathcal{M}). \tag{5.10}$$

The corresponding factor graph has two nodes (one for the likelihood function and one for the a-priori distribution) and a single edge ($\mathbf{X}$). Factor graphs give us no insight or computational advantage in solving this particular inference problem.

5.2.3.5   Problem 5–characteristics of the a-posteriori distribution of $\mathbf{X}$

Since $p(\mathbf{X}, \mathbf{Y} = \mathbf{y} | \mathcal{M})$ is hard to determine, finding its moments is usually also very hard. On the other hand, finding its *mode* turns out to be (rather surprisingly) feasible: let us think back to Section 4.7, where we showed how to use the max–sum algorithm to perform maximization over the $(\mathbb{R}, \max, +)$ semi-ring. This leads to the following solution.

- **Step 1.** Consider the *logarithm* of the joint distribution of the observations and the unknown parameters, evaluated in $\mathbf{Y} = \mathbf{y}$:

$$\log p(\mathbf{X}, \mathbf{Y} = \mathbf{y} | \mathcal{M}) = \log p(\mathbf{Y} = \mathbf{y} | \mathbf{X}, \mathcal{M}) + \log p(\mathbf{X} | \mathcal{M}). \tag{5.11}$$

  This is a function from $\mathcal{X}_1 \times \cdots \times \mathcal{X}_N \to \mathbb{R}$. The first term, $\log p(\mathbf{Y} = \mathbf{y} | \mathbf{X}, \mathcal{M})$, is the *log-likelihood function*.
- **Step 2.** We now factorize both the likelihood function and the a-priori distribution, so that we can write $\log p(\mathbf{Y} = \mathbf{y} | \mathbf{X}, \mathcal{M}) = \sum_k f_k(S_k)$, where $S_k \subseteq \{X_1, \ldots, X_N\}$, and $\log p(\mathbf{X} | \mathcal{M}) = \sum_{l=1}^{L} g_l(R_l)$, where $R_l \subseteq \{X_1, \ldots, X_N\}$. This factorization may require the introduction of additional variables and opening of nodes.
- **Step 3.** Create a cycle-free, connected factor graph of the factorization of $\log p(\mathbf{X}, \mathbf{Y} = \mathbf{y} | \mathcal{M})$. We now need to make a node for every term in the summation. Recall that $\mathbf{y}$ is fixed, and does not appear as a variable (edge) in the graph.

- **Step 4.** Perform the max–sum algorithm on this graph. This gives us the marginals $g_{X_k}(X_k)$, with

$$g_{X_k}(x_k) = \max_{\sim\{x_k\}} \log p(\mathbf{X} = \mathbf{x}, \mathbf{Y} = \mathbf{y}|\mathcal{M}) \qquad (5.12)$$

for $k = 1,\ldots,N$.
- **Step 5 – solution.** Take any k between 1 and N, then

$$\hat{x}_k = \arg\max_{x\in\mathcal{X}_k} g_{X_k}(x). \qquad (5.13)$$

Note that, to determine $\hat{x}_k$, we need to know the marginal $g_{X_k}(X_k)$ only up to an additive constant. As we argued in Section 4.7,

$$\hat{\mathbf{x}} = \left[\hat{x}_1,\ldots,\hat{x}_N\right] \qquad (5.14)$$
$$= \arg\max_{\mathbf{x}} \log p(\mathbf{X} = \mathbf{x}, \mathbf{Y} = \mathbf{y}|\mathcal{M}). \qquad (5.15)$$

Now, since the a-posteriori distribution and the joint distribution are related as

$$p(\mathbf{X}, \mathbf{Y} = \mathbf{y}|\mathcal{M}) = p(\mathbf{X}|\mathbf{Y} = \mathbf{y}, \mathcal{M})\, p(\mathbf{Y} = \mathbf{y}|\mathcal{M}), \qquad (5.16)$$

we find that

$$\hat{\mathbf{x}} = \arg\max_{\mathbf{x}} p(\mathbf{X} = \mathbf{x}|\mathbf{Y} = \mathbf{y}, \mathcal{M}), \qquad (5.17)$$

which is the desired result.

5.2.4   Examples

Example 5.1 (The burglar-alarm problem).
Problem. A famous example is the burglar-alarm problem from Pearl [56]. Consider three binary random variables: E represents the event of an earthquake occurring, B the event of a burglary taking place, and A the event of a burglar-alarm ringing. B and E are a-priori independent. We have a-priori knowledge, with $p(B = 1) = 1/100$, $p(E = 1) = 1/100$. We also know the distribution of A, conditioned on E and B:

$$p(A = 1|B = 0, E = 0) = 1/1000,$$
$$p(A = 1|B = 0, E = 1) = 1/10,$$
$$p(A = 1|B = 1, E = 0) = 7/10,$$
$$p(A = 1|B = 1, E = 1) = 9/10.$$

Suppose that we are at work, and get a call from the neighbor saying that our burglar-alarm is ringing. What is then the probability that a burglar is in our house, and what is the probability that an earthquake has occurred?

Solution. Our observation is the ringing of the burglar-alarm, so we can make the following associations:

$$Y \leftrightarrow A,$$

$$y \leftrightarrow 1,$$

$$X \leftrightarrow [B, E].$$

Our goal is to find the distributions $p(B|A = 1)$ and $p(E|A = 1)$. This corresponds to the second inference problem from Section 5.2.1. Let us work through the steps of the solution, as outlined in Section 5.2.3.2. In step 1, we factorize the joint distribution

$$p(B, E, A = 1) = p(A = 1|B, E) p(B, E).$$

In step 2, we factorize the likelihood function and the a-priori distribution. This leads to

$$p(B, E, A = 1) = p(A = 1|B, E) p(B) p(E).$$

In step 3, we create a factor graph of this factorization. This factor graph is shown in Fig. 5.1, where we have abbreviated $p(B)$ by $f_B(B)$, $p(E)$ by $f_E(E)$ and $p(A = 1|B, E)$ by $g(B, E)$. In step 4, we perform the SPA on this graph. We first send messages from the leaf nodes,

$$\mu_{f_B \rightarrow B}(b) = \begin{cases} 0.01 & b = 1 \\ 0.99 & b = 0 \end{cases}$$

and

$$\mu_{f_E \rightarrow E}(e) = \begin{cases} 0.01 & e = 1 \\ 0.99 & e = 0. \end{cases}$$

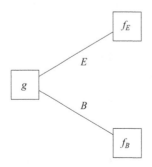

Figure 5.1. A factor graph for the burglar-alarm problem. We use the following abbreviations: $f_E(E) = p(E)$, $f_B(B) = p(B)$, and $g(B, E) = p(A = 1|B, E)$.

On the basis of these messages, we can now send messages from the node marked g:

$$\mu_{g\to B}(b) = \sum_{e=0}^{1} p(A = 1 | B = b, E = e)\,\mu_{f_E\to E}(e)$$

$$= \begin{cases} 0.70200 & b = 1 \\ 0.00199 & b = 0 \end{cases}$$

and

$$\mu_{g\to E}(e) = \sum_{b=0}^{1} p(A = 1 | B = b, E = e)\,\mu_{f_B\to B}(b)$$

$$= \begin{cases} 0.10800 & e = 1 \\ 0.00799 & e = 0. \end{cases}$$

We obtain the following marginals:

$$g_B(b) = p(A = 1, B = b)$$
$$= \mu_{g\to B}(b)\,\mu_{B\to g}(b)$$
$$= \begin{cases} 0.0070200 & b = 1 \\ 0.0019701 & b = 0 \end{cases}$$

and

$$g_E(e) = p(A = 1, E = e)$$
$$= \mu_{g\to E}(e)\,\mu_{E\to g}(e)$$
$$= \begin{cases} 0.0010800 & e = 1 \\ 0.0079101 & e = 0. \end{cases}$$

Finally in step 5, we find the normalization constants for B and E. They should be the same: for B, we find $1/0.0089901 \approx 111.2$, while for E we find $1/(0.0010800 + 0.0079101) \approx 111.2$, as we would expect. Multiplying the marginals by the normalization constant yields

$$p(B = b | A = 1) \approx \begin{cases} 0.7809 & b = 1 \\ 0.2191 & b = 0 \end{cases}$$

and

$$p(E = e | A = 1) \approx \begin{cases} 0.1201 & e = 1 \\ 0.8799 & e = 0. \end{cases}$$

So, even though a priori an earthquake is equally likely to occur as a burglary, the alarm ringing tells us that the probability that an earthquake has occurred is only 12%, while the probability of a burglary is almost 80%.

Example 5.2 (Repetition codes).

Problem. *Another basic example is from the world of digital communication. We wish to convey a single bit $b \in \mathbb{B}$ of information over an unreliable channel. The bit has the following a-priori distribution: $p(B = 0) = 1/2$. The channel is a binary asymmetric channel with the following transition probabilities (c is the input, y is the output):*

$$p(Y = 0|C = 0) = 0.5,$$

$$p(Y = 0|C = 1) = 0.1.$$

Note that this fully characterizes the channel. In order to protect the bit against this unreliable channel, we encode our bit b using a rate 1/4 repetition code. The transmitter then sends 4 bits $\mathbf{c} = [c_1, c_2, c_3, c_4] = [b, b, b, b]$ over the binary asymmetric channel. Suppose we receive $\mathbf{y} = [0, 0, 1, 1]$. How can we recover our original bit b?

Solution. *Let us create a factor graph of the joint distribution*

$$p(B, \mathbf{Y} = \mathbf{y}) = p(\mathbf{Y} = \mathbf{y}|B) p(B).$$

We will now introduce four additional variables and open up the node $p(\mathbf{Y} = \mathbf{y}|B)$. We replace the node $p(\mathbf{Y} = \mathbf{y}|B)$ by a factor graph of a factorization of $p(\mathbf{Y} = \mathbf{y}, \mathbf{C}|B)$. Note that this is a valid operation (as defined in Section 4.6.2) since

$$\sum_{\mathbf{c} \in \mathbb{B}^4} p(\mathbf{Y} = \mathbf{y}, \mathbf{C} = \mathbf{c}|B) = p(\mathbf{Y} = \mathbf{y}|B).$$

Now $p(\mathbf{Y} = \mathbf{y}, \mathbf{C}|B)$ factorizes as follows:

$$p(\mathbf{Y} = \mathbf{y}, \mathbf{C}|B) = \prod_{k=1}^{4} p(Y_k = y_k|C_k) \boxminus (C_k, B),$$

where $\boxminus(\cdot)$ is the equality function. A factor graph of the factorization of $p(\mathbf{Y} = \mathbf{y}, \mathbf{C}|B)$ is shown in Fig. 5.2, where $f_B(B)$ is a shorthand for $p(B)$, and $f_k(C_k)$ for $p(Y_k = y_k|C_k)$. It is clear that performing the SPA on this graph yields the marginal $p(B, \mathbf{Y} = \mathbf{y})$. The SPA is executed as follows. In the first phase, messages from the five leaf nodes are sent to the equality node:

$$\mu_{f_B \to B}(b) = \begin{cases} 0.5 & b = 1 \\ 0.5 & b = 0 \end{cases}$$

$$\mu_{f_k \to C_k}(c_k) = \begin{cases} 0.1 & c_k = 1 \\ 0.5 & c_k = 0 \end{cases}$$

for $k = 1, 2$ and

$$\mu_{f_k \to C_k}(c_k) = \begin{cases} 0.9 & c_k = 1 \\ 0.5 & c_k = 0 \end{cases}$$

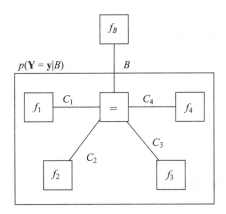

Figure 5.2. A factor graph of a repetition code. The node $p(\mathbf{Y} = \mathbf{y}|B)$ is opened up to reveal its structure. $f_B(B)$ is a shorthand for $p(B)$, and $f_k(C_k)$ for $p(Y_k = y_k|C_k)$.

for $k = 3, 4$. In the second phase, we compute the message from the equality node to the B-edge, as the pointwise multiplication of the incoming messages over the C-edges:

$$\mu_{\boxminus \to B}(B = b) = \sum_{c_1,c_2,c_3,c_4} \boxminus(b, c_1, c_2, c_3, c_4) \prod_{k=1}^{4} \mu_{g \to C_k}(c_k) \qquad (5.18)$$

$$= \prod_{k=1}^{4} \mu_{g \to C_k}(b)$$

so that

$$\mu_{\boxminus \to B}(B = 0) = (0.5)^4$$
$$= 0.0625$$

and

$$\mu_{\boxminus \to B}(B = 1) = (0.9)^2 \times (0.1)^2$$
$$= 0.0081.$$

We then find that

$$g_B(b) = p(B = b, \mathbf{Y} = \mathbf{y})$$

$$= \begin{cases} 0.00405 & b = 1 \\ 0.03125 & b = 0. \end{cases}$$

Normalization yields (with normalization constant ≈ 28.33):

$$p(B = b|\mathbf{Y} = \mathbf{y}) \approx \begin{cases} 0.11 & b = 1 \\ 0.89 & b = 0 \end{cases}$$

so that we can say with high confidence that the transmitted bit was equal to zero. We can also determine the likelihood of having received this particular observation **y**:

$$p(\mathbf{Y} = \mathbf{y}) = \sum_{b \in \mathbb{B}} g_B(b)$$

$$= 0.0353.$$

5.3   Messages and their representations

At this point the reader may very well wonder how messages can be represented in, say, a computer program. The examples in the previous sections indicate that for discrete variables a simple vector representation can be used. A problem that has also become apparent is that messages may have a very small dynamic range. In this section we will deal with some practical issues, such as message scaling and message representation.

This section is organized as follows. In Section 5.3.1, we show that messages can be scaled arbitrarily, without affecting the outcome of the SPA. To be more precise: in the sum–product algorithm we can multiply messages by constants, and in the max–sum algorithm we can add constants to messages. When we are interested in the first inference problem (computing the likelihood of the model), we need to keep track of all the scaling factors. When we are interested in any of the other inference problems, scaling factors/terms cancel out and can be forgotten. Section 5.3.2 deals with a particular type of scaling: normalization. Normalization allows us to interpret messages as distributions, which has some interesting consequences. We then move on to three representations of messages for *discrete variables* in Section 5.3.3: probability mass functions, log-likelihoods and log-likelihood ratios. In the same section, we also reveal a close link between the SPA and the max–sum algorithm. Finally, in Section 5.3.4 we present several ways of representing messages for *continuous variables*: quantization, and parametric and non-parametric representations.

5.3.1   Message scaling

5.3.1.1   Max–Sum algorithm
In the max–sum algorithm, any real number can be added to any message without affecting the outcome of the algorithm. Additive constants simply propagate through the max–sum algorithm. After running the max–sum algorithm, the marginals we obtain (say $\tilde{g}_{X_k}(X_k)$) are now equal to the true marginals ($g_{X_k}(X_k)$) up to an additive constant (say, C_k). We find that

$$\hat{x}_k = \arg\max_{x \in \mathcal{X}_k} \tilde{g}_{X_k}(X_k = x)$$

$$= \arg\max_{x \in \mathcal{X}_k} g_{X_k}(X_k = x)$$

so *we don't need to keep track of the the scaling terms C_k at all!* Adding constants to messages has some implementation advantages.

5.3.1.2 The sum–product algorithm

A common problem in message computation in the SPA for inference problems is that we multiply probabilities and likelihoods. As a result, messages are functions that tend to get smaller and smaller in magnitude as the SPA proceeds. When we want to compute the marginal a-posteriori distributions $p(X_k | \mathbf{Y} = \mathbf{y}, \mathcal{M})$, we will have to use large normalization factors. From a theoretical point of view this is not a problem. For computers, on the other hand, which have only a finite numerical precision, it is simply disastrous. To get around this problem, we can scale the messages so that they lie with in some predefined dynamic range. As long as we keep track of the scaling factors, we can still solve our inference problems exactly.

Consider the SPA computation rule:

$$\mu_{f_k \rightarrow X_m}(x_m) = \sum_{\sim \{x_m\}} f_k(x_1, \ldots, x_D) \prod_{n \neq m} \mu_{X_n \rightarrow f_k}(x_n). \tag{5.19}$$

Suppose that we scale the incoming messages $\mu_{X_n \rightarrow f_k}(x_n)$ by constants C_n, and denote $C_n \mu_{X_n \rightarrow f_k}(x_n)$ by $\tilde{\mu}_{X_n \rightarrow f_k}(x_n)$. When we apply the SPA computation rule to the scaled messages, we get

$$\tilde{\mu}_{f_k \rightarrow X_m}(x_m) = \sum_{\sim \{x_m\}} f_k(x_1, \ldots, x_D) \prod_{n \neq m} \tilde{\mu}_{X_n \rightarrow f_k}(x_n) \tag{5.20}$$

$$= \mu_{f_k \rightarrow X_m}(x_m) \prod_{n \neq m} C_n. \tag{5.21}$$

We can now scale $\tilde{\mu}_{f_k \rightarrow X_m}(X_m)$ by some constant C_m so that its values fall within some predefined dynamic range. Note that from the scaled message $\tilde{\mu}_{f_k \rightarrow X_m}(X_m)$ we can derive the original, unscaled message $\mu_{f_k \rightarrow X_m}(X_m)$ simply by dividing it by $\prod_n C_n$. When the SPA terminates, we multiply the two scaled messages over any edge and obtain scaled marginals for each of the variables

$$\tilde{g}_{X_n}(x_n) = \gamma g_{X_n}(x_n) \tag{5.22}$$

for some *known* constant γ.

Suppose that we are interested in determining the marginal a-posteriori probability $p(X_n | \mathbf{Y} = \mathbf{y}, \mathcal{M})$. We find that

$$p(X_1 = x_1 | \mathbf{Y} = \mathbf{y}, \mathcal{M}) = \frac{g_{X_n}(x_n)}{\sum_{x \in \mathcal{X}_n} g_{X_n}(x)} \tag{5.23}$$

$$= \frac{\tilde{g}_{X_n}(x_n)}{\sum_{x \in \mathcal{X}_n} \tilde{g}_{X_n}(x)}. \tag{5.24}$$

In other words, *to determine the marginal a-posteriori distributions, we don't need to keep track of the scaling factors at all!* We can simply normalize the scaled marginals.

Example 5.3 (The burglar-alarm problem). *Going back to our burglar-alarm problem, suppose that we scale* $\mu_{g \to B}(B)$ *by* $C_B = 100$, *so that*

$$\tilde{\mu}_{g \to B}(b) = \begin{cases} 70.2 & b = 1 \\ 0.199 & b = 0 \end{cases}$$

and we scale $\mu_{g \to E}(E)$ *by* $C_E = 10$, *so that*

$$\tilde{\mu}_{g \to E}(e) = \begin{cases} 1.0800 & e = 1 \\ 0.0799 & e = 0 \end{cases}$$

This leads to the following scaled marginals:

$$\tilde{g}_B(b) = \begin{cases} 0.702\,00 & b = 1 \\ 0.197\,01 & b = 0 \end{cases}$$

and

$$\tilde{g}_E(e) = \begin{cases} 0.010\,800 & e = 1 \\ 0.079\,101 & e = 0. \end{cases}$$

The normalization constant for B is now $1/(0.702\,00 + 0.197\,01) = 1/0.0899 \approx 11.12$, *and that for E,* $1/0.899 \approx 1.112$.

The original normalization constant (i.e., 111.2) can be obtained since C_E *and* C_B *are known. This is important for the first inference problem (see Section 5.2.3.1). On the other hand, to compute the marginal a-posteriori distributions, we don't need to keep track of the scaling factors* C_E *and* C_B, *since the marginal a-posteriori distributions can be obtained simply by normalizing the scaled marginals* $\tilde{g}_B(B)$ *and* $\tilde{g}_E(E)$. *For instance, without knowing* C_E *or* C_B, *we find that*

$$p(B = b|A = 1) = \frac{\tilde{g}_B(b)}{\tilde{g}_B(0) + \tilde{g}_B(1)}$$

$$\approx \begin{cases} 0.7809 & b = 1 \\ 0.2191 & b = 0, \end{cases}$$

which is consistent with our previous results.

5.3.2   Distributions as messages

From this point onward, we will focus almost exclusively on the SPA, rather than on the max–sum algorithm. Since messages can be scaled arbitrarily in the SPA, we can ask ourselves whether there is some smart way to scale them. One particularly attractive way of scaling is *normalization*: we first compute our message in the normal way, assuming that f_k has as variables $X_1, \ldots, X_D$:

$$\mu_{f_k \to X_m}(x_m) = \sum_{\sim\{x_m\}} f_k(x_1, \ldots, x_D) \prod_{n \neq m} \mu_{X_n \to f_k}(x_n) \tag{5.25}$$

and then, once we have determined $\mu_{f_k \to X_m}(x_m)$ for all $x_m \in \mathcal{X}_m$, we scale (multiply by some constant γ) the message such that

$$\sum_{x_m \in \mathcal{X}_m} \gamma \mu_{f_k \to X_m}(x_m) = 1. \tag{5.26}$$

Remember that we don't need to keep track of the normalization constants in order to determine the marginal a-posteriori distributions of X_m. However, when computing the likelihood of the model, $p(\mathbf{Y} = \mathbf{y}|\mathcal{M})$, normalization constants are important. Note that for continuous variables summations become integrals.

Interpretations
1. This normalization process has an important implication: messages can be interpreted as probability distributions (pmfs or pdfs). This explains the better-known name of the SPA: *belief propagation*.
2. We can write (5.25) as

$$\mu_{f_k \to X_m}(x_m) \propto \mathbb{E}\{f_k(X_1, \ldots, X_D)|X_m = x_m\} \tag{5.27}$$

when we interpret the variables X_n, $n \neq m$, as being *independent* random variables with *a-priori distributions* $p(X_n) = \mu_{X_n \to f_k}(X_n)$. Here "$\propto$" refers to equality up to a multiplicative constant.
3. Since $\mu_{f_k \to X_m}(x_m)$ is also a distribution, we can go one step further and interpret (5.25) as

$$\mu_{f_k \to X_m}(x_m) = p(X_m = x_m) \tag{5.28}$$

$$= \sum_{\sim\{x_m\}} p\big(X_m = x_m, \{X_n = x_n\}_{n \neq m}\big) \tag{5.29}$$

$$= \sum_{\sim\{x_m\}} \underbrace{p\big(X_m = x_m \,\big|\, \{X_n = x_n\}_{n \neq m}\big)}_{\propto f_k(x_1, \ldots, x_D)} \prod_{n \neq m} \underbrace{p(X_n = x_n)}_{\mu_{X_n \to f_k}(x_n)}. \tag{5.30}$$

In other words, we can interpret $f_k(x_1, \ldots, x_D)$ as being proportional to a conditional distribution

$$p(X_m = x_m|\{X_n = x_n\}_{n \neq m}).$$

These interpretations will allow us to attach meaning to messages and compute them efficiently, capitalizing on the rich structure of probability theory.

5.3.3   Representation of messages for discrete variables

While we now know that we can scale messages and attach a meaning to them, we are still left with the question of how to represent them efficiently. In this section we will describe three common ways to achieve this. We again focus mainly on the SPA. As we have seen, messages in the max-sum algorithm can be represented by vectors and can be modified by adding constants without affecting the final result.

5.3.3.1   Probability mass functions

When a variable X_n is defined over a finite domain $\mathcal{X}_n$, the normalized messages are essentially probability mass functions (pmfs). As such, they can be represented as a vector of size $|\mathcal{X}_n|$. So, instead of writing the function $\mu_{f_k \to X_m}(X_m)$, we can write a vector $\mathbf{p}_{f_k \to X_m}$. For notational convenience, we will use the following notation to access the elements: $\mathbf{p}_{f_k \to X_m}(x_m) = \mu_{f_k \to X_m}(x_m)$. The SPA computation rule (5.25) now becomes

$$\mathbf{p}_{f_k \to X_m}(x_m) \propto \sum_{\sim\{x_m\}} f_k(x_1, \ldots, x_D) \prod_{n \neq m} \mathbf{p}_{X_n \to f_k}(x_n). \tag{5.31}$$

The normalization constant is found by computing the right-hand side of (5.31) for every $x_m \in \mathcal{X}_n$ and then normalizing so that $\sum_{x_m \in \mathcal{X}_n} \mathbf{p}_{f_k \to X_m}(x_m) = 1$.

5.3.3.2   Log-likelihoods

Suppose that we have a message $\mu_{f_k \to X_m}(X_m)$. We introduce a $|\mathcal{X}_n| \times 1$ vector $\mathbf{L}_{f_k \to X_m}$ (known as a log-domain representation or log-likelihood):

$$\mathbf{L}_{f_k \to X_m}(x_m) = \log \mu_{f_k \to X_m}(x_m). \tag{5.32}$$

Note that the original message can be recovered by exponentiating the vector $\mathbf{L}_{f_k \to X_m}$. This transformation has the benefit of converting the range $[0, +\infty]$ of $\mu_{f_k \to X_m}(X_m)$ to the range $[-\infty, +\infty]$ of $\mathbf{L}_{f_k \to X_m}$, effectively increasing the dynamic range of messages. Now (5.25) becomes

$$\mathbf{L}_{f_k \to X_m}(x_m) = \log\left(\sum_{\sim\{x_m\}} f_k(x_1, \ldots, x_D) \prod_{n \neq m} e^{\mathbf{L}_{X_n \to f_k}(x_n)} \right). \tag{5.33}$$

In the conventional SPA, the marginals are given by

$$g_{X_m}(x_m) = \mu_{f_k \to X_m}(x_m) \, \mu_{X_m \to f_k}(x_m). \tag{5.34}$$

In the log-domain this becomes

$$\log g_{X_m}(x_m) = \mathbf{L}_{f_k \to X_m}(x_m) + \mathbf{L}_{X_m \to f_k}(x_m). \tag{5.35}$$

We can add any constant to the message $\mathbf{L}_{f_k \to X_m}$ without affecting the SPA.

- When determining the likelihood of the model $p(\mathbf{Y} = \mathbf{y}|\mathcal{M})$ we need to keep track of all constants added to the messages.
- When determining the marginal a-posteriori distribution $p(X_m|\mathbf{Y} = \mathbf{y}, \mathcal{M})$, the constants are irrelevant, since determining

$$p(X_m = x_m|\mathbf{Y} = \mathbf{y}, \mathcal{M}) \propto \exp\left(\mathbf{L}_{f_k \to X_m}(x_m) + \mathbf{L}_{X_m \to f_k}(x_m)\right) \tag{5.36}$$

is not affected by adding any constant to $\mathbf{L}_{f_k \to X_m}$ or $\mathbf{L}_{X_m \to f_k}$.

In (5.33), we must repeatedly exponentiate messages and take logarithms. This is a rather costly affair, so we would rather perform message computation directly in the log-domain. To this end, it is helpful to introduce the *Jacobian logarithm*.

DEFINITION 5.1 (Jacobian logarithm). *For any $L \in \mathbb{N}_0$, the Jacobian logarithm is a function $\mathbb{M} : \mathbb{R}^L \to \mathbb{R}$ defined according to the following recursive rule:*

$$\mathbb{M}(L_1, \ldots, L_L) = \mathbb{M}(L_1, \mathbb{M}(L_2, \ldots, L_L)), \tag{5.37}$$

where

$$\mathbb{M}(L_1, L_2) = \max(L_1, L_2) + \log\left(1 + e^{-|L_1 - L_2|}\right) \tag{5.38}$$

and

$$\mathbb{M}(L_1) = L_1. \tag{5.39}$$

Similarly to the summation operator Σ, we will often abbreviate $\mathbb{M}(L_1, \ldots, L_L)$ as $\mathbb{M}_{i=1}^L(L_i)$. The core computation of the Jacobian logarithm (5.38) is usually implemented using a maximization and a table look-up, requiring a short table as a function of $|L_1 - L_2|$. This avoids explicit exponentiation and taking of logarithms, and allows the Jacobian algorithm to be implemented very efficiently.

The Jacobian logarithm has the following important property [64]:

$$\mathbb{M}(L_1, \ldots, L_L) = \log\left(\sum_{l=1}^{L} e^{L_l}\right). \tag{5.40}$$

This property allows us to express (5.33) as

$$\mathbf{L}_{f_k \to X_m}(x_m) = \mathbb{M}_{\sim\{x_m\}}\left(\log f_k(x_1, \ldots, x_D) + \sum_{n \neq m} \mathbf{L}_{X_n \to f_k}(x_n)\right), \tag{5.41}$$

where the Jacobian logarithm goes over all configurations $[x_1, \ldots, x_D]$ with mth entry equal to x_m. The low complexity of the Jacobian logarithm allows us to perform the entire SPA on log-domain messages without resorting to computationally demanding exponentiations and logarithms.

Example 5.4 (Repetition code). *Let us go back to our repetition-code example, with a factor graph shown in Fig. 5.2. We first write the messages from the leaves to the equality node as normalized vectors, so that*

$$\mathbf{p}_{f_B \to B} = \gamma [0.5 \quad 0.5]^{\mathsf{T}}$$

$$= \begin{bmatrix} \dfrac{1}{2} & \dfrac{1}{2} \end{bmatrix}^{\mathsf{T}}$$

and, for $k \in \{1,2\}$

$$\mathbf{p}_{f_k \to C_k} = \gamma [0.5 \quad 0.1]^{\mathsf{T}}$$

$$= \begin{bmatrix} \dfrac{5}{6} & \dfrac{1}{6} \end{bmatrix}^{\mathsf{T}}$$

and, for $k \in \{3,4\}$,

$$\mathbf{p}_{f_k \to C_k} = \gamma [0.5 \quad 0.9]^{\mathsf{T}}$$

$$= \begin{bmatrix} \dfrac{5}{14} & \dfrac{9}{14} \end{bmatrix}^{\mathsf{T}}.$$

Taking the logarithm yields $\mathbf{L}_{f_B \to B} = [-0.69 \ -0.69]^{\mathsf{T}}$, $\mathbf{L}_{f_k \to C_k} = [-0.18 \ -1.79]^{\mathsf{T}}$ for $k \in \{1,2\}$ and $\mathbf{L}_{f_k \to C_k} = [-1.03 \ -0.44]^{\mathsf{T}}$ for $k \in \{3,4\}$. Using (5.41), we transform (5.18) and compute the message $\mathbf{L}_{\boxminus \to B}$ as follows, for $b \in \{0,1\}$:

$$\mathbf{L}_{\boxminus \to B}(b) = \mathbb{M}_{c_1,c_2,c_3,c_4} \left(\log \boxminus (b, c_1, c_2, c_3, c_4) + \sum_{n=1}^{4} \mathbf{L}_{C_n \to \boxminus}(c_n) \right).$$

Since $\log \boxminus (b, c_1, c_2, c_3, c_4) = 0$ when $c_1 = c_2 = c_3 = c_4 = b$ and $-\infty$ otherwise, we find that

$$\mathbf{L}_{\boxminus \to B}(b) = \mathbb{M} \left(-\infty, -\infty, \ldots, -\infty, 0 + \sum_{n=1}^{4} \mathbf{L}_{C_n \to \boxminus}(b) \right)$$

$$= \sum_{n=1}^{4} \mathbf{L}_{C_n \to \boxminus}(b)$$

$$\approx \begin{cases} -4.46 & b = 1 \\ -2.42 & b = 0. \end{cases}$$

The marginals in the log-domain are then given by (up to an unknown additive constant)

$$\log g_B(b) = \mathbf{L}_{\boxminus \to B}(b) + \mathbf{L}_{B \to \boxminus}(b)$$

$$= \begin{cases} -5.16 & b = 1 \\ -3.12 & b = 0. \end{cases}$$

Since $\log g_B(1) < \log g_B(0)$, we can conclude that the transmitted bit was most likely a zero. If we wish to determine $p(B = b | \mathbf{Y} = \mathbf{y})$, we take the exponential of $\log g_B(B)$ and then normalize, giving us

$$p(B = b | \mathbf{Y} = \mathbf{y}) \approx \begin{cases} 0.11 & b = 1 \\ 0.89 & b = 0, \end{cases}$$

which is identical to the result from Section 5.2.4.

5.3.3.3 A link between sum–product and max–sum

Before we move on to a third representation of messages for discrete variables, let us make a small digression and reveal a link between the max–sum algorithm (which operates on the logarithm of a distribution) and log-likelihood representations in the SPA. Suppose that we have a factorization of $p(\mathbf{X}, \mathbf{Y} = \mathbf{y} | \mathcal{M})$, where $f_k(X_1, \dots, X_D)$ appears as a factor. Let us perform the SPA where messages are represented by log-likelihoods. We have the following rule:

$$\mathbf{L}_{f_k \to X_m}(x_m) = \mathbb{M}_{\sim \{x_m\}} \left(\log f_k(x_1, \dots, x_D) + \sum_{n \neq m} \mathbf{L}_{X_n \to f_k}(x_n) \right). \tag{5.42}$$

Since

$$\mathbb{M}(L_1, L_2) = \max(L_1, L_2) + \log \left(1 + e^{-|L_1 - L_2|} \right) \tag{5.43}$$

$$\approx \max(L_1, L_2) \tag{5.44}$$

when $|L_1 - L_2|$ is large, we can approximate $\mathbf{L}_{f_k \to X_m}(x_m)$ as

$$\mathbf{L}_{f_k \to X_m}(x_m) \approx \max_{\sim \{x_m\}} \left(\log f_k(x_1, \dots, x_D) + \sum_{n \neq m} \mathbf{L}_{X_n \to f_k}(x_n) \right). \tag{5.45}$$

If we were now to perform the max–sum algorithm on the factorization of $\log p(\mathbf{X}, \mathbf{Y} = \mathbf{y} | \mathcal{M})$, then the function $\log f_k(X_1, \dots, X_D)$ will appear as a factor. From the definition of the max–sum algorithm in Section 4.7 from Chapter 4, we know that the message-computation rule is given by

$$\mu_{f_k \to X_m}(x_m) = \max_{\sim \{x_m\}} \left(\log f_k(x_1, \dots, x_D) + \sum_{n \neq m} \mu_{X_n \to f_k}(x_n) \right). \tag{5.46}$$

On comparing (5.45) and (5.46), we notice that they are exactly the same. Performing the SPA with log-domain messages, combined with an approximation of the Jacobian logarithm, results in the max–sum algorithm.

5.3.3.4   Log-likelihood ratios

When we use log-likelihoods, we can add or subtract any real number to a message before transmitting it over an edge. One particular choice of such a number is the log-likelihood of a fixed reference element in $\mathcal{X}_m$. Let us order the elements in $\mathcal{X}_m$ as $\mathcal{X}_m = \{a_m^{(1)}, a_m^{(2)}, \ldots, a_m^{(|\mathcal{X}_m|)}\}$ and define a vector

$$\lambda_{f_k \to X_m} = \mathbf{L}_{f_k \to X_m} - \mathbf{L}_{f_k \to X_m}\left(a_m^{(1)}\right). \tag{5.47}$$

By subtracting a fixed entry, the vector will always have a zero in a fixed position (in this case, the first position). This means that we might as well not send this entry (since the receiving node knows that it is zero anyway). Since $\mathcal{X}_m$ contains $|\mathcal{X}_m|$ elements, this results in a memory saving of $(|\mathcal{X}_m| - 1)/|\mathcal{X}_m|$. This is useful only when $|\mathcal{X}_m|$ is small, for instance when X_m is a binary variable.

Note the following relation between the message represented as a pmf $\mathbf{p}_{f_k \to X_m}$ and the message as $\lambda_{f_k \to X_m}$,

$$\lambda_{f_k \to X_m}\left(a_m^{(n)}\right) = \log\left(\frac{\mathbf{p}_{f_k \to X_m}\left(a_m^{(n)}\right)}{\mathbf{p}_{f_k \to X_m}\left(a_m^{(1)}\right)}\right), \tag{5.48}$$

so the message $\lambda_{f_k \to X_m}$ is usually called a *log-likelihood ratio* (LLR). Since LLRs are a special case of log-likelihoods, all comment regarding computation rules and marginals remain valid.

Example 5.5 (Repitition code).   *From Section 5.3.3.2, we know that* $\mathbf{L}_{f_B \to B} \approx [-0.69 - 0.69]^T$, $\mathbf{L}_{f_k \to C_k} \approx [-0.18 - 1.79]^T$ *for* $k \in \{1, 2\}$ *and* $\mathbf{L}_{f_k \to C_k} \approx [-1.03 - 0.44]^T$ *for* $k \in \{3, 4\}$. *We take as reference element in the binary domain the element* 0, *and introduce the LLRs* $\lambda_{f_B \to B} = \mathbf{L}_{f_B \to B}(1) - \mathbf{L}_{f_B \to B}(0) = 0$, $\lambda_{f_k \to C_k} \approx -1.61$, *for* $k \in \{1, 2\}$, *and* $\lambda_{f_k \to C_k} \approx 0.59$, *for* $k \in \{3, 4\}$. *Since*

$$\mathbf{L}_{\boxminus \to B}(b) = \sum_{n=1}^{4} \mathbf{L}_{C_n \to \boxminus}(b)$$

we have also

$$\lambda_{\boxminus \to B} = \sum_{n=1}^{4} \lambda_{C_n \to \boxminus}$$

$$\approx -2.04.$$

Finally, the LLR of the marginal is given by

$$\lambda_B = \lambda_{\boxminus \to B} + \lambda_{B \to \boxminus}$$
$$= -2.04$$
$$= \log\left(\frac{p(B = 1|\mathbf{Y} = \mathbf{y})}{p(B = 0|\mathbf{Y} = \mathbf{y})}\right).$$

Since $\lambda_B < 0$, we can conclude that the transmitted bit was most likely a zero. If we wish to determine $p(B = b|\mathbf{Y} = \mathbf{y})$, we take the exponential and then normalize, giving us the same result as before:

$$p(B = 0|\mathbf{Y} = \mathbf{y}) = \frac{1}{1 + e^{\lambda_B}} \approx 0.89,$$
$$p(B = 1|\mathbf{Y} = \mathbf{y}) = \frac{e^{\lambda_B}}{1 + e^{\lambda_B}} \approx 0.11.$$

5.3.4   Representation of messages for continuous variables

While messages for discrete variables can be represented easily by means of a vector, this is no longer true for continuous variables. We focus solely on the SPA. When a variable X_n is defined over a continuous domain $\mathcal{X}_n$, the normalized messages in the SPA are essentially probability density functions (pdfs), and the sum–product rule now involves multi-dimensional integrations. This poses two problems.

1. How should the messages be represented?
2. How do we implement the sum–product rule?

In these sections, we will describe various ways of describing messages, and how to implement the sum–product rule. For the sake of clarity, we will focus on a node with three variables, $f(X, Y, Z)$, resulting in the following sum–product rule:

$$\mu_{f \to Z}(z) = \gamma \int_x \int_y f(x, y, z) \, \mu_{X \to f}(x) \, \mu_{Y \to f}(y) dx \, dy \qquad (5.49)$$

where γ is a normalization constant, chosen such that

$$\int_z \mu_{f \to Z}(z) dz = 1. \qquad (5.50)$$

The incoming messages over edges X and Y, $\mu_{X \to f}(X)$ and $\mu_{Y \to f}(Y)$, can be interpreted as pdfs $p_X(x)$ and $p_Y(y)$, respectively. Similarly, the outgoing message $\mu_{f \to Z}(Z)$ is a pdf $p_Z(z)$. As before, we also have the following interpretation of (5.49):

$$p_Z(z) = \int \int p_{Z|X,Y}(z|x, y) \, p_X(x) \, p_Y(y) dx \, dy \qquad (5.51)$$

with $p_{Z|X,Y}(z|x,y)$ equal to $\gamma f(x,y,x)$. In other words, when computing the message $\mu_{f \to Z}(z)$, we can use (5.51) and treat X and Y as independent random variables with known a-priori distributions. The conditional distribution of z, $p_{Z|X,Y}(z|x,y)$, is known only up to a constant.

In general, representing the pdfs exactly is impossible, and, even if they can be represented in closed form, computing the integrals required by the SPA may be very hard. For these reasons, we often look for approximate representations of the pdfs. We must (a) take care that the approximations are sufficiently accurate for our purpose, and (b) make the integrals tractable (either analytically or numerically).

We will describe three common representations: quantization, parametric representations, and non-parametric representations.

5.3.4.1   Quantization

The most natural way to circumvent the above problems is by quantization of the domains of X, Y, and Z. This leads to quantized versions of the functions, as well as the messages. Esentially, we approximate the pdf by a pmf. In this case, the techniques from Section 5.3.3 can be applied. The main drawback of such techniques is that the complexity and memory requirements scale exponentially in the dimensionality of the variables involved.

5.3.4.2   Parametric representations

The idea behind parametric representations is that the distributions can be captured entirely by a finite set of parameters. A classic example is the Gaussian mixture density (GMD). Suppose that $\mathbf{X}$ is an L-dimensional variable distributed according to a GMD, then the pdf $p_{\mathbf{X}}(\mathbf{x})$ is given by

$$p_{\mathbf{X}}(\mathbf{x}) = \sum_{l=1}^{L} \alpha_l \mathcal{N}_{\mathbf{x}}(\mathbf{m}_l, \Sigma_l) \qquad (5.52)$$

where α_l is the weight of the lth mixture component (with $\sum_l \alpha_l = 1$). In many applications, such a GMD is a good approximation of the true density. The message from a node is then given by the list of L mixture weights, L means, and L covariance matrices.

When the factors in the factorization have a convenient structure (an example will be given in Section 6.4), the SPA message-computation rule then reverts to transforming incoming GMD messages into an output GMD message.

5.3.4.3   Non-parametric representations

In non-parametric representations, we resort to particle representations. We recall from Chapter 3 that we have a particle representation $\mathcal{R}_L(p_X(\cdot))$ of a distribution $p_X(\cdot)$ with

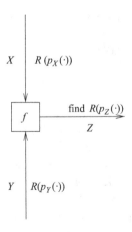

Figure 5.3. Non-parametric representations of messages. The function $f(x, y, z)$ is proportional to $p_{Z|X,Y}(z|x, y)$.

L properly weighted samples $\{(w_l, x_l)\}_{l=1}^{L}$ when, for any integrable function $f(x)$,

$$\int f(x) \, p_X(x) \mathrm{d}x \approx \sum_{l=1}^{L} \int f(x) \, w_l \boxminus (x_l, x) \mathrm{d}x \tag{5.53}$$

$$= \sum_{l=1}^{L} w_l f(x_l). \tag{5.54}$$

Solving (5.51) now becomes equivalent to solving the following problem (see Fig. 5.3): *given two independent random variables X and Y, with $\mathcal{R}_L(p_X(\cdot)) = \{(w_l, x_l)\}_{l=1}^{L}$, and $\mathcal{R}_L(p_Y(\cdot)) = \{(v_l, y_l)\}_{l=1}^{L}$, as well as a conditional pdf $p_{Z|X,Y}(z|x, y)$, known only up to a constant, find a particle representation of $p_Z(z)$, $\mathcal{R}_L(p_Z(\cdot)) = \{(u_l, z_l)\}_{l=1}^{L}$.* We will describe three possible solutions: importance sampling, mixture sampling, and regularization.

Approach 1 – importance sampling
We go through the following steps.

- Let us draw L iid samples from $p_X(x)$: $x^{(k)} \sim p_X(x)$, and L iid samples from $p_Y(y)$: $y^{(k)} \sim p_Y(y)$. Since, for the purpose of computing the message $p_Z(z)$, we can treat X and Y as independent, this results in L samples $\{(x^{(k)}, y^{(k)})\}_{k=1}^{L}$ from the joint distribution $p_{X,Y}(x, y) = p_X(x)p_Y(y)$.
- We now use importance sampling (see Section 3.3.2) to generate samples from $p_{X,Y,Z}(x, y, z)$. For every couple $(x^{(k)}, y^{(k)})$, we draw one sample $z^{(k)} \sim q_{Z|X,Y}(z|x^{(k)}, y^{(k)})$, where $q_{Z|X,Y}(z|x^{(k)}, y^{(k)})$ is a well-chosen sampling distribution. The sample $(x^{(k)}, y^{(k)}, z^{(k)})$ is thus taken from the distribution

$q_{Z|X,Y}(z|x,y)p_X(x)p_Y(y)$. We set the weighting of the sample $z^{(k)}$ to

$$u^{(k)} = \frac{p_{Z|X,Y}\left(z^{(k)}\big|x^{(k)},y^{(k)}\right)p_X\left(x^{(k)}\right)p_Y\left(y^{(k)}\right)}{q_{Z|X,Y}\left(z^{(k)}\big|x^{(k)},y^{(k)}\right)p_X\left(x^{(k)}\right)p_Y\left(y^{(k)}\right)} \tag{5.55}$$

$$\propto \frac{f\left(x^{(k)},y^{(k)},z^{(k)}\right)}{q_{Z|X,Y}\left(z^{(k)}\big|x^{(k)},y^{(k)}\right)}. \tag{5.56}$$

We remind the reader that $f(x,y,z)$ can be evaluated for any x,y,z and is proportional to $p_{Z|X,Y}(z|x,y)$. We normalize the weights such that $\sum_k u^{(k)} = 1$. The L weighted samples $\{(u^{(k)},(x^{(k)},y^{(k)},z^{(k)}))\}_{k=1}^L$ form a particle representation of the joint distribution $p_{X,Y,Z}(x,y,z) = p_{Z|X,Y}(z|x,y)p_X(x)p_Y(y)$.

• Retaining only the third component in the samples results in a list of L properly weighted samples $\{(u^{(k)},z^{(k)})\}_{k=1}^L$, which form a particle representation of $p_Z(z)$, the desired distribution (see Section 3.3.1).

The algorithm is shown in Algorithm 5.1. In some applications it is necessary to resample from the particle representation of $p_Z(z)$ in order to avoid degeneration problems [65]: as messages propagate over the factor graph, it can happen that all the weight gets concentrated on a single sample. Resampling at every message update removes this problem.

Algorithm 5.1 The sum–product rule for continuous variables – importance sampling

1: *input:* $\mathcal{R}_L(p_X(x)) = \{(w^{(l)},x^{(l)})\}_{l=1}^L$ and $\mathcal{R}_L(p_Y(y)) = \{(v^{(l)},y^{(l)})\}_{l=1}^L$
2: **for** $k = 1$ to L **do**
3:     draw $x^{(k)} \sim p_X(x)$
4:     draw $x^{(k)} \sim p_Y(y)$
5:     draw $z^{(k)} \sim q_{Z|X,Y}(z|x^{(k)},y^{(k)})$
6:     set $u^{(k)} = f(x^{(k)},y^{(k)},z^{(k)})/q_{Z|X,Y}(z^{(k)}|x^{(k)},y^{(k)})$
7: **end for**
8: normalize weights
9: *output:* $\mathcal{R}_L(p_Z(z)) = \{(u^{(k)},z^{(k)})\}_{k=1}^L$

Approach 2 – mixture sampling

On substituting the particle representations of $p_X(x)$ and $p_Y(y)$ into (5.51), we can write

$$p_Z(z) \approx \int \int p_{Z|X,Y}(z|x,y) \left\{ \sum_{l_1=1}^L w^{(l_1)}\boxminus\left(x,x^{(l_1)}\right) \right\} \left\{ \sum_{l_2=1}^L w^{(l_2)}\boxminus\left(y,y^{(l_2)}\right) \right\} dx\,dy \tag{5.57}$$

$$= \sum_{l_1=1}^L w^{(l_1)} \sum_{l_2=1}^L v^{(l_2)} p_{Z|X,Y}\left(z\big|x^{(l_1)},y^{(l_2)}\right) \tag{5.58}$$

so that we can interpret $p_Z(z)$ as a mixture density with L^2 mixture components. We will now draw a weighted sample from every mixture component: for every combination $(x^{(l_1)}, y^{(l_2)})$, we draw a sample $z^{(l_1,l_2)} \sim q_{Z|X,Y}(z|x^{(l_1)}, y^{(l_2)})$ and set the corresponding weight to

$$u^{(l_1,l_2)} = w^{(l_1)} v^{(l_2)} \frac{f\left(x^{(l_1)}, y^{(l_2)}, z^{(l_1,l_2)}\right)}{q_{Z|X,Y}\left(z^{(l_1,l_2)} \middle| x^{(l_1)}, y^{(l_2)}\right)}. \tag{5.59}$$

We end up with L^2 samples (in general, for a node of degree D, we end up with L^{D-1} samples). We normalize the weights, resulting in a particle representation $\{(u^{(k)}, z^{(k)})\}_{k=1}^{L^2}$ of $p_Z(z)$ with L^2 samples. The number of samples can be reduced by resampling. The algorithm is shown in Algorithm 5.2.

Algorithm 5.2 Sum–product rule for continuous variables – mixture sampling

1: *input:* $\mathcal{R}_L(p_X(\cdot)) = \{(w^{(l)}, x^{(l)})\}_{l=1}^L$ and $\mathcal{R}_L(p_Y(\cdot)) = \{(v^{(l)}, y^{(l)})\}_{l=1}^L$
2: **for** $l_1 = 1$ to L **do**
3:     **for** $l_2 = 1$ to L **do**
4:         draw $z^{(l_1,l_2)} \sim q_{Z|X,Y}(z|x^{(l_1)}, y^{(l_2)})$
5:         set $u^{(l_1,l_2)} = w^{(l_1)} v^{(l_2)} f(x^{(l_1)}, y^{(l_2)}, z^{(l_1,l_2)}) / q_{Z|X,Y}(z^{(l_1,l_2)} | x^{(l_1)}, y^{(l_2)})$
6:     **end for**
7: **end for**
8: normalize weights
9: *output:* $\mathcal{R}_L(p_Z(\cdot)) = \{(u^{(k)}, z^{(k)})\}_{k=1}^{L^2}$

Approach 3 – regularization

The above techniques suffer from one important drawback: in some cases the function $f(x^{(l_1)}, y^{(l_2)}, z)$, as a function of z, is zero almost everywhere. This occurs for instance when $f(\cdot)$ corresponds to an equality node: $f(x^{(l_1)}, y^{(l_2)}, z) = \boxminus(x^{(l_1)}, y^{(l_2)}, z)$. Since the sample from $p_X(x)$ and the sample from $p_Y(y)$ will generally not be the same, $f(x^{(l_1)}, y^{(l_2)}, z) = 0$, $\forall z$. This problem can be circumvented by regularization (see Section 3.3.1): we approximate the particle representation by a mixture of Gaussian densities. The process requires two steps.

1. Convert the particle representations $\mathcal{R}_L(p_Y(\cdot))$ and $\mathcal{R}_L(p_X(\cdot))$ into an appropriate Gaussian mixture. Starting from $\mathcal{R}_L(p_X(\cdot)) = \{(w^{(l)}, x^{(l)})\}_{l=1}^L$, we approximate $p_X(x)$ as

$$p_X(x) \approx \sum_{l=1}^L w^{(l)} \mathcal{N}_x\left(x^{(l)}, \sigma^2\right), \tag{5.60}$$

where, for scalar X, $\sigma^2 = (4/(3L))^{1/5}$ [45]. We do the same for $p_Y(y)$. Now we have approximations of the distributions for all x and for all y.

2. Draw L properly weighted samples from $p_Z(z)$. There are many ways to do this, depending on the structure of $f(x, y, z)$.

   (a) In the particular case of an equality node $(f(x, y, z) = \boxed{=}(x, y, z))$, $p_Z(z)$ is also a Gaussian mixture with L^2 components:

   $$p_Z(z) = \gamma p_X(z) p_Y(z). \tag{5.61}$$

   Samples from this Gausssian mixture can be obtained by sampling from $z^{(k)} \sim p_X(z)$, and setting the importance weight $u^{(k)}$ to be $p_Y(z_k)$ (or vice versa). An alternative technique based on Gibbs sampling is described in [66].

   (b) For a more general $f(x, y, z)$, we can sample from some appropriate joint sampling distribution $q_{XYZ}(x, y, z) = \psi(x, y, z)$, and then set the weight $u^{(k)}$ of sample $(x^{(k)}, y^{(k)}, z^{(k)})$ to be $u^{(k)} \propto p_X(x^{(k)}) p_Y(y^{(k)}) f(x^{(k)}, y^{(k)}, z^{(k)}) / \psi(x^{(k)}, y^{(k)}, z^{(k)})$. Retaining only the third component of every sample results in a particle representation $\{(u^{(k)}, z^{(k)})\}_{k=1}^{L}$ of $p_Z(z)$.

5.4   Loopy inference

We now know how to solve inference problems on factor graphs, and how to represent and compute messages both for discrete and for continuous variables. Throughout this chapter, we have managed to avoid factor graphs with cycles. However, many inference problems do not lend themselves well to cycle-free graphs. What happens when the factor graph contains cycles? We have seen in Section 4.6.6 that the SPA needs to be modified to take into account cyclic dependencies. This is easily achieved as follows.

- A selected set of messages within the graph is set to uniform distributions over the corresponding domains.
- After a certain (predefined or dynamically decided) number of iterations, the SPA is halted, and *approximate* marginal a-posteriori distributions are computed. These approximate distributions are sometimes referred to as *beliefs* (in order to distinguish them from the true a-posteriori distributions).

This at least allows us to execute the SPA on cyclic graphs. The SPA now becomes iterative (or loopy). This leads to the terms *loopy inference, loopy belief propagation, iterative processing, and turbo processing*. In Section 4.6.6, we showed that messages tend to zero or to infinity with progressive iterations. In inference problems, we normalize messages so that this problem is removed: messages may possibly converge to reasonable distributions. But how should we interpret these distributions? Is there any relationship between the marginals obtained and the true marginal a-posteriori distributions? Do the messages converge at all? These questions are hard to answer in general, and in fact it is fair to say that the answer is not yet fully known. The interested reader is directed to [61, 67, 68] and references therein for more information. In particular [61] deals with generalizations of the SPA that may give better performance and reveals an important

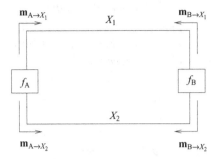

Figure 5.4. A factor graph of the distribution $p(\mathbf{Y} = \mathbf{y}|X_1, X_2)p(X_1X_2)$, with $f_A(X_1, X_2) = p(\mathbf{Y} = \mathbf{y}|X_1, X_2)$ and $f_B(X_1, X_2) = p(X_1, X_2)$.

link between the fixed points of the SPA and free-energy minimization in statistical physics.

This incomplete understanding doesn't mean that we should forget about loopy inference! Indeed, in most practical scenarios where we are interested in determining marginal a-posteriori distributions it turns out that implementing the SPA for inference on cyclic factor graphs gives excellent empirical results. Similarly, the likelihood of the model can be approximated by applying the techniques from [61], where it is shown that $\log p(\mathbf{Y} = \mathbf{y}|\mathcal{M}) = -F_H$, where F_H is the Helmholtz free energy, and that F_H can be approximated by the Bethe free energy F_{Bethe}, which in turn is a simple function of the beliefs obtained in the SPA[1].

Let us study the simplest of examples: a factor graph with a single cycle and two nodes. Extension to more general factor graphs with a single cycle is straightforward. As we will see, there is a nice relationship between the SPA marginals (the beliefs) and the true marginal a-posteriori distributions, and loopy inference may give good results.

Example 5.6. *We have two variables, X_1 and X_2, defined over a finite domain $\Omega = \{a_1, a_2, \ldots, a_L\}$ with $L > 1$ elements. After observing $\mathbf{Y} = \mathbf{y}$, we write the joint distribution of $\mathbf{X} = [X_1X_2]$ and $\mathbf{Y} = \mathbf{y}$:*

$$p(\mathbf{X}, \mathbf{Y} = \mathbf{y}) = p(\mathbf{Y} = \mathbf{y}|X_1, X_2)\, p(X_1X_2)$$
$$= p(\mathbf{Y} = \mathbf{y}|X_1, X_2)\, p(X_1)\, p(X_2|X_1).$$

When neither the likelihood function nor the a-priori distribution has a convenient factorizations, we end up with a factorization of the form $f(X_1, Y_2) = f_A(X_1, X_2)f_B(X_1, X_2)$, which is exactly the situation we encountered in Section 4.6.6. The factor graph is shown in Fig. 5.4. As we did in Chapter 4, we will represent the messages as vectors so that $\mu_{f_A \to X_1}(X_1) = \mu_{X_1 \to f_B}(X_1)$ becomes $\mathbf{m}_{A \to X_1}$. Similarly, $\mu_{f_B \to X_1}(X_1)$ becomes $\mathbf{m}_{B \to X_1}$, $\mu_{f_B \to X_2}(X_2)$ becomes $\mathbf{m}_{B \to X_2}$, and $\mu_{f_A \to X_2}(X_2)$ becomes

---

$\mathbf{m}_{A \to X_2}$. *We evaluate* $f_A(x_1, x_2)$, $\forall (x_1, x_2) \in \Omega^2$ *and place the result in an* $L \times L$ *matrix* **A**. *Similarly,* $f_B(x_1, x_2)$, $\forall (x_1, x_2) \in \Omega^2$ *is placed into an* $L \times L$ *matrix* **B**. *Using these notations, we see that the sum–product computation rules now revert to two disjunct set of rules:* $\mathbf{m}_{A \to X_1}$ *and* $\mathbf{m}_{B \to X_2}$ *are updated clockwise (see Fig. 5.4):*

$$
\begin{aligned}
\mathbf{m}_{B \to X_2} &\propto \mathbf{B} \mathbf{m}_{A \to X_1}, \\
\mathbf{m}_{A \to X_1} &\propto \mathbf{A}^{\mathsf{T}} \mathbf{m}_{B \to X_2},
\end{aligned}
$$

while the messages $\mathbf{m}_{A \to X_2}$ *and* $\mathbf{m}_{B \to X_1}$ *are updated counterclockwise:*

$$
\begin{aligned}
\mathbf{m}_{A \to X_2} &\propto \mathbf{A}^{\mathsf{T}} \mathbf{m}_{B \to X_1}, \\
\mathbf{m}_{B \to X_1} &\propto \mathbf{B} \mathbf{m}_{A \to X_2}.
\end{aligned}
$$

If we initialize $\mathbf{m}_{A \to X}^{(0)}$ *to some (e.g., random or uniform) distribution, then the clockwise messages can be updated iteratively. We do the same for* $\mathbf{m}_{B \to X_1}^{(0)}$, *enabling iterative updating of the counterclockwise messages. After n iterations,* $\mathbf{m}_{A \to X_1}$ *becomes*

$$
\mathbf{m}_{A \to X_1}^{(n)} = \gamma^{(n)} \mathbf{C}^n \mathbf{m}_{A \to X_1}^{(0)}
$$

where $\mathbf{C} = \mathbf{A}\mathbf{B}^{\mathsf{T}}$ *and* $\gamma^{(n)}$ *is a normalization constant. Suppose that we can find a set of eigenvectors of* **C** *and corresponding eigenvalues* $\{(\lambda_k, \mathbf{v}_k)\}_{k=1}^{L}$ *that span* $\mathbb{R}^L$, *ordered such that* $|\lambda_k| \geq |\lambda_{k-1}|$. *We assume that there is a single dominant eigenvector* λ_1. *Introducing* $\mathbf{V} = [\mathbf{v}_1, \ldots, \mathbf{v}_L]$, *and* $\Lambda = \mathrm{diag}\{\lambda_1, \ldots, \lambda_L\}$, *we know that* $\mathbf{C}\mathbf{V} = \mathbf{V}\Lambda$, *so that* $\mathbf{C}^{\mathsf{T}}(\mathbf{V}^{\mathsf{T}})^{-1} = (\mathbf{V}^{\mathsf{T}})^{-1}\Lambda$. *In other words, the columns in the matrix* $(\mathbf{V}^{\mathsf{T}})^{-1} = \mathbf{W}$ *are ordered eigenvectors of* $\mathbf{C}^{\mathsf{T}}$.

Going back to our messages, we can now express $\mathbf{m}_{A \to X_1}^{(0)}$ *as a function of the eigenvectors:*

$$
\mathbf{m}_{A \to X_1}^{(0)} = \sum_{k=1}^{L} \alpha_k \mathbf{v}_k
$$

for some coefficients $\alpha_1, \ldots, \alpha_L$. *It then follows that*

$$
\mathbf{m}_{A \to X_1}^{(n)} = \gamma^{(n)} \sum_{k=1}^{L} \alpha_k \lambda_k^n \mathbf{v}_k.
$$

Assuming that $\alpha_k \neq 0$, $\forall k$, *as* $n \to +\infty$, $\mathbf{m}_{A \to X_1}^{(n)}$ *will lie along the dominant eigenvector (corresponding to the eigenvalue with the largest magnitude) of* **C**. *Similarly,* $\mathbf{m}_{B \to X_1}^{(n)}$ *will lie along the dominant eigenvector of* $\mathbf{C}^{\mathsf{T}}$. *In both cases, we see that the convergence rate is proportional to* λ_1/λ_2: *when the dominant eigenvalue is much larger than the other eigenvalues, messages will converge faster (in fewer iterations). In other words,* $\mathbf{m}_{A \to X_1}^{(+\infty)} = \gamma_1 \mathbf{v}_1$ *and* $\mathbf{m}_{B \to X}^{(+\infty)} = \gamma_2 \mathbf{w}_1$, *where* $\mathbf{w}_1$ *is the first column of* **W**, *and* γ_1 *and* γ_2

are normalization constants. The normalized marginals are then given by

$$g_{X_1}(a_i) = [\mathbf{v}_1]_i [\mathbf{w}_1]_i$$

$$= [\mathbf{V}]_{i,1} \left[\mathbf{V}^{-1}\right]_{1,i}.$$

It is easily verified that $\sum_{a \in \Omega} g_{X_1}(a) = 1$. On the other hand, the true marginal a-posteriori distribution of X_1 is given by

$$p(X_1 = a_i | \mathbf{Y} = \mathbf{y}) = \frac{C_{ii}}{\sum_{j=1}^{L} C_{jj}}.$$

We re-write $\sum_{j=1}^{L} C_{jj} = \sum_{j=1}^{L} \lambda_j$, and $C_{ii} = \mathbf{e}_i^{\mathsf{T}} C \mathbf{e}_i$, where $\mathbf{e}_i$ is a column vector with all zeros, except a one on index i. We find the following relationship between $p(X_1 = a_i | \mathbf{Y} = \mathbf{y})$ and $g_{X_1}(a_i)$:

$$\underbrace{p(X_1 = a_i | \mathbf{Y} = \mathbf{y})}_{\text{true a-post. distribution}} = \frac{\mathbf{e}_i^{\mathsf{T}} C \mathbf{e}_i}{\sum_{j=1}^{L} \lambda_j}$$

$$= \frac{\mathbf{e}_i^{\mathsf{T}} \mathbf{V} \Lambda \mathbf{V}^{-1} \mathbf{e}_i}{\sum_{j=1}^{L} \lambda_j}$$

$$= \frac{\sum_{n=1}^{L} \lambda_n [\mathbf{V}]_{i,n} \left[\mathbf{V}^{-1}\right]_{n,i}}{\sum_{j=1}^{L} \lambda_j}$$

$$= \frac{\lambda_1}{\sum_{j=1}^{L} \lambda_j} g_{X_1}(a_i) + \sum_{n=2}^{L} \frac{\lambda_n [\mathbf{V}]_{i,n} \left[\mathbf{V}^{-1}\right]_{n,i}}{\sum_{j=1}^{L} \lambda_j}$$

$$= (1 - \varepsilon) \underbrace{g_{X_1}(a_i)}_{\text{belief}} + \varepsilon \sum_{n=2}^{L} \frac{\lambda_n [\mathbf{V}]_{i,n} \left[\mathbf{V}^{-1}\right]_{n,i}}{\sum_{j=2}^{L} \lambda_j},$$

where

$$\varepsilon = 1 - \frac{\lambda_1}{\sum_{j=1}^{L} \lambda_j}.$$

This leads to the interpretation that the loopy SPA gives us the correct marginal a-posteriori distributions up to an additive error. This error is small when the largest eigenvalue is much larger (in magnitude) than the other eigenvalues (i.e., when ε is close to 0).

5.5   Main points

In statistical inference we are interested in obtaining information regarding certain variables $\mathbf{X}$, on the basis of an observation $\mathbf{Y} = \mathbf{y}$. We have seen how factor graphs can help in solving the following inference problems.

(1) What is $p(\mathbf{Y} = \mathbf{y}|\mathcal{M})$, the likelihood of the model $\mathcal{M}$?
(2) What is the a-posteriori distribution $p(X_k|\mathbf{Y} = \mathbf{y}, \mathcal{M})$ of X_k, given a certain model $\mathcal{M}$? What are the characteristics of this distribution (its mode, its moments)?
(3) What is $\hat{\mathbf{x}} = \arg\max_\mathbf{x} p(\mathbf{X} = \mathbf{x}|\mathbf{Y} = \mathbf{y}, \mathcal{M})$, the mode of the a-posteriori distribution of $\mathbf{X}$, given a certain model $\mathcal{M}$?

To solve the first two problems, the general idea is to create a cycle-free factor graph of a factorization of the joint distribution $p(\mathbf{X}, \mathbf{Y} = \mathbf{y}|\mathcal{M})$ and implement the SPA on this graph. The last problem is solved by working in the $(\mathbb{R}, \max, +)$ semi-ring and implementing the max–sum algorithm on the cycle-free factor graph of $\log p(\mathbf{X}, \mathbf{Y} = \mathbf{y}|\mathcal{M})$.

We have described how messages can be scaled without affecting the outcome of the SPA. For discrete variables, messages can be represented by vectors of three types: probability mass functions, log-likelihoods, and log-likelihood ratios. For continuous variables, we must resort to quantization, parametric representations, or particle representations.

While the SPA guarantees to give the correct marginals only for cycle-free factor graphs, the most exciting and promising applications are precisely on cyclic factor graphs. This naturally leads to iterative inference techniques. While not much is known regarding convergence behavior, it is generally accepted that the marginals obtained (the beliefs) are approximations of the true marginal a-posteriori distributions. A wealth of empirical evidence supports this claim.

6 State-space models

6.1 Introduction

State-space models (SSMs) are a mathematical abstraction of many real-life dynamic systems. They have been proven to be useful in a wide variety of fields, including robot tracking, speech processing, control systems, stock prediction, and bio-informatics, basically anywhere there is a dynamic system [70–75]. These models are not only of great practical relevance, but also a good illustration of the power of factor graphs and the SPA. The central idea behind an SSM is that the system at any given time can be described by a *state*, belonging to a state space. The state space can be either discrete or continuous. The state changes dynamically over time according to a known statistical rule. We cannot observe the state directly; the state is said to be hidden. Instead we observe another quantity (the *observation*), which has a known statistical relationship with the state. Once we have collected a sequence of observations, our goal is to infer the corresponding sequence of states.

This chapter is organized as follows.

- In **Section 6.2** we will describe the basic concepts of SSMs, create an appropriate factor graph, and show how the sum–product and max–sum algorithms can be executed on this factor graph. Then, we will consider three cases of SSM in detail.
- In **Section 6.3**, we will cover models with discrete state spaces, known as hidden Markov models (HMMs), where we reformulate the well-known forward–backward and Viterbi algorithms using factor graphs.
- Then, in **Section 6.4** we go on to linear Gaussian models, and will derive a version of the Kalman filter and Kalman smoother.
- Lastly, we will consider general SSM with continuous state spaces in **Section 6.5**, and show how Monte Carlo techniques can be combined with factor graphs to perform inference. This leads to the so-called particle filter and variations thereof.

While the concept of an SSM is an important topic in and by itself, for the purpose of iterative receiver design, the reader can restrict his/her attention to Sections 6.2 and 6.3. Sections 6.4 and 6.5 are provided merely for the sake of completeness.

6.2   State-space models

6.2.1   Definition

In discrete-time SSMs, the variable $X_k \in \mathcal{X}_k$ is referred to as the *state* at *time instant* k. The variable $Y_k \in \mathcal{Y}_k$ is the *output* at time k. We will abbreviate $[X_l, X_{l+1}, \ldots, X_k]$ by $\mathbf{X}_{l:k}$, and $[Y_l, \ldots, Y_k]$ by $\mathbf{Y}_{l:k}$, for $l \leq k$. The states form a first-order Markov chain, such that

$$p\big(X_k = x_k \big| \mathbf{X}_{0:k-1} = \mathbf{x}_{0:k-1}, \mathcal{M}\big) = p\big(X_k = x_k \big| X_{k-1} = x_{k-1}, \mathcal{M}\big). \qquad (6.1)$$

There are two types of SSM, depending on how the output depends on the states.

- **Type 1 – transition-emitting SSMs:** the output at time k depends solely on the states at times $k - 1$ and k,

$$p(Y_k = y_k | \mathbf{X}_{0:N} = \mathbf{x}_{0:N}, \mathcal{M}) = p\big(Y_k = y_k \big| X_{k-1} = x_{k-1}, X_k = x_k, \mathcal{M}\big). \qquad (6.2)$$

- **Type 2 – state-emitting SSMs:** the output at time k depends solely on the state at time k,

$$p(Y_k = y_k | \mathbf{X}_{0:N} = \mathbf{x}_{0:N}, \mathcal{M}) = p(Y_k = y_k | X_k = x_k, \mathcal{M}). \qquad (6.3)$$

The probabilities $p(X_k | X_{k-1}, \mathcal{M})$ and $p(Y_k | X_{k-1}, X_k, \mathcal{M})$ are known as the transition probabilities and the output probabilities, respectively. In the world of finite state machines (where the state space is discrete and finite), a state-emitting SSM is known as a Moore machine, while a transition-emitting SSM is a Mealy machine. From their definitions, it is clear that a state-emitting SSM is a particular case of a transition-emitting SSM. The role of the model $\mathcal{M}$ in this context can be the detailed description of the transition and output probabilities, the number of states, etc.

Higher-order Markov models
In an Lth-order Markov chain, the states satisfy

$$p\big(X_k = x_k \big| \mathbf{X}_{0:k-1} = \mathbf{x}_{0:k-1}, \mathcal{M}\big) = p\big(X_k = x_k \big| X_{k-1} = x_{k-1}, \ldots, X_{k-L} = x_{k-L}, \mathcal{M}\big). \qquad (6.4)$$

This can be transformed into a first-order Markov chain by introducing the super-states

$$S_k = [X_k, X_{k-1}, \ldots, X_{k-L+1}] \qquad (6.5)$$

such that

$$p\big(S_k = s_k \big| \mathbf{S}_{0:k-1} = \mathbf{s}_{0:k-1}, \mathcal{M}\big) = p\big(S_k = s_k \big| S_{k-1} = s_{k-1}, \mathcal{M}\big). \qquad (6.6)$$

This implies that, without any loss in generality, we can focus on first-order Markov models.

Example 6.1 (Inferring human emotions). *We are performing a psychological experiment on a human being called Nick. By observing Nick's facial expressions, we hope to infer his underlying emotions. The state space consists of two emotions $\mathcal{X} = \{h, s\}$, where h stands for "happy" and s for "sad." Let us assume this to be a reasonable model that captures the possible emotions Nick can have. We observe Nick every minute (this corresponds to the discrete time index k) and take note of his facial expression. There are three possible facial expression Nick is able to display $\mathcal{Y} = \{l, c, b\}$, where l stands for "laughing," c for "crying," and b for "blank." Nick's facial expression depends on his emotion at that time, with*

$$p(Y_k = l \mid X_k = h) = 0.3,$$
$$p(Y_k = c \mid X_k = h) = 0.1,$$
$$p(Y_k = b \mid X_k = h) = 0.6,$$

and

$$p(Y_k = l \mid X_k = s) = 0.1,$$
$$p(Y_k = c \mid X_k = s) = 0.2,$$
$$p(Y_k = b \mid X_k = s) = 0.7.$$

We also know the dynamic model of how Nick's mood changes over time:

$$p(X_k = h \mid X_{k-1} = s) = 0.1,$$
$$p(X_k = s \mid X_{k-1} = h) = 0.9,$$

as well as the a-priori probabilities of Nick's mood before we start our observations:

$$p(X_0 = h) = 0.1$$
$$p(X_0 = s) = 0.9.$$

Taking note of Nick's facial expressions, we find that $\mathbf{y} = [l \, b]$. Our goal is to determine Nick's emotions during those 2 minutes. It is clear that the emotions and facial expressions satisfy the conditions of a state-emitting SSM. We will see how to infer Nick's underlying emotions in the next sections.

6.2.2   Factor-graph representation

The joint distribution of the state sequence $\mathbf{X} = X_{0:N}$ and the observation $\mathbf{Y} = \mathbf{y} = y_{0:N}$ can be factorized as follows, for a transition-emitting SSM,

$$p(\mathbf{X}, \mathbf{Y} = \mathbf{y}|\mathcal{M}) = p(X_0|\mathcal{M}) \prod_{k=1}^{N} \underbrace{p\big(Y_k = y_k|X_k, X_{k-1}, \mathcal{M}\big) p\big(X_k|X_{k-1}, \mathcal{M}\big)}_{f_k(X_{k-1}, X_k)}, \quad (6.7)$$

and for the particular case of a state-emitting SSM,

$$p(\mathbf{X}, \mathbf{Y} = \mathbf{y}|\mathcal{M}) = p(X_0|\mathcal{M}) \prod_{k=1}^{N} \underbrace{p(Y_k = y_k|X_k, \mathcal{M}) p\big(X_k|X_{k-1}, \mathcal{M}\big)}_{f_k(X_{k-1}, X_k)}. \quad (6.8)$$

The factor graphs of (6.7) and (6.8) are shown in Fig. 6.1 and Fig. 6.2, respectively. In a state-emitting SSM, we can open up (as described in Section 4.6.2) the nodes f_k, for $k > 0$. The result is depicted in Fig. 6.2. Note that opening the node f_k cannot be applied to the more general transition-emitting SSM, since this would result in a factor graph with cycles.

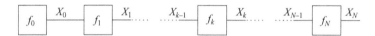

Figure 6.1. A factor graph for a transition-emitting SSM, with $f_0(X_0) = p(X_0|\mathcal{M})$ and $f_k(X_{k-1}, X_k) = p(Y_k = y_k|X_k, X_{k-1}, \mathcal{M})p\,(X_k|X_{k-1}, \mathcal{M})$.

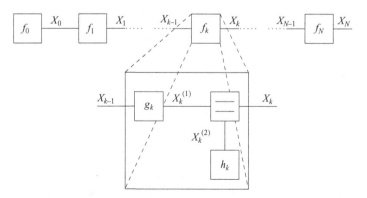

Figure 6.2. A factor graph for a state-emitting SSM, with $f_0(X_0) = p(X_0)$ and $f_k(X_{k-1}, X_k) = p(Y_k = y_k|X_k)p(X_k|X_{k-1})$. The node f_k is opened to reveal its structure, with $g_k(X_{k-1}, X_k^{(1)}) = p(X_k^{(1)}|X_{k-1})$ and $h_k(X_k^{(2)}) = p(Y_k = y_k|X_k^{(2)})$, and an equality node.

6.2.3   The sum–product algorithm for state-space models

6.2.3.1   General solution

In this section we will see how to solve inference on SSMs in a general way, and how this leads to sequential processing. From Figs. 6.1 and 6.2, we immediately see (i) that the factor graph has no cycles, and (ii) how to perform the SPA. Messages are propagated from left to right, and, at the same time, from right to left. These two phases are known as the forward and backward phases, respectively. Both phases are depicted in Fig. 6.3.

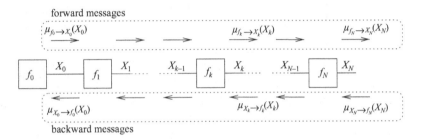

Figure 6.3. The sum–product algorithm on a state-space model with forward and backward phases.

- **Forward phase**: initially, a message $\mu_{f_0 \to X_0}(X_0)$ is sent at the left-most leaf of the graph. This allows us to compute $\mu_{f_1 \to X_1}(X_1)$, and then $\mu_{f_2 \to X_2}(X_2)$, and so forth up until $\mu_{f_N \to X_N}(X_N)$. The messages are known as forward messages (since they are propagated from left to right, from the past to the future).
- **Backward phase**: at the same time, the half-edge X_N can send a message (a constant) to f_N: $\mu_{X_N \to f_N}(X_N)$, this allows us to compute $\mu_{X_{N-1} \to f_{N-1}}(X_{N-1})$, and then $\mu_{X_{N-2} \to f_{N-2}}(X_{N-2})$, and so forth up until $\mu_{X_0 \to f_0}(X_0)$. The messages are known as backward messages (since they are propagated from right to left, from the future to the past).

Once both phases have been completed, the marginals can be computed by pointwise multiplication of the forward and the backward message on every edge.

Example 6.2 (Inferring human emotions). *Returning to our human test subject, we depict the factor graph of $p(\mathbf{X}, \mathbf{Y} = \mathbf{y} | \mathcal{M})$ in Fig. 6.4. Since the graph is so small, we will not bother normalizing the messages. The reader can verify that normalizing gives exactly the same result (the same marginal a-posteriori distributions).*

Forward phase

The forward message from node f_0 over X_0 is given by the a-priori distribution of X_0: $\mu_{f_0 \to X_0}(s) = 0.9$ and $\mu_{f_0 \to X_0}(h) = 0.1$. We can then compute $\mu_{g_1 \to X_1^{(1)}}(X_1^{(1)})$

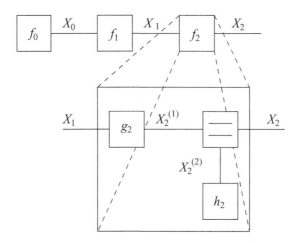

Figure 6.4. A factor graph for an SSM of Nick with four observations, with $f_0(X_0) = p(X_0)$ and $f_k(X_{k-1}, X_k) = p(Y_k = y_k | X_k, X_{k-1}) p(X_k | X_{k-1})$. The node f_k is opened to reveal its structure, with $g_k(X_{k-1}, X_k^{(1)}) = p(X_k^{(1)} | X_{k-1})$ and $h_k(X_k^{(2)}) = p(Y_k = y_k | X_k^{(2)})$, and an equality node.

as follows:

$$\mu_{g_1 \to X_1^{(1)}}\left(x_1^{(1)}\right) = \sum_{x_0} p\left(X_k^{(1)} = x_1^{(1)} | X_0 = x_0\right) \mu_{f_0 \to X_0}(x_0)$$

$$= p\left(x_1^{(1)} | s\right) 0.9 + p\left(x_1^{(1)} | h\right) 0.1,$$

so that $\mu_{g_1 \to X_1^{(1)}}(h) = 0.1$ *and* $\mu_{g_1 \to X_1^{(1)}}(s) = 0.9$. *Next, the upward message on the edge* $X_1^{(2)}$ *is given by* $h_1(X_1^{(2)}) = p(Y_1 = 1 | X_1^{(2)})$, *so that*

$$\mu_{h_1 \to X_1^{(2)}}(s) = 0.1,$$

$$\mu_{h_1 \to X_1^{(2)}}(h) = 0.3.$$

Pointwise multiplication of $\mu_{g_1 \to X_1^{(1)}}(X_1^{(1)})$ *by* $\mu_{h_1 \to X_1^{(2)}}(X_1^{(2)})$ *leads to* $\mu_{f_1 \to X_1}(X_1)$:

$$\mu_{f_1 \to X_1}(s) = 0.09,$$

$$\mu_{f_1 \to X_1}(h) = 0.03.$$

The remaining forward messages are given by

$$\mu_{g_2 \to X_2^{(1)}}(s) = 0.108,$$

$$\mu_{g_2 \to X_2^{(1)}}(h) = 0.012,$$

and

$$\mu_{f_2 \to X_2}(\text{s}) = 0.0756,$$
$$\mu_{f_2 \to X_2}(\text{h}) = 0.0072.$$

Backward phase

In parallel with the forward phase, we can start the backward phase: since X_2 is a half-edge, $\mu_{X_2 \to f_2}(\text{s}) = \mu_{X_2 \to f_2}(\text{h}) = 1$. We continue computing messages from right to left, leading to

$$\mu_{X_2^{(1)} \to g_2}(\text{s}) = 0.7,$$
$$\mu_{X_2^{(1)} \to g_2}(\text{h}) = 0.6,$$

and $\mu_{g_2 \to X_1}(\text{s}) = \mu_{g_2 \to X_1}(\text{h}) = 0.69$. Then

$$\mu_{X_1^{(1)} \to g_1}(\text{s}) = 0.069,$$
$$\mu_{X_1^{(1)} \to g_1}(\text{h}) = 0.207,$$

and finally

$$\mu_{X_0 \to f_0}(\text{s}) = 0.0828,$$
$$\mu_{X_0 \to f_0}(\text{h}) = 0.0828.$$

Marginals

The marginal of X_1 can now be obtained as

$$p(X_1 = \text{s}, \mathbf{Y} = [\text{l b}]|\mathcal{M}) = \mu_{g_2 \to X_1}(\text{s}) \times \mu_{X_1 \to g_2}(\text{s})$$
$$= 0.0621,$$
$$p(X_1 = \text{h}, \mathbf{Y} = [\text{l b}]|\mathcal{M}) = \mu_{g_2 \to X_1}(\text{h}) \times \mu_{X_1 \to g_2}(\text{h})$$
$$= 0.0207.$$

Summing out X_1 tells us that $p(\mathbf{Y} = [\text{l b}]|\mathcal{M}) = 0.0820$. Normalizing gives us $p(X_1 = \text{s}|\mathbf{Y} = [\text{l b}], \mathcal{M}) = 0.75$ and $p(X_1 = \text{h}|\mathbf{Y} = [\text{l b}], \mathcal{M}) = 0.25$. We can do the same for X_2, resulting in

$$p(X_2 = \text{s}|\mathbf{Y} = [\text{l b}], \mathcal{M}) \approx 0.91,$$
$$p(X_2 = \text{h}|\mathbf{Y} = [\text{l b}], \mathcal{M}) \approx 0.09.$$

At the end of the experiment, we can say with high confidence that Nick is sad. Suppose that we now note a third observation $Y_3 = \text{c}$. In order to perform the forward phase of the SPA, we can re-use the already computed message $\mu_{f_2 \to X_2}(X_2)$. This allows *the forward phase to process the observations as they are generated. The backward phase, however, has to wait until all observations are available. This brings us seamlessly to the next section.*

6.2.3.2   Sequential processing

In some applications, N may be very large (or even infinite), and we wish to track the state of the system as observations become available. This is known as *sequential processing*, online processing, or filtering, depending on the context. In such cases, we implement only the forward phase of the SPA. At each time instant k we compute the message $\mu_{f_k \to X_k}(X_k)$, on the basis of $\mu_{f_{k-1} \to X_{k-1}}(X_{k-1})$ and the observation y_k. When messages are normalized, they have the following meaning: $\mu_{f_k \to X_k}(X_k)$ is the a-posteriori distribution of X_k, given all observations up until time instant k:

$$\mu_{f_k \to X_k}(x_k) = p(X_k = x_k | \mathbf{Y}_{1:k} = \mathbf{y}_{1:k}, \mathcal{M}). \tag{6.9}$$

In the case of a state-emitting SSM (see Fig. 6.2), the messages $\mu_{g_k \to X_k^{(1)}}(X_k^{(1)})$ also have an interesting interpretation: they are the a-posteriori distribution of X_k, given all past observations:

$$\mu_{g_k \to X_k^{(1)}}(x_k) = p\left(X_k = x_k | \mathbf{Y}_{1:k-1} = \mathbf{y}_{1:k-1}, \mathcal{M}\right). \tag{6.10}$$

In other words, this message is a *prediction* of the state at time instant k, before the observation y_k is made.

In certain applications, once all the observations $\mathbf{y}_{1:N}$ are available, and all the forward messages have been computed, we can combine the backward phase with the computation of the marginal a-posteriori distributions $p(X_k | \mathbf{Y}_{1:N} = \mathbf{y}_{1:N}, \mathcal{M})$. This is known as *smoothing*. Note that now the backward phase is performed *after* the forward phase (rather than in parallel).

6.2.4   Three types of state-space model

Hidden Markov models

Those SSMs with discrete state spaces are commonly known as hidden Markov models (HMMs). They will turn out to be important in the design of iterative receivers, and will be treated in Section 6.3. Since the state space is discrete, we can represent messages by vectors. Our example from the previous section was a HMMs with two states. For HMMs, we will solve the following inference problems: how to determine the likelihood of the model $\mathcal{M}$, the marginal a-posteriori distributions $p(X_k | \mathbf{Y}_{1:N} = \mathbf{y}_{1:N}, \mathcal{M})$, and the mode of the joint a-posteriori distribution $p(\mathbf{X}_{0:N} | \mathbf{Y}_{1:N} = \mathbf{y}_{1:N}, \mathcal{M})$.

Linear Gaussian models

In linear Gaussian models, the state at time instant k is a linear function of the state at time $k - 1$ with additive Gaussian noise. Similarly, the output at time k is a linear function of the state at time k with additive Gaussian noise. Although these models are not used later on in this book, they are quite important for other applications, and we will treat them in detail in Section 6.4. For linear Gaussian models, it turns out that all the messages are Gaussian distributions, and can thus be represented by a mean and a covariance matrix. We will describe how to determine the likelihood of the model $\mathcal{M}$, the marginal a-posteriori distributions, and the mode of the joint a-posteriori distribution.

Arbitrary SSMs

Finally, in Section 6.5 we will deal with more general state-emitting SSMs with continuous state spaces. In general, exact inference is no longer possible since the messages cannot be represented exactly. We look to Monte Carlo techniques to perform *approximate inference*. Similarly to the linear Gaussian models, approximate inference for SSM will not be used in later chapters of this book, but is of importance in many applications and is worthy of our attention. We will show how to determine the marginal a-posteriori distributions and the likelihood of the model $\mathcal{M}$.

6.3 Hidden Markov models

6.3.1 Introduction

We will consider only transition-emitting SSMs, since they are the more general case, and leave the case of state-emitting SSMs to the reader. We remind the reader that, since the spaces over which the variables are defined are discrete, messages can be represented by vectors (see also Section 5.3.3). We will first implement the SPA. This allows us to determine the marginal a-posteriori distributions $p(X_k|\mathbf{Y} = \mathbf{y}, \mathcal{M})$, as well as the likelihood of the model $\mathcal{M}$. We will then use the max–sum algorithm to determine the mode of the joint a-posteriori distribution $p(\mathbf{X}|\mathbf{Y} = \mathbf{y}, \mathcal{M})$. For more information on HMMs, the reader is referred to the excellent tutorial [74].

6.3.2 Determining the marginal a-posteriori distributions

Direct implementation

Let us apply the SPA on the factor graph from Fig. 6.1. We start from the half-edge X_N and the node of degree 1, f_0. We can then compute messages from left to right (forward messages) sequentially. At the same time, we can compute messages from right to left (backward messages) sequentially. At every step we normalize the messages such that $\sum_{x_k \in \mathcal{X}_k} \mu_{f_k \to X_k}(x_k) = 1$ and $\sum_{x_k \in \mathcal{X}_k} \mu_{f_k \to X_k}(x_k) = 1$. We will denote the normalization constants for the forward messages γ_k, and those for the backward messages ρ_k. The entire algorithm is shown in Algorithm 6.1. At the end we have the marginals $g_{X_k}(X_k) = p(X_k, \mathbf{Y} = \mathbf{y}|\mathcal{M})$, for $k = 0, \ldots, N$. Note that we must take into account the normalization constants.

Algorithm 6.1 Hidden Markov models: sum–product algorithm with message normalization

---

1: initialization,

$$\mu_{f_0 \to X_0}(x_0) = \gamma_0 p(x_0), \forall x_0 \in \mathcal{X}_0$$
$$\mu_{X_N \to f_N}(x_N) = \rho_N, \forall x_N \in \mathcal{X}_N$$

2: **for** $k = 1$ to N **do**

3:     compute forward message, $\forall x_k \in \mathcal{X}_k$:

$$\mu_{f_k \to X_k}(x_k) = \gamma_k \sum_{x_{k-1} \in \mathcal{X}_{k-1}} f_k(x_{k-1}, x_k) \, \mu_{f_{k-1} \to X_{k-1}}(x_{k-1})$$

4:     compute backward message $\forall x_{N-k} \in \mathcal{X}_k$: set $l = N - k$

$$\mu_{X_l \to f_l}(x_l) = \rho_l \sum_{x_{l+1} \in \mathcal{X}_{l+1}} f_{l+1}(x_l, x_{l+1}) \, \mu_{X_{l+1} \to f_{l+1}}(x_{l+1})$$

5: **end for**

6: **for** $k = 0$ to N **do**

7:     introduce $C_k = \prod_{l=0}^{k} \gamma_l \prod_{n=k}^{N} \rho_n$

8:     marginal of X_k:

$$g_{X_k}(x_k) = \frac{1}{C_k} \mu_{X_k \to f_k}(x_k) \, \mu_{f_k \to X_k}(x_k)$$

9: **end for**

Vector–matrix implementation

In discrete state spaces, the messages can be represented by vectors. Suppose that we index the elements in $\mathcal{X}_k$ as $\mathcal{X}_k = \{a_k^{(1)}, \ldots, a_k^{(|\mathcal{X}_k|)}\}$. Let us represent $\mu_{f_k \to X_k}(X_k)$ by the vector $\mathbf{p}_k^{(\mathrm{F})}$ and $\mu_{X_k \to f_k}(X_k)$ by the vector $\mathbf{p}_k^{(\mathrm{B})}$. Both are $|\mathcal{X}_k| \times 1$ column vectors, with

$$\left[\mathbf{p}_k^{(\mathrm{F})}\right]_i = \mu_{f_k \to X_k}\left(a_k^{(i)}\right) \tag{6.11}$$

and

$$\left[\mathbf{p}_k^{(\mathrm{B})}\right]_i = \mu_{X_k \to f_k}\left(a_k^{(i)}\right) \tag{6.12}$$

for $i = 1, \ldots, |\mathcal{X}_k|$. The transition probabilities $p(X_k = x_k | X_{k-1} = x_{k-1}, \mathcal{M})$, combined with the output probabilities $p(Y_k = y_k | X_{k-1} = x_{k-1}, X_k = x_k, \mathcal{M})$, can be represented as matrices: $\mathbf{A}_k$ is an $|\mathcal{X}_{k-1}| \times |\mathcal{X}_k|$ matrix, with

$$[\mathbf{A}_k]_{i,j} = p\left(Y_k = y_k \middle| X_k = a_k^{(j)}, X_{k-1} = a_{k-1}^{(i)}, \mathcal{M}\right) p\left(X_k = a_k^{(j)} \middle| X_{k-1} = a_{k-1}^{(i)}, \mathcal{M}\right). \tag{6.13}$$

In that case, the SPA can be implemented as described in Algorithm 6.2.

Algorithm 6.2 Hidden Markov models: sum–product algorithm with message normalization using vector and matrix representation

---

1: initialization,
$$[\mathbf{p}_0^{(F)}]_i = \gamma_0 p(a_0^{(i)}), \forall i \in \{1, \ldots, |\mathcal{X}_0|\}.$$
$$[\mathbf{p}_N^{(B)}]_i = \rho_N, \forall i \in \{1, \ldots, |\mathcal{X}_N|\}$$
2: **for** $k = 1$ to N **do**
3:    compute forward message:

$$\mathbf{p}_k^{(F)} = \gamma_k \mathbf{A}_k^{\mathrm{T}} \mathbf{p}_{k-1}^{(F)}$$

4:    compute backward message:

$$\mathbf{p}_{N-k}^{(B)} = \rho_{N-k} \mathbf{A}_{N-k+1} \mathbf{p}_{N-k+1}^{(B)}$$

5: **end for**
6: **for** $k = 0$ to N **do**
7:    introduce $C_k = \prod_{l=0}^{k} \gamma_l \prod_{n=k}^{N} \rho_n$
8:    marginal of X_k:

$$g_{X_k}\left(a_k^{(i)}\right) = \frac{1}{C_k}\left[\mathbf{p}_k^{(F)}\right]_i \left[\mathbf{p}_k^{(B)}\right]_i$$

9: **end for**

Marginal a-posteriori distributions

The marginal a-posteriori distributions $p(X_k|\mathbf{Y} = \mathbf{y}, \mathcal{M})$ are given by

$$p(X_k = x_k|\mathbf{Y} = \mathbf{y}, \mathcal{M}) = \frac{g_{X_k}(x_k)}{\sum_{x \in \mathcal{X}_k} g_{X_k}(x)}. \tag{6.14}$$

Note that to find these distributions, keeping track of the normalization constants γ_k and ρ_k is not necessary, since they will cancel out in (6.14).

6.3.3 Determining the likelihood of the model

From the same marginals $g_{X_k}(X_k)$, we can now determine the likelihood of the model $p(\mathbf{Y} = \mathbf{y}|\mathcal{M})$ as

$$p(\mathbf{Y} = \mathbf{y}|\mathcal{M}) = \sum_{x_k \in \mathcal{X}_k} g_{X_k}(x_k) \tag{6.15}$$

for any $k \in \{0, \ldots, N\}$. In contrast to the case in this previous section, we are now required to keep track of the normalization constants γ_k and ρ_k.

6.3.4   Determining the mode of the joint a-posteriori distribution

We will work in the $(\mathbb{R}, \max, +)$ semi-ring, based on the factorization of $\log p(\mathbf{X}, \mathbf{Y} = \mathbf{y} | \mathcal{M})$:

$$\log p(\mathbf{X}, \mathbf{Y} = \mathbf{y} | \mathcal{M}) = \log p(X_0 | \mathcal{M}) + \sum_{k=1}^{N} (\log p(Y_k = y_k | X_k, X_{k-1}, \mathcal{M}))$$

$$+ \log p(X_k | X_{k-1}, \mathcal{M})). \tag{6.16}$$

The corresponding factor graph is again the one from Fig. 6.1, where now $f_0(X_0) = \log p(X_0 | \mathcal{M})$, and $f_k(X_{k-1}, X_k) = \log p(Y_k = y_k | X_k, X_{k-1}, \mathcal{M}) + \log p(X_k | X_{k-1}, \mathcal{M})$. The details of the max–sum algorithm are described in Algorithm 6.3. Notice the similarities to the SPA in the $(\mathbb{R}^+, +, \times)$ semi-ring. For max–sum, we generally don't introduce scaling. However, adding a real number to a message will not affect the outcome of the max–sum algorithm, in a sense that the marginals will be correct, up to an additive constant. From the marginals $g_{X_k}(X_k)$, we can now determine the mode of the a-posteriori distributions $p(\mathbf{X} = \mathbf{x} | \mathbf{Y} = \mathbf{y}, \mathcal{M})$, given by $\hat{\mathbf{x}} = [\hat{x}_0, \ldots, \hat{x}_N]$ with

$$\hat{x}_k = \arg \max_{x_k \in \mathcal{X}_k} g_{X_k}(x_k). \tag{6.17}$$

Observe that adding a constant to $g_{X_k}(X_k)$ does not affect $\hat{x}_k$.

Algorithm 6.3 Hidden Markov models: max–sum algorithm

1: initialization:
$$\mu_{f_0 \rightarrow X_0}(x_0) = \log p(x_0), \forall x_0 \in \mathcal{X}_0$$
$$\mu_{X_N \rightarrow f_N}(x_N) = 0, \forall x_N \in \mathcal{X}_N$$

2: **for** $k = 1$ to N **do**

3:     compute forward message, $\forall x_k \in \mathcal{X}_k$:

$$\mu_{f_k \rightarrow X_k}(x_k) = \max_{x_{k-1} \in \mathcal{X}_{k-1}} \left\{ f_k(x_{k-1}, x_k) + \mu_{f_{k-1} \rightarrow X_{k-1}}(x_{k-1}) \right\}$$

4:     compute backward message $\forall x_{N-k} \in \mathcal{X}_k$: set $l = N - k$

$$\mu_{X_l \rightarrow f_l}(x_l) = \max_{x_{l+1} \in \mathcal{X}_{l+1}} \left\{ f_{l+1}(x_l, x_{l+1}) + \mu_{X_{l+1} \rightarrow f_{l+1}}(x_{l+1}) \right\}$$

5: **end for**

6: **for** $k = 0$ to N **do**

7:     marginal of X_k:

$$g_{X_k}(x_k) = \mu_{X_k \rightarrow f_k}(x_k) + \mu_{f_k \rightarrow X_k}(x_k)$$

8: **end for**

6.3.5 Concluding remarks

The algorithms described above are well known in the technical literature. The SPA is more commonly known as the *forward–backward algorithm* [74]. The max–sum algorithm is equivalent to the *Viterbi algorithm* [52]. Without knowledge of factor graphs, deriving both algorithms is more involved. From Section 5.3.3.3 we now understand that the Viterbi algorithm can also be obtained by implementing the forward–backward algorithm with log-domain messages, combined with an approximation of the Jacobian logarithm.

We have solved the following inference problems:

(1) determining the a-posteriori distribution of X_k, $p(X_k|\mathbf{Y} = \mathbf{y}, \mathcal{M})$,
(2) determining the likelihood of the model $\mathcal{M}$, $p(\mathbf{Y} = \mathbf{y}|\mathcal{M})$, and
(3) determining the mode of the joint a-posteriori distribution, $p(\mathbf{X}|\mathbf{Y} = \mathbf{y}, \mathcal{M})$.

We have seen that, when the state spaces are discrete, inference problems can be solved exactly. The complexity is linear in N and in the number of states. When the state space is continuous, all summations need to be replaced by integrals, and messages can no longer be represented by vectors. Generally, exact inference is impossible. There is an important exception, as we will see in the next section.

At this point the reader is free to skip ahead to the next chapter. The remaining two sections in this chapter deal with types of SSM that are important, but are not necessary in order to understand the rest of the book. These sections are provided mainly for the sake of completeness.

6.4 Linear Gaussian models

6.4.1 Introduction

In real[1] linear Gaussian models, we have the following state-emitting SSM. The state at time k is given by a $K \times 1$ vector $\mathbf{X}_k$, the output $\mathbf{Y}_k$ is an $M \times 1$ vector. With $\mathbf{x}_0 \sim \mathcal{N}_{\mathbf{x}_0}(\mathbf{m}_0, \boldsymbol{\Sigma}_0)$, the system is fully described by the following two equations:

$$\mathbf{x}_k = \mathbf{A}_k \mathbf{x}_{k-1} + \mathbf{B}_k \mathbf{v}_k, \tag{6.18}$$

$$\mathbf{y}_k = \mathbf{C}_k \mathbf{x}_k + \mathbf{D}_k \mathbf{w}_k, \tag{6.19}$$

where, for all $k \in \{1, \ldots, N\}$, $\mathbf{A}_k$ and $\mathbf{B}_k$ are known $K \times K$ matrices, $\mathbf{C}_k$ is a known $M \times K$ matrix, and $\mathbf{D}_k$ is a known $M \times M$ matrix. Furthermore, $\mathbf{v}_k$ and $\mathbf{w}_k$ are independent zero-mean Gaussian noise processes with $\mathbf{v}_k \sim \mathcal{N}_{\mathbf{v}_k}(\mathbf{0}, \mathbf{I}_K)$ and $\mathbf{w}_k \sim \mathcal{N}_{\mathbf{w}_k}(\mathbf{0}, \mathbf{I}_M)$. Here, we use bold capital letters both for matrices and for vector random variables. This should not cause too much confusion. Only the quantities $\mathbf{A}_k, \mathbf{B}_k, \mathbf{C}_k, \mathbf{D}_k$, and covariance matrices are matrices, all the rest are vectors. For simplicity, we will assume that $\mathbf{A}_k, \mathbf{D}_k$, and $\mathbf{B}_k$

---

[1] The extension to complex models is straightforward.

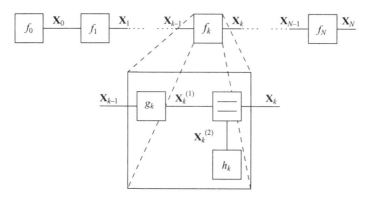

Figure 6.5. A factor graph for a state-emitting SSM, with $f_0(\mathbf{X}_0) = p(\mathbf{X}_0)$ and $f_k(\mathbf{X}_{k-1}, \mathbf{X}_k) = p(\mathbf{Y}_k = \mathbf{y}_k|\mathbf{X}_k)p(\mathbf{X}_k|\mathbf{X}_{k-1})$. The node f_k is opened to reveal its structure, with $g_k(\mathbf{X}_{k-1}, \mathbf{X}_k^{(1)}) = p(\mathbf{X}_k^{(1)}|\mathbf{X}_{k-1})$ and $h_k(\mathbf{X}_k^{(2)}) = p(\mathbf{Y}_k = \mathbf{y}_k|\mathbf{X}_k^{(2)})$, and an equality node.

are invertible. For more general cases, the reader is referred to the texts that form the inspiration to this section [76,77].

We will first transform the general factor graph from Fig. 6.5 by opening the various nodes. We then execute the SPA and determine the marginal a-posteriori distributions. We will also show how to determine the likelihood of the model and the mode of the joint a-posteriori distribution. For notational convenience, we will omit the conditioning on model $\mathcal{M}$, except in Section 6.4.3.

The factor graph

The joint distribution of $\mathbf{X} = \mathbf{X}_{0:N}$ and the observation $\mathbf{Y} = \mathbf{Y}_{1:N}$ factorizes as for any state-emitting SSM as

$$p(\mathbf{X}, \mathbf{Y} = \mathbf{y}|\mathcal{M}) = p(\mathbf{X}_0) \prod_{k=1}^{N} \underbrace{p(\mathbf{Y}_k = \mathbf{y}_k|\mathbf{X}_k)p(\mathbf{X}_k|\mathbf{X}_{k-1})}_{f_k(\mathbf{X}_{k-1}, \mathbf{X}_k)} \qquad (6.20)$$

with the factor graph shown in Fig. 6.5. As before, we have opened the nodes to reveal their structure, with $g_k(\mathbf{X}_{k-1}, \mathbf{X}_k^{(1)}) = p(\mathbf{X}_k^{(1)}|\mathbf{X}_{k-1})$ and $h_k(\mathbf{X}_k^{(2)}) = p(\mathbf{Y}_k = \mathbf{y}_k|\mathbf{X}_k^{(2)})$. This leads to the introduction of two additional variables, $\mathbf{X}_k^{(1)}$ and $\mathbf{X}_k^{(2)}$ for every time instant $k > 0$.

Opening the node g_k

Because of the model (6.18) and (6.19), the nodes g_k and h_k can be opened further. Let us introduce the intermediate variables $\mathbf{s}_k = \mathbf{B}_k \mathbf{v}_k$ and $\mathbf{z}_k = \mathbf{A}_k \mathbf{x}_{k-1}$ so that $\mathbf{x}_k^{(1)} = \mathbf{z}_k + \mathbf{s}_k$. We know from Section 5.2.2 that we can replace the node corresponding to $p(\mathbf{x}_k^{(1)}|\mathbf{x}_{k-1})$

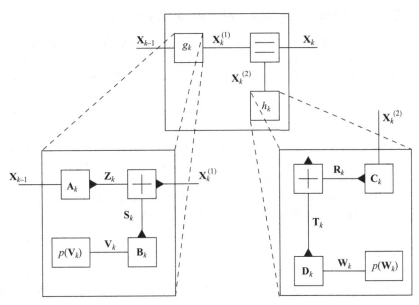

Figure 6.6. A factor graph with $g_k(\mathbf{X}_{k-1}, \mathbf{X}_k^{(1)}) = p(\mathbf{X}_k^{(1)} \mid \mathbf{X}_{k-1})$ and $h_k(\mathbf{X}_k^{(2)}) = p(\mathbf{Y}_k = \mathbf{y}_k | \mathbf{X}_k^{(2)})$. The nodes g_k and h_k can again be opened to reveal their structure.

by the factor graph of a factorization of $p(\mathbf{x}_k^{(1)}, \mathbf{z}_k, \mathbf{v}_k, \mathbf{s}_k | \mathbf{x}_{k-1})$. Clearly,

$$p(\mathbf{x}_k^{(1)}, \mathbf{z}_k, \mathbf{v}_k, \mathbf{s}_k | \mathbf{x}_{k-1}) = \boxed{=}(\mathbf{x}_k^{(1)}, \mathbf{z}_k + \mathbf{s}_k)\boxed{=}(\mathbf{z}_k, \mathbf{A}_k\mathbf{x}_{k-1})\boxed{=}(\mathbf{s}_k, \mathbf{B}_k\mathbf{v}_k)p(\mathbf{v}_k). \quad (6.21)$$

This leads to the factor graph shown in Fig. 6.6 (lower left part). The nodes marked $\mathbf{A}_k$ and $\mathbf{B}_k$ are defined as $\boxed{=}(\mathbf{z}_k, \mathbf{A}_k\mathbf{x}_{k-1})$ and $\boxed{=}(\mathbf{s}_k, \mathbf{B}_k\mathbf{v}_k)$, respectively. The node marked with a $\boxplus$ is defined as $\boxed{=}(\mathbf{x}_k^{(1)}, \mathbf{z}_k + \mathbf{s}_k)$. The arrows on those three nodes are a representation of the output of the operation (matrix multiplication and addition). This enables us to interpret the nodes in the factor graph using a simple notation. For instance, $\mathbf{s}_k$ is obtained by multiplying $\mathbf{v}_k$ by the matrix $\mathbf{B}_k$, so the arrow on the node points toward the edge $\mathbf{S}_k$.

Opening the node h_k

We can do the same for $p(\mathbf{Y}_k = \mathbf{y}_k | \mathbf{X}_k)$: upon introducing $\mathbf{t}_k = \mathbf{D}_k\mathbf{w}_k$ and $\mathbf{r}_k = \mathbf{C}_k\mathbf{x}_k^{(2)}$, this gives us the factorization

$$p(\mathbf{y}_k, \mathbf{r}_k, \mathbf{w}_k, \mathbf{t}_k | \mathbf{x}_{k-1}^{(2)}) = \boxed{=}(\mathbf{y}_k, \mathbf{r}_k + \mathbf{t}_k)\boxed{=}(\mathbf{r}_k, \mathbf{C}_k\mathbf{x}_k^{(2)})\boxed{=}(\mathbf{t}_k, \mathbf{D}_k\mathbf{w}_k)p(\mathbf{w}_k) \quad (6.22)$$

leading to the factor graph in Fig. 6.6 (lower right part). The nodes marked $\mathbf{C}_k$ and $\mathbf{D}_k$ are defined as $\boxed{=}(\mathbf{r}_k, \mathbf{C}_k\mathbf{x}_k^{(2)})$ and $\boxed{=}(\mathbf{t}_k, \mathbf{D}_k\mathbf{w}_k)$. The node marked with a $\boxplus$ is now defined as $\boxed{=}(\mathbf{y}_k, \mathbf{r}_k + \mathbf{t}_k)$. Notice that the arrow is not pointing toward an edge since $\mathbf{y}_k$ is not a variable in the factor graph.

6.4.2   Determining the marginal a-posteriori distributions

As for any SSM, the SPA consists of a forward phase with messages from left to right in Fig. 6.5, and a backward phase with messages from right to left. It will become apparent that all the messages in the factor graph are Gaussian distributions. Note that not all the messages in the graph need to be computed. For instance, the message $\mu_{\boxplus \to S_k}(S_k)$ (see Fig. 6.6) serves no purpose in the inference problem, since we are not interested in the marginal of S_k.

We will proceed as follows. We will first concentrate on some basic building blocks and show how messages should be computed. We can then perform the forward and backward phases in parallel exploiting these basic building blocks. In some cases, it is preferred to perform the backward phase after the forward phase (rather than in parallel); this computation will be treated at the end of this section.

6.4.2.1   Building blocks

Looking at the factor graph in Fig. 6.6, we notice a number of building blocks, which are repeated at every time instant k. We will investigate how the SPA behaves in each of the building blocks separately. We can distinguish the following types of nodes (see Fig. 6.7):

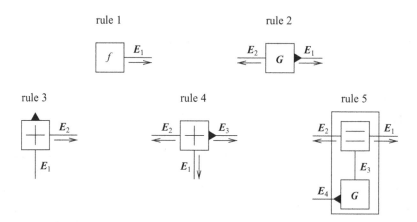

Figure 6.7. Five building blocks for linear Gaussian models.

1. $f(\mathbf{e}) = \mathcal{N}_{\mathbf{e}}(\mathbf{m}, \boldsymbol{\Sigma})$,
2. $f(\mathbf{e}_1, \mathbf{e}_2) = \boxed{=}(\mathbf{e}_1, \mathbf{G}\mathbf{e}_2)$, with $\mathbf{G}$ square and invertible,
3. $f(\mathbf{e}_1, \mathbf{e}_2) = \boxed{=}(\mathbf{y}, \mathbf{e}_1 + \mathbf{e}_2)$, with $\mathbf{y}$ a fixed vector,
4. $f(\mathbf{e}_1, \mathbf{e}_2, \mathbf{e}_3) = \boxed{=}(\mathbf{e}_3, \mathbf{e}_1 + \mathbf{e}_2)$, and
5. $f(\mathbf{e}_1, \mathbf{e}_2, \mathbf{e}_3, \mathbf{e}_4) = \boxed{=}(\mathbf{e}_1, \mathbf{e}_2, \mathbf{e}_3)\boxed{=}(\mathbf{e}_4, \mathbf{G}\mathbf{e}_3)$ with $\mathbf{G}$ non-square.

The sum–product rule for these nodes is then given as follows.

1. **RULE 1:** $\mu_{f \to \mathbf{E}}(\mathbf{e}) = \mathcal{N}_{\mathbf{e}}(\mathbf{m}, \boldsymbol{\Sigma})$.

2. **RULE 2:** Given $\mu_{E_1 \to f}(e_1) = \mathcal{N}_{e_1}(m_1, \Sigma_1)$ and $\mu_{E_2 \to f}(e_2) = \mathcal{N}_{e_2}(m_2, \Sigma_2)$, we find that

$$\mu_{f \to E_1}(e_1) = \mathcal{N}_{e_1}\left(Gm_2, G\Sigma_2 G^T\right),$$

$$\mu_{f \to E_2}(e_2) = \mathcal{N}_{e_2}\left(G^{-1}m_1, G^{-1}\Sigma_1\left(G^{-1}\right)^T\right).$$

3. **RULE 3:** Given $\mu_{E_1 \to f}(e_1) = \mathcal{N}_{e_1}(0, \Sigma_1)$, then

$$\mu_{f \to E_2}(e_2) = \mathcal{N}_{e_2}(y, \Sigma_1).$$

4. **RULE 4:** Given $\mu_{E_1 \to f}(e_1) = \mathcal{N}_{e_1}(m_1, \Sigma_1)$, $\mu_{E_2 \to f}(e_2) = \mathcal{N}_{e_2}(m_2, \Sigma_2)$, and $\mu_{E_3 \to f}(e_3) = \mathcal{N}_{e_3}(m_3, \Sigma_3)$, then

$$\mu_{f \to E_3}(e_3) = \mathcal{N}_{e_3}(m_1 + m_2, \Sigma_1 + \Sigma_2),$$

$$\mu_{f \to E_1}(e_1) = \mathcal{N}_{e_1}(m_3 - m_2, \Sigma_3 + \Sigma_2).$$

5. **RULE 5:** Given $\mu_{E_2 \to f} = \mathcal{N}_{e_2}(m_2, \Sigma_2)$, and $\mu_{E_4 \to f}(e_4) = \mathcal{N}_{e_4}(m_4, \Sigma_4)$, let us determine $\mu_{f \to E_1}(e_1)$:

$$\mu_{f \to E_1}(e_1) \propto \mu_{E_2 \to f}(e_1) \mu_{E_3 \to \boxminus}(e_1), \tag{6.23}$$

where

$$\mu_{E_3 \to \boxminus}(e_3) \propto \int \boxminus(e_4, Ge_3) \mu_{E_4 \to f}(e_4) de_4. \tag{6.24}$$

Substituting (6.24) into (6.23) yields

$$\mu_{f \to E_1}(e_1) \propto \mu_{E_2 \to f}(e_1) \mu_{E_4 \to f}(Ge_1).$$

Since $\mu_{E_2 \to f}(e_2) = \mathcal{N}_{e_2}(m_2, \Sigma_2)$, we find (and this requires a minimum of effort, to multiply two Gaussian distributions) that $\mu_{f \to E_1}(e_1) = \mathcal{N}_{e_1}(m_1, \Sigma_1)$, where

$$\Sigma_1 = \left(\Sigma_2^{-1} + G^T \Sigma_4^{-1} G\right)^{-1}$$

and

$$m_1 = \Sigma_1 \left(\Sigma_2^{-1} m_2 + G^T \Sigma_4^{-1} m_4\right).$$

Using the matrix-inversion lemma[2], we can re-write Σ_1 as

$$\Sigma_1 = (I - KG)\Sigma_2,$$

[2] One form of the matrix-inversion lemma states that
$(A^{-1} + C^T B^{-1} C)^{-1} = (I - AC^T(B + CAC^T)^{-1}C)A$.

where $\mathbf{K} = \boldsymbol{\Sigma}_2 \mathbf{G}^\mathrm{T} (\boldsymbol{\Sigma}_4 + \mathbf{G}\boldsymbol{\Sigma}_2\mathbf{G}^\mathrm{T})^{-1}$. This leads to

$$\mathbf{m}_1 = \mathbf{m}_2 + \mathbf{K}(\mathbf{m}_4 - \mathbf{G}\mathbf{m}_2).$$

In the particular case that the entries in $\boldsymbol{\Sigma}_2$ tend to infinity, then we have $\boldsymbol{\Sigma}_1 = (\mathbf{G}^\mathrm{T}\boldsymbol{\Sigma}_4^{-1}\mathbf{G})^{-1}$ and take over $\mathbf{m}_1 = \boldsymbol{\Sigma}_1 \mathbf{G}^\mathrm{T}\boldsymbol{\Sigma}_4^{-1}\mathbf{m}_4$.

Although there are other types of messages in the graph, they are irrelevant since we are interested only in the marginals of the states $\mathbf{X}_k$. Now that the basic building blocks are defined, performing the SPA is fairly straightforward.

6.4.2.2   Notation

It is important to note that, in the previous section, *all messages are Gaussian distributions*. This means that all messages can be represented by a mean and a covariance matrix (this is an example of a parametric representation, as described in Section 5.3.4). We will use the following notations for the messages. The forward messages are denoted

$$\mu_{f_{k-1} \to \mathbf{X}_{k-1}}(\mathbf{x}_{k-1}) = \mathcal{N}_{\mathbf{x}_{k-1}}\left(\mathbf{m}_{k-1|k-1}, \mathbf{P}_{k-1|k-1}\right), \tag{6.25}$$

$$\mu_{\mathbf{X}_k^{(1)} \to \boxminus}\left(\mathbf{x}_k^{(1)}\right) = \mathcal{N}_{\mathbf{x}_k^{(1)}}\left(\mathbf{m}_{k|k-1}, \mathbf{P}_{k|k-1}\right). \tag{6.26}$$

The backward messages are denoted

$$\mu_{\mathbf{X}_k \to f_k}(\mathbf{x}_k) = \mathcal{N}_{\mathbf{x}_k}\left(\mathbf{n}_{k|k}, \mathbf{Q}_{k|k}\right), \tag{6.27}$$

$$\mu_{\boxminus \to \mathbf{X}_k^{(1)}}\left(\mathbf{x}_k^{(1)}\right) = \mathcal{N}_{\mathbf{x}_{k-1}}\left(\mathbf{n}_{k|k-1}, \mathbf{Q}_{k|k-1}\right). \tag{6.28}$$

The marginals will be written as

$$p(\mathbf{X}_k = \mathbf{x}_k | \mathbf{Y}_{1:N} = \mathbf{y}_{1:N}) = \mathcal{N}_{\mathbf{x}_k}\left(\mathbf{m}_{k|N}, \mathbf{P}_{k|N}\right).$$

6.4.2.3   The forward phase

The forward phase starts with the message $\mu_{f_0 \to \mathbf{X}_0}(\mathbf{X}_0)$, at time $k = 0$. The messages in the forward phase are depicted in Fig. 6.8, for node f_k. The numbers indicate which of the five rules was used to compute a specific message. At a given time instant k, the incoming message to node f_k is $\mu_{f_{k-1} \to \mathbf{X}_{k-1}}(\mathbf{X}_{k-1})$. The outgoing message is $\mu_{f_k \to \mathbf{X}_k}(\mathbf{X}_k)$. It is easily verified that

$$\mathbf{m}_{k|k-1} = \mathbf{A}_k \mathbf{m}_{k-1|k-1} \tag{6.29}$$

and

$$\mathbf{P}_{k|k-1} = \mathbf{A}_k \mathbf{P}_{k-1|k-1} \mathbf{A}_k^\mathrm{T} + \mathbf{B}_k \mathbf{B}_k^\mathrm{T}. \tag{6.30}$$

We also find that

$$\mathbf{m}_{k|k} = \mathbf{m}_{k|k-1} + \mathbf{K}_k \left(\mathbf{y}_k - \mathbf{C}_k \mathbf{m}_{k|k-1}\right) \tag{6.31}$$

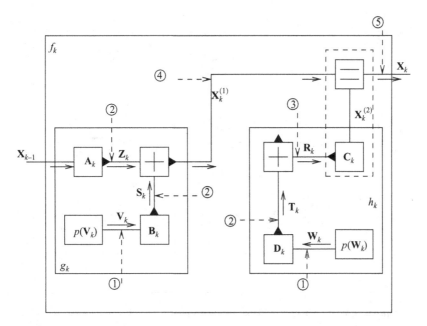

Figure 6.8. Linear Gaussian models: forward phase. The circled numbers indicate which rule is used to generate the message.

and

$$\mathbf{P}_{k|k} = (\mathbf{I} - \mathbf{K}_k \mathbf{C}_k)\, \mathbf{P}_{k|k-1}, \tag{6.32}$$

where $\mathbf{K}_k = \mathbf{P}_{k|k-1}\mathbf{C}_k^{\mathsf{T}}(\mathbf{D}_k \mathbf{D}_k^{\mathsf{T}} + \mathbf{C}_k \mathbf{P}_{k|k-1}\mathbf{C}_k^{\mathsf{T}})^{-1}$.

6.4.2.4 Backward phase in parallel with forward phase

In a similar vein, the backward phase is executed, starting from time instant $N-1$. The messages in the backward phase are depicted in Fig. 6.9, for node f_k. The incoming message is $\mu_{\mathbf{X}_k \to f_k}(\mathbf{X}_k)$. The outgoing message is $\mu_{\mathbf{X}_{k-1} \to f_{k-1}}(\mathbf{x}_{k-1})$. We see that

$$\mathbf{n}_{k|k-1} = \mathbf{n}_{k|k} + \mathbf{L}_k\big(\mathbf{y}_k - \mathbf{C}_k \mathbf{n}_{k|k}\big) \tag{6.33}$$

and

$$\mathbf{Q}_{k|k-1} = (\mathbf{I} - \mathbf{L}_k \mathbf{C}_k)\, \mathbf{Q}_{k|k} \tag{6.34}$$

where $\mathbf{L}_k = \mathbf{Q}_{k|k}\mathbf{C}_k^{\mathsf{T}}(\mathbf{D}_k \mathbf{D}_k^{\mathsf{T}} + \mathbf{C}_k \mathbf{Q}_{k|k}\mathbf{C}_k^{\mathsf{T}})^{-1}$. It is also clear that

$$\mathbf{n}_{k-1|k-1} = \mathbf{A}_k^{-1}\mathbf{n}_{k|k-1}$$

and

$$\mathbf{Q}_{k-1|k-1} = \mathbf{A}_k^{-1}\big(\mathbf{B}_k \mathbf{B}_k^{\mathsf{T}} + \mathbf{Q}_{k|k-1}\big)\big(\mathbf{A}_k^{-1}\big)^{\mathsf{T}}.$$

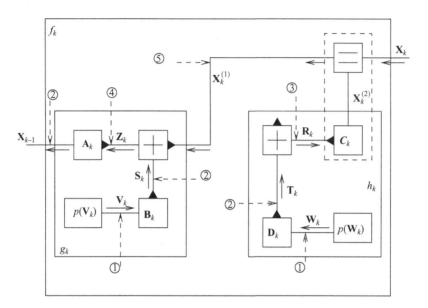

Figure 6.9. Linear Gaussian models: backward phase. The circled numbers indicate which rule is used to generate the message.

Note that in the forward and backward phases there are quite a few messages in common (for instance those on the edges $\mathbf{V}_k$, $\mathbf{S}_k$, $\mathbf{W}_k$, $\mathbf{T}_k$, and $\mathbf{R}_k$). This means that certain results can be re-used.

Marginals

The marginals $p(\mathbf{X}_k = \mathbf{x}_k | \mathbf{Y}_{1:N} = \mathbf{y}_{1:N})$ can finally be obtained as follows. For any $\mathbf{X}_k$, there is a forward message $\mu_{f_k \to \mathbf{x}_k}(\mathbf{x}_k) = \mathcal{N}_{\mathbf{x}_k}(\mathbf{m}_{k|k}, \mathbf{P}_{k|k})$ and a backward message $\mu_{\mathbf{X}_k \to f_k}(\mathbf{x}_k) = \mathcal{N}_{\mathbf{x}_k}(\mathbf{n}_{k|k}, \mathbf{Q}_{k|k})$. The marginal is obtained by multiplying these messages, followed by normalization. This yields

$$p(\mathbf{X}_k = \mathbf{x}_k | \mathbf{Y}_{1:N} = \mathbf{y}_{1:N}) = \mathcal{N}_{\mathbf{x}_k}(\mathbf{m}_{k|N}, \mathbf{P}_{k|N}), \tag{6.35}$$

where

$$\mathbf{m}_{k|N} = \mathbf{m}_{k|k} + \mathbf{K}_{k|N}(\mathbf{n}_{k|k} - \mathbf{m}_{k|k}) \tag{6.36}$$

and

$$\mathbf{P}_{k|N} = (\mathbf{I} - \mathbf{K}_{k|N})\mathbf{P}_{k|k} \tag{6.37}$$

with $\mathbf{K}_{k|N} = \mathbf{P}_{k|k}(\mathbf{P}_{k|k} + \mathbf{Q}_{k|k})^{-1}$.

6.4.2.5 Smoothing: backward phase after forward phase

As we mentioned in Section 6.2.3, in some situations the observations become available only one at a time. In this case, we can start the forward phase as observations become available, and postpone the backward phase. This is known as *filtering*. After time instant $k = N$, all observations are available, and all forward messages have been computed. The message $\mu_{f_N \to \mathbf{X}_N}(\mathbf{x}_N) = \mathcal{N}_{\mathbf{x}_N}(\mathbf{m}_{N|N}, \mathbf{P}_{N|N})$ is equal to the a-posteriori distribution of $\mathbf{X}_N$, given $\mathbf{y}_{1:N}$: $p(\mathbf{X}_N = \mathbf{x}_N | \mathbf{Y}_{1:N} = \mathbf{y}_{1:N}) = \mathcal{N}_{\mathbf{x}_N}(\mathbf{m}_{N|N}, \mathbf{P}_{N|N})$.

Rather than performing the backward phase and then determining the marginal a-posteriori distributions as we did in the previous section, we can combine these two steps. This is known as *smoothing*: given the forward messages and $\mathcal{N}_{\mathbf{x}_k}(\mathbf{m}_{k|N}, \mathbf{P}_{k|N})$, we compute $\mathcal{N}_{\mathbf{x}_{k-1}}(\mathbf{m}_{k-1|N}, \mathbf{P}_{k-1|N})$ for $k = N, N-1, \ldots, 1$. We know that

$$p(\mathbf{x}_{k-1}|\mathbf{x}_k, \mathbf{y}_{1:N}) = p(\mathbf{x}_{k-1}|\mathbf{y}_{1:k-1}) \frac{p(\mathbf{x}_k|\mathbf{x}_{k-1})}{p(\mathbf{x}_k|\mathbf{y}_{1:k-1})} \tag{6.38}$$

$$= \mathcal{N}_{\mathbf{x}_{k-1}}(\mathbf{m}_{k-1|k-1}, \mathbf{P}_{k-1|k-1}) \frac{\mathcal{N}_{\mathbf{x}_k}(\mathbf{A}_k \mathbf{x}_{k-1}, \mathbf{B}_k \mathbf{B}_k^{\mathsf{T}})}{\mathcal{N}_{\mathbf{x}_k}(\mathbf{m}_{k|k-1}, \mathbf{P}_{k|k-1})} \tag{6.39}$$

$$\propto \mathcal{N}_{\mathbf{x}_{k-1}}(\mu(\mathbf{x}_k), \Sigma) \tag{6.40}$$

with

$$\mu(\mathbf{x}_k) = \mathbf{m}_{k-1|k-1} + \mathbf{P}_{k-1|k-1} \mathbf{A}_k^{\mathsf{T}} \mathbf{P}_{k|k-1}^{-1}(\mathbf{x}_k - \mathbf{A}_k \mathbf{m}_{k-1|k-1}), \tag{6.41}$$

$$\Sigma = \mathbf{P}_{k-1|k-1} - \mathbf{P}_{k-1|k-1} \mathbf{A}_k^{\mathsf{T}} \mathbf{P}_{k|k-1}^{-1} \mathbf{A}_k \mathbf{P}_{k-1|k-1}. \tag{6.42}$$

We have used the fact that $\mathbf{P}_{k|k-1} = \mathbf{A}_k \mathbf{P}_{k-1|k-1} \mathbf{A}_k^{\mathsf{T}} + \mathbf{B}_k \mathbf{B}_k^{\mathsf{T}}$. Let us recall the well-known law of iterated expectation:

$$\mathbb{E}_{X_1}\{f(X_1)\} = \mathbb{E}_{X_2}\{\mathbb{E}_{X_1}\{f(X_1)|X_2\}\}. \tag{6.43}$$

It then follows that

$$\mathbf{m}_{k-1|N} = \mathbb{E}_{\mathbf{X}_{k-1}}\{\mathbf{X}_{k-1}|\mathbf{y}_{1:N}\} \tag{6.44}$$

$$= \mathbb{E}_{\mathbf{X}_k}\{\mathbb{E}_{\mathbf{X}_{k-1}}\{\mathbf{X}_{k-1}|\mathbf{X}_k, \mathbf{y}_{1:N}\}|\mathbf{y}_{1:N}\} \tag{6.45}$$

$$= \mathbb{E}_{\mathbf{X}_k}\{\mu(\mathbf{X}_k)|\mathbf{y}_{1:N}\} \tag{6.46}$$

$$= \mathbf{m}_{k-1|k-1} + \mathbf{P}_{k-1|k-1} \mathbf{A}_k^{\mathsf{T}} \mathbf{P}_{k|k-1}^{-1}(\mathbf{m}_{k|N} - \mathbf{A}_k \mathbf{m}_{k-1|k-1}) \tag{6.47}$$

and, similarly,

$$\mathbf{P}_{k-1|N} = \mathbb{E}_{\mathbf{X}_{k-1}}\{(\mathbf{X}_{k-1} - \mathbf{m}_{k-1|N})(\mathbf{X}_{k-1} - \mathbf{m}_{k-1|N})^{\mathsf{T}}|\mathbf{y}_{1:N}\} \tag{6.48}$$

$$= \mathbb{E}_{\mathbf{X}_k}\{\mathbb{E}_{\mathbf{X}_{k-1}}\{(\mathbf{X}_{k-1} - \mathbf{m}_{k-1|N})(\mathbf{X}_{k-1} - \mathbf{m}_{k-1|N})^{\mathsf{T}}|\mathbf{X}_k, \mathbf{y}_{1:N}\}|\mathbf{y}_{1:N}\} \tag{6.49}$$

$$= \mathbf{P}_{k-1|k-1} + \mathbf{P}_{k-1|k-1} \mathbf{A}_k^{\mathsf{T}} \mathbf{P}_{k|k-1}^{-1}(\mathbf{P}_{k|N} - \mathbf{P}_{k|k-1}) \mathbf{P}_{k|k-1}^{-1} \mathbf{A}_k \mathbf{P}_{k-1|k-1}. \tag{6.50}$$

6.4.3   Determining the likelihood of the model

Since we have not kept track of any normalization constants in the previous section, it may seem hard to determine $p(\mathbf{Y}_{1:N} = \mathbf{y}_{1:N}|\mathcal{M})$. However, we note that the logarithm of the likelihood of the model can be written as follows:

$$\log p(\mathbf{Y}_{1:N} = \mathbf{y}_{1:N}|\mathcal{M}) = \log p(\mathbf{Y}_1 = \mathbf{y}_1|\mathcal{M})$$

$$+ \sum_{k=2}^{N} \log p(\mathbf{Y}_k = \mathbf{y}_k|\mathbf{Y}_{1:k-1} = \mathbf{y}_{1:k-1}, \mathcal{M}). \qquad (6.51)$$

It is easy to see that $p(\mathbf{Y}_k|\mathbf{Y}_{1:k-1} = \mathbf{y}_{1:k-1}, \mathcal{M})$ is a Gaussian distribution with mean, say $\mathbf{m}_k^{(y)}$, and covariance matrix, say $\boldsymbol{\Sigma}_k^{(y)}$. Using iterated expectation, we can determine $\mathbf{m}_k^{(y)}$ as

$$\mathbf{m}_k^{(y)} = \mathbb{E}_{\mathbf{Y}_k}\{\mathbf{Y}_k|\mathbf{Y}_{1:k-1} = \mathbf{y}_{1:k-1}\} \qquad (6.52)$$

$$= \mathbb{E}_{\mathbf{X}_k}\{\mathbb{E}_{\mathbf{Y}_k}\{\mathbf{Y}_k|\mathbf{X}_k, \mathbf{Y}_{1:k-1} = \mathbf{y}_{1:k-1}\}|\mathbf{Y}_{1:k-1} = \mathbf{y}_{1:k-1}\} \qquad (6.53)$$

$$= \mathbb{E}_{\mathbf{X}_k}\{\mathbb{E}_{\mathbf{Y}_k}\{\mathbf{Y}_k|\mathbf{X}_k\}|\mathbf{Y}_{1:k-1} = \mathbf{y}_{1:k-1}\} \qquad (6.54)$$

$$= \mathbb{E}_{\mathbf{X}_k}\{\mathbf{C}_k\mathbf{X}_k|\mathbf{Y}_{1:k-1} = \mathbf{y}_{1:k-1}\} \qquad (6.55)$$

$$= \mathbf{C}_k\mathbf{m}_{k|k-1}. \qquad (6.56)$$

Similarly,

$$\boldsymbol{\Sigma}_k^{(y)} = \mathbb{E}_{\mathbf{Y}_k}\left\{\left(\mathbf{Y}_k - \mathbf{m}_k^{(y)}\right)\left(\mathbf{Y}_k - \mathbf{m}_k^{(y)}\right)^{\mathrm{T}}|\mathbf{Y}_{1:k-1} = \mathbf{y}_{1:k-1}\right\} \qquad (6.57)$$

$$= \mathbb{E}_{\mathbf{X}_k}\left\{\mathbb{E}_{\mathbf{Y}_k}\left\{\left(\mathbf{Y}_k - \mathbf{m}_k^{(y)}\right)\left(\mathbf{Y}_k - \mathbf{m}_k^{(y)}\right)^{\mathrm{T}}|\mathbf{X}_k\right\}|\mathbf{Y}_{1:k-1} = \mathbf{y}_{1:k-1}\right\} \qquad (6.58)$$

$$= \mathbf{D}_k\mathbf{D}_k^{\mathrm{T}} + \mathbf{C}_k\mathbf{P}_{k|k-1}\mathbf{C}_k^{\mathrm{T}}. \qquad (6.59)$$

This yields

$$\log p(\mathbf{Y} = \mathbf{y}|\mathcal{M})$$

$$= \sum_{k=1}^{N}\left\{-\frac{1}{2}\log\left(2\pi \det \boldsymbol{\Sigma}_k^{(y)}\right) - \frac{1}{2}\left(\mathbf{y}_k - \mathbf{m}_k^{(y)}\right)^{\mathrm{T}}\left(\boldsymbol{\Sigma}_k^{(y)}\right)^{-1}\left(\mathbf{y}_k - \mathbf{m}_k^{(y)}\right)\right\}. \qquad (6.60)$$

6.4.4   Determining the mode of the joint a-posteriori distribution

It is easy to show that $p(\mathbf{X}|\mathbf{Y} = \mathbf{y}, \mathcal{M})$ is a Gaussian distribution. Gaussian distributions have a great deal of useful properties, such as the fact that the mode of a (multivariate) Gaussian distribution corresponds to the concatenation of the modes of the marginals. This implies that the mode of $p(\mathbf{X}|\mathbf{Y} = \mathbf{y}, \mathcal{M})$ coincides with the concatenation of the modes of the marginals:

$$\hat{\mathbf{x}}_{0:N} = \arg\max_{\mathbf{x}} p(\mathbf{X}_{0:N} = \mathbf{x}|\mathbf{Y}_{1:N} = \mathbf{y}_{1:N}) \qquad (6.61)$$

$$= \left[\hat{\mathbf{x}}_0^{\mathrm{T}}, \ldots, \hat{\mathbf{x}}_N^{\mathrm{T}}\right]^{\mathrm{T}}, \qquad (6.62)$$

where $\hat{\mathbf{x}}_k$ is the mode of $p(\mathbf{X}_k|\mathbf{Y} = \mathbf{y})$. In other words, $\hat{\mathbf{x}}_k = \mathbf{m}_{k|N}$.

6.4.5 Concluding remarks

In the technical literature, the forward phase of the SPA is known as Kalman filtering, while the backward phase after the forward phase is known as Kalman smoothing [70].
    We have solved the following inference problems:

1. determining the a-posteriori distribution of X_k, $p(X_k|\mathbf{Y} = \mathbf{y}, \mathcal{M})$,
2. determining the likelihood of the model $\mathcal{M}$, $p(\mathbf{Y} = \mathbf{y}|\mathcal{M})$, and
3. determining the mode of the joint a-posteriori distribution, $p(\mathbf{X}|\mathbf{Y} = \mathbf{y}, \mathcal{M})$.

6.5 Approximate inference for state-space models

6.5.1 Introduction

While the above techniques are useful for linear Gaussian models, most practical systems will be neither Gaussian nor linear. Let us focus on state-emitting SSMs with continuous state spaces, where both the states X_k and the outputs Y_k are scalars. Extension to vector models is straightforward. The joint distribution of the states and the observation is given by

$$p(\mathbf{X}, \mathbf{Y} = \mathbf{y}|\mathcal{M}) = p(X_0|\mathcal{M}) \prod_{k=1}^{N} \underbrace{p(Y_k = y_k|X_k, \mathcal{M})p(X_k|X_{k-1}, \mathcal{M})}_{f_k(X_k, X_{k-1})}. \qquad (6.63)$$

The factor graph of the factorization (6.63) is depicted in Fig. 6.10. Again, we have opened the node f_k to reveal its structure. The reader should note that we have used a slightly different notation for some of the variables with respect to previous factor graphs (such as Figs. 6.2 and 6.5).
    Building on the techniques described in Section 5.3.4, we will describe how to execute the SPA by using particle representations of the messages. This allows us to determine, at least approximately, the marginal a-posteriori distributions $p(X_k|\mathbf{Y} = \mathbf{y}, \mathcal{M})$. In general, finding the mode of the joint a-posteriori distribution $p(\mathbf{X}|\mathbf{Y} = \mathbf{y}, \mathcal{M})$ is hard. On the other hand, it turns out to be possible to determine the likelihood $p(\mathbf{Y} = \mathbf{y}|M)$ of the model $\mathcal{M}$. For notational convenience, we will drop the conditioning on the model $\mathcal{M}$, except in Section 6.5.3. More information regarding particle methods in the context of state-space models can be found in [48, 65, 73, 78, 79].

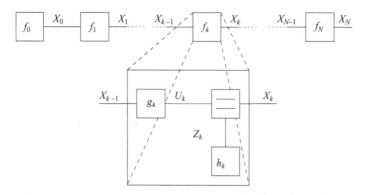

Figure 6.10. A factor graph for a state-emitting SSM, with $f_0(X_0) = p(X_0)$, and $f_k(X_{k-1}, X_k) = p(Y_k = y_k | X_k)p(X_k | X_{k-1})$. The node f_k is opened to reveal its structure, with $g_k(X_{k-1}, U_k) = p(U_k | X_{k-1})$ and $h_k(Z_k) = p(Y_k = y_k | Z_k)$, and an equality node.

6.5.2   Determining the marginal a-posteriori distributions

6.5.2.1   Notations

In the forward phase we will use the following particle representations for the messages:

$$\mathcal{R}_L\big(\mu_{f_{k-1}\to X_{k-1}}(X_{k-1})\big) = \left\{\left(w^{(l)}_{k-1|k-1}, x^{(l)}_{k-1|k-1}\right)\right\}^L_{l=1} \tag{6.64}$$

and

$$\mathcal{R}_L\big(\mu_{g_k\to U_k}(U_k)\big) = \left\{\left(w^{(l)}_{k|k-1}, x^{(l)}_{k|k-1}\right)\right\}^L_{l=1}. \tag{6.65}$$

For the backward phase, we will use

$$\mathcal{R}_L\big(\mu_{X_k\to f_k}(X_k)\big) = \left\{\left(\tilde{w}^{(l)}_{k|k}, \tilde{x}^{(l)}_{k|k}\right)\right\}^L_{l=1} \tag{6.66}$$

and

$$\mathcal{R}_L\big(\mu_{U_k\to g_k}(U_k)\big) = \left\{\left(\tilde{w}^{(l)}_{k|k-1}, \tilde{x}^{(l)}_{k|k-1}\right)\right\}^L_{l=1}. \tag{6.67}$$

6.5.2.2   The forward phase

We start from a particle representation of $p(X_0)$, $\mathcal{R}_L(p(X_0)) = \{(w^{(l)}_{0|0}, x^{(l)}_{0|0})\}^L_{l=1}$, which also serves as a representation of $\mu_{f_0\to X_0}(X_0)$. Let us assume that we have a particle representation of $\mu_{f_{k-1}\to X_{k-1}}(X_{k-1})$: $\mathcal{R}_L(\mu_{f_{k-1}\to X_{k-1}}(X_{k-1})) = \{(w^{(l)}_{k-1|k-1}, x^{(l)}_{k-1|k-1})\}^L_{l=1}$. The forward phase consists of two steps.

1. determine a particle representation of $\mu_{g_k\to U_k}(U_k)$ based on the particle representation of $\mu_{f_{k-1}\to X_{k-1}}(X_{k-1})$.
2. determine a particle representation of $\mu_{f_k\to X_k}(X_k)$ based on the particle representation of $\mu_{g_k\to U_k}(U_k)$.

Both steps are described in Algorithm 6.4.

Algorithm 6.4 The forward phase of sum–product algorithm on a state-space model using particle representations

---

1: initialization: $\mathcal{R}_L(p(X_0)) = \{(w_{0|0}^{(l)}, x_{0|0}^{(l)})\}_{l=1}^L$

2: **for** $k = 1$ to N **do**

3:     **for** $l = 1$ to L **do**

4:         draw $x_{k|k-1}^{(l)} \sim q(U_k = u_k | X_{k-1} = x_{k-1|k-1}^{(l)})$

5:         set importance weight

$$w_{k|k-1}^{(l)} = w_{k-1|k-1}^{(l)} \frac{p\left(X_k = x_{k|k-1}^{(l)} \middle| X_{k-1} = x_{k-1|k-1}^{(l)}\right)}{q\left(U_k = x_{k|k-1}^{(l)} \middle| X_{k-1} = x_{k-1|k-1}^{(l)}\right)}$$

6:     **end for**

7:     normalize weights to obtain $\mathcal{R}_L(\mu_{g_k \to U_k}(U_k)) = \{(w_{k|k-1}^{(l)}, x_{k|k-1}^{(l)})\}_{l=1}^L$

8:     **for** $l = 1$ to N **do**

9:         set $x_{k|k}^{(l)} = x_{k|k-1}^{(l)}$

10:        set importance weight $w_{k|k}^{(l)} = p(Y_k = y_k | X_k = x_{k|k-1}^{(l)}) w_{k|k-1}^{(l)}$

11:     **end for**

12:     normalize weights to obtain $\mathcal{R}_L(\mu_{f_k \to X_k}(X_k)) = \{(w_{k|k}^{(l)}, x_{k|k}^{(l)})\}_{l=1}^L$

13: **end for**

Step 1

We determine a representation of $\mu_{g_k \to U_k}(U_k)$ using *mixture sampling* (as described in Section 5.3.4). In other words, we approximate $\mu_{g_k \to U_k}(u_k)$ as a mixture density by substituting the particle representation of $\mu_{f_{k-1} \to X_{k-1}}(x_{k-1})$ into the sum–product rule for $\mu_{g_k \to U_k}(U_k)$:

$$\mu_{g_k \to U_k}(u_k) \propto \int p(X_k = u_k | X_{k-1} = x_{k-1}) \mu_{f_{k-1} \to X_{k-1}}(x_{k-1}) dx_{k-1} \qquad (6.68)$$

$$\approx \sum_{l=1}^L w_{k-1|k-1}^{(l)} p\left(X_k = u_k \middle| X_{k-1} = x_{k-1|k-1}^{(l)}\right). \qquad (6.69)$$

Now, for every $x_{k-1|k-1}^{(l)}$, and for a suitable importance sampling function $q(U_k | X_{k-1} = x_{k-1|k-1}^{(l)})$, we draw a single sample (say, $x_{k|k-1}^{(l)}$), with weight

$$w_{k|k-1}^{(l)} = w_{k-1|k-1}^{(l)} \frac{p\left(X_k = x_{k|k-1}^{(l)} \middle| X_{k-1} = x_{k-1|k-1}^{(l)}\right)}{q\left(U_k = x_{k|k-1}^{(l)} \middle| X_{k-1} = x_{k-1|k-1}^{(l)}\right)}. \qquad (6.70)$$

Once L samples have been drawn, we normalize the weights, and end up with a particle representation $\mathcal{R}_L(\mu_{g_k \to U_k}(U_k)) = \{(w_{k|k-1}^{(l)}, x_{k|k-1}^{(l)})\}_{l=1}^L$. As a special case, we can set

$q(U_k = u_k | X_{k-1} = x_{k-1}) = p(X_k = u_k | X_{k-1} = x_{k-1})$, in which case the weights are unchanged: $w_{k|k-1}^{(l)} = w_{k-1|k-1}^{(l)}$.

Step 2

In the second step in the forward phase, we determine the message $\mu_{f_k \to X_k}(X_k)$. Simply plugging the particle representation into the sum–product rule gives us

$$\mu_{f_k \to X_k}(x_k) \propto \mu_{Z_k \to \boxminus}(x_k)\, \mu_{g_k \to U_k}(x_k)$$

$$\propto p(Y_k = y_k | X_k = x_k)\mu_{g_k \to U_k}(x_k)$$

$$\approx \sum_{l=1}^{L} \underbrace{p(Y_k = y_k | X_k = x_k) w_{k|k-1}^{(l)}}_{\propto w_{k|k}^{(l)}}\, \boxminus\left(x_k, x_{k|k-1}^{(l)}\right).$$

We find that $\mathcal{R}_L(\mu_{f_k \to X_k}(X_k)) = \{(w_{k|k}^{(l)}, x_{k|k}^{(l)})\}_{l=1}^{L}$ with

$$x_{k|k}^{(l)} = x_{k|k-1}^{(l)} \tag{6.71}$$

$$w_{k|k}^{(l)} \propto p\left(Y_k = y_k \middle| X_k = x_{k|k-1}^{(l)}\right) w_{k|k-1}^{(l)}. \tag{6.72}$$

To avoid degeneration, it may be necessary to resample at this point [65].

6.5.2.3 Backward phase in parallel with forward phase

If the backward phase is executed in parallel with the forward phase, it is an almost perfect mirror-image of the forward phase. We start with a representation of $\mu_{X_N \to f_N}(X_N)$, as $\{(\tilde{w}_{N|N}^{(l)}, \tilde{x}_{N|N}^{(l)})\}_{l=1}^{L}$, obtained by uniform sampling.

Step 1

Suppose that we have available a particle representation of $\mu_{X_k \to f_k}(X_k)$: $\{(\tilde{w}_{k|k}^{(l)}, \tilde{x}_{k|k}^{(l)})\}_{l=1}^{L}$. Since

$$\mu_{U_k \to g_k}(u_k) \propto p(Y_k = y_k | X_k = x_k)\mu_{X_k \to \boxminus}(u_k) \tag{6.73}$$

we can apply the same reasoning as in the forward phase to find that $\mu_{U_k \to g_k}(u_k)$ can be represented by $\{(\tilde{x}_{k|k-1}^{(l)}, \tilde{w}_{k|k-1}^{(l)})\}_{l=1}^{L}$, where

$$\tilde{x}_{k|k-1}^{(l)} = \tilde{x}_{k|k}^{(l)} \tag{6.74}$$

$$\tilde{w}_{k|k-1}^{(l)} \propto \tilde{w}_{k|k}^{(l)} p\left(Y_k = y_k \middle| X_k = \tilde{x}_k^{(l)}\right). \tag{6.75}$$

Step 2
In the second phase, we again perform mixture sampling:

$$\mu_{g_k \to X_{k-1}}(x_{k-1}) \propto \int p(X_k = u_k | X_{k-1} = x_{k-1}) \mu_{U_k \to g_k}(u_k) du_k \tag{6.76}$$

$$\approx \sum_{l=1}^{L} \tilde{w}_{k|k-1}^{(l)} p\left(X_k = \tilde{x}_{k|k-1}^{(l)} \middle| X_{k-1} = x_{k-1}\right). \tag{6.77}$$

Now, for every $\tilde{x}_{k|k-1}^{(l)}$, and a suitable importance sampling function $\tilde{q}(X_{k-1} = x_{k-1} | \tilde{x}_{k|k-1}^{(l)})$, draw a single sample $\tilde{x}_{k-1|k-1}^{(l)}$, with weight

$$\tilde{w}_{k-1|k-1}^{(l)} = \tilde{w}_{k|k-1}^{(l)} \frac{p\left(X_k = \tilde{x}_{k|k-1}^{(l)} \middle| X_{k-1} = \tilde{x}_{k-1|k-1}^{(l)}\right)}{\tilde{q}\left(X_{k-1} = \tilde{x}_{k-1|k-1}^{(l)} \middle| \tilde{x}_{k|k-1}^{(l)}\right)}. \tag{6.78}$$

Once L samples have been drawn, we normalize the weights, and end up with $\mathcal{R}_L(\mu_{x_{k-1} \to f_{k-1}}(X_{k-1}))$ as $\{(\tilde{w}_{k-1|k-1}^{(l)}, \tilde{x}_{k-1|k-1}^{(l)})\}_{l=1}^{L}$. To avoid degeneration, it may be necessary to resample at this point.

Marginals
The marginals can then by found by regularization of the forward and backward messages. This regularization process is rather cumbersome and will not be pursued here. In practice, we commonly prefer to perform the backward phase in combination with determining the marginals, once all the forward messages have been computed (this is similar to the smoothing technique for linear Gaussian models described in Section 6.4.2.5). This is explained in the next section.

6.5.2.4   Smoothing: backward phase after forward phase
As with the linear Gaussian model, we can compute the marginal a-posteriori distributions by performing the backward after the forward phase. At the end of the forward phase, we have a particle representation of the message $\mu_{f_N \to X_N}(X_N) = p(X_N | \mathbf{Y}_{1:N} = \mathbf{y}_{1:N})$. Suppose now that we have a particle representation of $p(X_k | \mathbf{Y}_{1:N} = \mathbf{y}_{1:N})$ as $\{(w_{k|N}^{(l)}, \mathbf{x}_{k|N}^{(l)})\}_{l=1}^{L}$. Let us try to express $p(X_{k-1} | \mathbf{Y}_{1:N} = \mathbf{y}_{1:N})$ as a function of $p(X_k | \mathbf{Y}_{1:N} = \mathbf{y}_{1:N})$ and the forward messages [78]. Using a notational shorthand, we find that

$$p(x_{k-1} | \mathbf{y}_{1:N}) = p(x_{k-1} | \mathbf{y}_{1:k-1}) \int p(x_k | \mathbf{y}_{1:N}) \frac{p(x_k | x_{k-1})}{p(x_k | \mathbf{y}_{1:k-1})} dx_k. \tag{6.79}$$

Substitution of the particle representations of $p(x_k|\mathbf{y}_{1:N})$ and $p(x_{k-1}|\mathbf{y}_{1:k-1})$ yields

$$p(x_{k-1}|\mathbf{y}_{1:N}) \approx \underbrace{\sum_{l_1=1}^{L} w_{k-1|k-1}^{(l_1)} \sum_{l_2=1}^{L} w_{k|N}^{(l_2)} \frac{p\left(x_{k|N}^{(l_2)}\Big|x_{k-1|k-1}^{(l_1)}\right)}{p\left(x_{k|N}^{(l_2)}\Big|\mathbf{y}_{1:k-1}\right)}}_{\propto w_{k-1|N}^{(l_1)}} \equiv \left(x_{k-1}, x_{k-1|k-1}^{(l_1)}\right).$$

(6.80)

Now we make use of

$$p\left(x_{k|N}^{(l_2)}\Big|\mathbf{y}_{1:k-1}\right) = \int p\left(x_{k|N}^{(l_2)}\Big|x_{k-1}\right) p\left(x_{k-1}|\mathbf{y}_{1:k-1}\right) dx_{k-1} \quad (6.81)$$

$$\approx \sum_{l_3=1}^{L} w_{k-1|k-1}^{(l_3)} p\left(x_{k|N}^{(l_2)}\Big|x_{k-1|k-1}^{(l_3)}\right). \quad (6.82)$$

After substitution of (6.82) into (6.80), we find that $p(x_{k-1}|\mathbf{y}_{1:N})$ can be represented by a list of L samples $\{\{(w_{k-1|N}^{(l)}, x_{k-1|N}^{(l)})\}_{l=1}^{L}\}$, where

$$x_{k-1|N}^{(l)} = x_{k-1|k-1}^{(l)} \quad (6.83)$$

and

$$w_{k-1|N}^{(l)} = w_{k-1|k-1}^{(l)} \sum_{l_2=1}^{L} w_{k|N}^{(l_2)} \frac{p\left(x_{k|N}^{(l_2)}\Big|x_{k-1|k-1}^{(l)}\right)}{\sum_{l_3=1}^{L} w_{k-1|k-1}^{(l_3)} p\left(x_{k|N}^{(l_2)}\Big|x_{k-1|k-1}^{(l_3)}\right)}. \quad (6.84)$$

Observe that the computational complexity per sample is proportional to L^2.

6.5.3   Determining the likelihood of the model

As with linear Gaussian models (see Section 6.4.3), the likelihood of the model $p(\mathbf{Y}_{1:N} = \mathbf{y}_{1:N}|\mathcal{M})$ can be obtained after the forward phase. We will work again in the log domain and compute $\log p(\mathbf{Y}_{1:N} = \mathbf{y}_{1:N}|\mathcal{M})$. We know that

$$\log p(\mathbf{Y}_{1:N} = \mathbf{y}_{1:N}|\mathcal{M})$$

$$= \log p(Y_1 = y_1|\mathcal{M}) + \sum_{k=2}^{N} \log p(Y_k = y_k|\mathbf{Y}_{1:k-1} = \mathbf{y}_{1:k-1}, \mathcal{M}), \quad (6.85)$$

where

$$p(Y_1 = y_1 | \mathcal{M}) = \int p(Y_1 = y_1, X_1 = x_1 | \mathcal{M}) \, dx_1 \tag{6.86}$$

$$= \int p(Y_1 = y_1 | X_1 = x_1, \mathcal{M}) p(X_1 = x_1 | \mathcal{M}) \, dx_1 \tag{6.87}$$

$$\approx \sum_{l=1}^{L} w_{1|0}^{(l)} p\left(Y_1 = y_1 \Big| X_1 = x_{1|0}^{(l)}, \mathcal{M}\right) \tag{6.88}$$

and similarly

$$p(Y_k = y_k | \mathbf{Y}_{1:k-1} = \mathbf{y}_{1:k-1}, \mathcal{M})$$

$$= \int p(Y_k = y_k, X_k = x_k | \mathbf{Y}_{1:k-1} = \mathbf{y}_{1:k-1}, \mathcal{M}) \, dx_k \tag{6.89}$$

$$\approx \sum_{l=1}^{L} w_{k|k-1}^{(l)} p\left(Y_k = y_k | X_k = x_{k|k-1}^{(l)}, \mathcal{M}\right). \tag{6.90}$$

Note that $p(Y_1 = y_1 | \mathcal{M})$ and $p(Y_k = y_k | \mathbf{Y}_{1:k-1} = \mathbf{y}_{1:k-1}, \mathcal{M})$ are scalars, not distributions.

6.5.4   Concluding remarks

In the technical literature, the forward phase is known as particle filtering, while the backward phase is called particle smoothing. When $q(U_k = u_k | X_{k-1} = x_{k-1}) = p(X_k = u_k | X_{k-1} = x_{k-1})$, the forward phase is known as the bootstrap filter [80]. Performing Monte Carlo-based approximate inference for SSM with continuous state spaces allows us to solve the following inference problems:

(1) determining approximately the a posteriori distribution of X_k, $p(X_k | \mathbf{Y} = \mathbf{y}, \mathcal{M})$, and
(2) determining approximately the likelihood of the model $\mathcal{M}$, $p(\mathbf{Y} = \mathbf{y} | \mathcal{M})$.

6.6   Main points

In this chapter, we have covered inference for state-space models (SSMs). They are important in many applications in which dynamically changing systems are modeled, and are used in a wide variety of engineering problems. We have given a detailed overview of inference on three important state-space models: hidden Markov models, linear Gaussian models, and general models with continuous state spaces. Creating a factor graph and executing the SPA leads to reformulations of several well-known algorithms from the technical literature, including the forward–backward algorithm, the Viterbi algorithm, the Kalman filter, and the particle filter.

This ends our digression into factor graphs and statistical inference. We now have all the necessary tools to go back to our problem from Chapter 2: designing receivers for digital communication.

7 Factor graphs in digital communication

7.1 Introduction

Let us now return to our original problem from Chapter 2: receiver design. Our ultimate goal is to recover (in an optimal manner) the transmitted information bits **b** from the received waveform $r(t)$. In this chapter, we will formulate an inference problem that will enable us to achieve this task. In the first stage the received waveform is converted into a suitable observation **y**. We then create a factor graph of the distribution $p(\mathbf{B}, \mathbf{Y} = \mathbf{y} | \mathcal{M})$. This factor graph will contain three important nodes, expressing the relationship between information bits and coded bits (the *decoding* node), between coded bits and coded symbols (the *demapping* node), and between coded symbols and the observation (the *equalization* node). Correspondingly, the inference problem breaks down into three sub-problems: decoding, demapping, and equalization.

Decoding will be covered in Chapter 8, where we will describe some state-of-the-art iterative decoding take over schemes, including turbo codes, RA codes, and LDPC codes. The demapping problem will be considered in Chapter 9 for two common modulation techniques: bit-interleaved coded modulation and trellis-coded modulation. Equalization highly depends on the specific digital communication scheme. In Chapter 10, we will derive several general-purpose equalization strategies. In the three subsequent chapters, we will then show how these general-purpose strategies can be applied to the digital communication schemes from Chapter 2. In Chapter 11 the focus is on single-user, single-antenna communication. This is then extended to multi-antenna and multi-user communication in Chapters 12 and 13, respectively.

This short chapter is organized as follows.

- In **Section 7.2**, we will describe how designing a receiver can be cast in the factor-graph framework.
- In **Section 7.3** we will then open various nodes to reveal their structure.

7.2 The general principle

7.2.1 Inference problems for digital receivers

Before we design a receiver, we will first formulate an inference problem and create a factor graph. As we have seen in Chapter 2, a sequence of N_b information bits (**b**) is first encoded to a sequence of N_c coded bits ($\mathbf{c}$), with $\mathbf{c} = f_c(\mathbf{b})$. The coded bits are converted into a sequence of N_s complex coded symbols ($\mathbf{a}$, with each symbol belonging to a constellation), with $\mathbf{a} = f_a(\mathbf{c})$. The functions $f_a(\cdot)$ and $f_c(\cdot)$ are deterministic and known both to the transmitter and to the receiver. The symbol sequence $\mathbf{a}$ is further processed to give rise to the transmitted equivalent baseband signal $s(t)$. This signal propagates through the equivalent baseband channel, resulting in a received baseband signal $r(t)$ with vector representation $\mathbf{r}$.

Sequence detection

Since the final goal of the receiver is to recover the information sequence $\mathbf{b}$ from the observation $\mathbf{r}$, the optimal way to proceed is to determine the information sequence that maximizes the a-posteriori probability:

$$\hat{\mathbf{b}} = \arg \max_{\mathbf{b} \in \mathbb{B}^{N_b}} p(\mathbf{B} = \mathbf{b}|\mathbf{R} = \mathbf{r}, \mathcal{M}), \tag{7.1}$$

where the model $\mathcal{M}$ can encapsulate the channel parameters, code type, modulation format, etc. We know from Section 3.2 that solving (7.1) is optimal in the sense that it minimizes the probability of error, which in this context is known as the word error rate or the frame error rate (FER).

In most practical receivers, the received waveform is first converted into a suitable observation. This can be through filtering, sampling, projection, transformation to a different domain, etc. As we go through the various transmission techniques in Chapters 11–13, we will show how exactly this conversion is achieved. The observation is denoted $\mathbf{y}$. When $p(\mathbf{B}|\mathbf{R} = \mathbf{r}, \mathcal{M}) = p(\mathbf{B}|\mathbf{Y} = \mathbf{y}, \mathcal{M})$, $\mathbf{Y}$ is known as a *sufficient statistic* for $\mathbf{B}$. In this book, we generally don't care whether or not we are dealing with sufficient statistics.

Once we have determined a suitable observation $\mathbf{y}$, we can find the MAP information sequence by creating a factor graph of $\log p(\mathbf{B}, \mathbf{Y} = \mathbf{y}|\mathcal{M})$ and applying the max–sum algorithm.

Bit-by-bit detection

Alternatively, we can also create a factor graph of $p(\mathbf{B}, \mathbf{Y} = \mathbf{y}|\mathcal{M})$ and implement the SPA. This leads to the marginal a-posteriori distributions $p(B_k|\mathbf{Y} = \mathbf{y}, \mathcal{M})$, from which we can make optimal bit-by-bit decisions:

$$\hat{b}_k = \arg \max_{b_k \in \{0,1\}} p(B_k = b_k|\mathbf{Y} = \mathbf{y}, \mathcal{M}). \tag{7.2}$$

This approach is no longer optimal in the sense of minimizing the word error rate. It is, however, optimal in the sense of minimizing the error probability of the individual bits, namely the bit error rate (BER). In many receivers, due to the cycles in the factor graph, sequence detection is not possible, and bit-by-bit detection is the only feasible approach. Unless mentioned explicitly otherwise, we will always perform bit-by-bit detection.

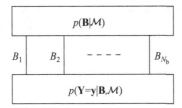

Figure 7.1. A factor graph of the distribution $p(\mathbf{B}, \mathbf{Y} = \mathbf{y}|\mathcal{M})$.

7.2.2 Factor graphs

Factorizing $p(\mathbf{B}, \mathbf{Y} = \mathbf{y}|\mathcal{M})$ leads to

$$p(\mathbf{B}, \mathbf{Y} = \mathbf{y}|\mathcal{M}) = p(\mathbf{Y} = \mathbf{y}|\mathbf{B}, \mathcal{M})p(\mathbf{B}|\mathcal{M}), \tag{7.3}$$

where $p(\mathbf{Y} = \mathbf{y}|\mathbf{B}, \mathcal{M})$ is the likelihood function and $p(\mathbf{B}|\mathcal{M})$ is the a-priori distribution. We then create a factor graph (shown in Fig. 7.1). On the basis of relations among the information bits, coded bits, and coded symbols, we will open the nodes further in the next section. In most cases the factor graph will have cycles, so that we end up with an iterative version of the SPA. This raises some additional issues.

- The marginals we compute are beliefs, and are considered approximations of the true marginals $p(B_k|\mathbf{Y} = \mathbf{y}, \mathcal{M})$.
- The presence of cycles forces us to initialize some messages with uniform distributions.
- We also have to deal with scheduling, i.e., the order in which messages are computed. Different scheduling strategies may have different performances. Some strategies may be preferred from an implementation point of view.
- Finally, at some point the SPA needs to be terminated.

7.3 Opening nodes

7.3.1 Principles

In this section, we will open the nodes $p(\mathbf{B}|\mathcal{M})$ and $p(\mathbf{Y} = \mathbf{y}|\mathbf{B}, \mathcal{M})$. We remind the reader that we can open a node representing the function $f(X_1, \ldots, X_D)$ by replacing it by the factorization of another function, $g(X_1, \ldots, X_D, U_1, \ldots, U_K)$, as long as the following conditions are met:

- the variables $U_1, \ldots, U_K$ appear nowhere else in the graph, and
- the functions f and g satisfy

$$\sum_{u_1, \ldots, u_K} g(x_1, \ldots, x_D, u_1, u_2, \ldots, u_K) = f(x_1, \ldots, x_D). \tag{7.4}$$

It is easy to see that this replacement will not affect the messages computed on the edges $X_1, \ldots, X_D$ as long as the factor graph of the factorization of $g(X_1, \ldots, X_D, U_1, \ldots, U_K)$ has no cycles.

7.3.2   Opening the a-priori node

In this book, we will restrict ourselves to information bits that are independent with known a-priori probabilities, so that

$$p(\mathbf{B} = \mathbf{b}|\mathcal{M}) = \prod_{k=1}^{N_b} p(B_k = b_k|\mathcal{M}). \tag{7.5}$$

7.3.3   Opening the likelihood node

The likelihood function $p(\mathbf{Y} = \mathbf{y}|\mathbf{B}, \mathcal{M})$ can be factorized by introducing suitable additional variables, say $\mathbf{D}$. We can replace the node $p(\mathbf{Y} = \mathbf{y}|\mathbf{B}, \mathcal{M})$ by a factorization of $p(\mathbf{Y} = \mathbf{y}, \mathbf{D}|\mathbf{B}, \mathcal{M})$. The question is, of course, what should $\mathbf{D}$ be? Let us look at an example to get some insight.

Example 7.1. *In a simple communication scheme, the N_b information bits $\mathbf{b}$ are converted into a sequence of N_c coded bits $\mathbf{c}$, with $\mathbf{c} = f_c(\mathbf{b})$, for some deterministic mapping $f_c(\cdot)$. Then, the N_c coded bits are mapped to $N_s = N_c$ BPSK symbols, corresponding to the constellation $\Omega = \{-1, +1\}$. The mapping works as follows: $a_n = 2c_n - 1$, for $n = 1, \ldots, N_c$. This yields a sequence of N_s BPSK symbols $\mathbf{a}$, with $\mathbf{a} = f_a(\mathbf{c})$. Let us assume that $\mathbf{y}$ is related to $\mathbf{a}$ by*

$$\mathbf{y} = \alpha \mathbf{a} + \mathbf{n},$$

where $\alpha \in \mathbb{R}$ is the channel amplitude and $\mathbf{n} \sim \mathcal{N}_{\mathbf{n}}(\mathbf{0}, \sigma^2 \mathbf{I}_{N_s})$, a natural choice for $\mathbf{D} = [\mathbf{A} \quad \mathbf{C}]$. We then open the node $p(\mathbf{Y} = \mathbf{y}|\mathbf{B}, \mathcal{M})$ and replace it by a factorization of $p(\mathbf{Y} = \mathbf{y}, \mathbf{A}, \mathbf{C}|\mathbf{B}, \mathcal{M})$. Upon applying Bayes' rule, and taking into account the appropriate conditional dependencies, we find the following factorization:

$$p(\mathbf{Y} = \mathbf{y}, \mathbf{A}, \mathbf{C}|\mathbf{B}, \mathcal{M}) = \underbrace{p(\mathbf{Y} = \mathbf{y}|\mathbf{A}, \mathbf{C}, \mathbf{B}, \mathcal{M})}_{=p(\mathbf{Y}=\mathbf{y}|\mathbf{A},\mathcal{M})} p(\mathbf{A}, \mathbf{C}|\mathbf{B}, \mathcal{M})$$

$$= p(\mathbf{Y} = \mathbf{y}|\mathbf{A}, \mathcal{M}) \underbrace{p(\mathbf{A}|\mathbf{C}, \mathbf{B}, \mathcal{M})}_{=p(\mathbf{A}|\mathbf{C},\mathcal{M})} p(\mathbf{C}|\mathbf{B}, \mathcal{M})$$

$$= \prod_{l=1}^{N_s} p(Y_l = y_l|A_l, \mathcal{M}) \boxdot (\mathbf{A}, f_a(\mathbf{C})) \boxdot (\mathbf{C}, f_c(\mathbf{B}))$$

$$\propto \prod_{l=1}^{N_s} \left\{ \exp\left(-\frac{1}{2\sigma^2} (y_l - A_l)^2\right) \boxdot (A_l, 2C_l - 1) \right\} \boxdot (\mathbf{C}, f_c(\mathbf{B}))$$

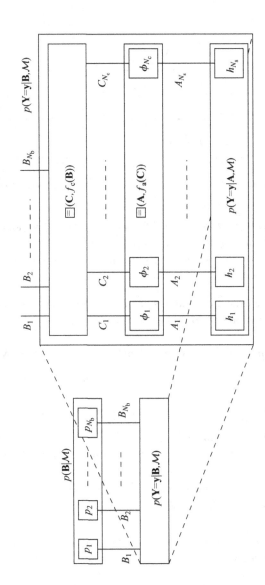

Figure 7.2. A factor graph of the joint distribution $p(\mathbf{B}, \mathbf{Y} = \mathbf{y}|\mathcal{M})$ with independent information bits. The node $p(\mathbf{Y} = \mathbf{y}|\mathbf{B}, \mathcal{M})$ is opened to reveal its structure, with $\phi(c_l, a_l) = \boxminus(a_l, 2c_l - 1)$ and $h_l(a_l) = \exp[-(y_l - a_l)^2/(2\sigma^2)]$.

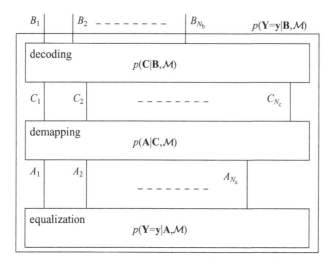

Figure 7.3. A factor graph of $p(\mathbf{Y} = \mathbf{y}|\mathbf{B}, \mathcal{M})$. The node is opened to reveal its structure.

with the factor graph shown in Fig. 7.2. The node $\boxminus(\mathbf{C}, f_c(\mathbf{B}))$ can again be opened if we know the structure of a code. Implementing the SPA on the graph from Fig. 7.2 will give us (approximations of) the desired a-posteriori probabilities $p(B_k|\mathbf{Y} = \mathbf{y}, \mathcal{M})$.

The above example illustrates that it is a good idea to place $\mathbf{C}$ (the coded bits) and $\mathbf{A}$ (the coded symbols) in $\mathbf{D}$. In general, this leads to

$$p(\mathbf{Y} = \mathbf{y}, \mathbf{A}, \mathbf{C}|\mathbf{B}, \mathcal{M}) = p(\mathbf{Y} = \mathbf{y}|\mathbf{A}, \mathbf{C}, \mathbf{B}, \mathcal{M})p(\mathbf{A}, \mathbf{C}|\mathbf{B}, \mathcal{M})$$
$$= p(\mathbf{Y} = \mathbf{y}|\mathbf{A}, \mathcal{M})p(\mathbf{A}|\mathbf{C}, \mathcal{M})p(\mathbf{C}|\mathbf{B}, \mathcal{M}).$$

The corresponding factor graph is shown in Fig. 7.3. We discern three nodes.

1. The node on the top represents $p(\mathbf{C}|\mathbf{B}, \mathcal{M})$, the relation between the information bits and the coded bits. We will name the process of executing the SPA on this node *decoding*.
2. The node in the middle represents $p(\mathbf{A}|\mathbf{C}, \mathcal{M})$, the relation between the coded bits and the coded symbols. We will name the process of executing the SPA on this node *demapping*.
3. Finally, the bottom-most node represents the function $p(\mathbf{Y} = \mathbf{y}|\mathbf{A}, \mathcal{M})$, the relation between the coded symbols and the observation. The SPA on this node is called *equalization*.

We have seen in Chapter 4 that factor graphs inherently allow functional decomposition. This will be of great help as we develop our receivers, since we need focus only on the various nodes in Fig. 7.3 separately. In Chapter 8 and Chapter 9 we will focus on the decoding and demapping aspects, respectively. The four subsequent chapters will deal with equalization.

7.4   Main points

In this chapter we have seen how (at least in principle) the problem of recovering
information bits from a received signal can be resolved by first converting the received
waveform into a suitable observation and then creating a factor graph of the joint
distribution of the information bits and the observation. Opening nodes has given us
insight into the receiver's operation. We have exposed three critical functions/nodes:
decoding, *demapping*, and *equalization*. These functions will be covered in that order in
the next few chapters.

8   Decoding

8.1   Introduction

Error-correcting codes are a way to protect a binary information sequence against adverse channel effects by adding a certain amount of redundancy. This is known as encoding. The receiver can then try to recover the original binary information sequence, using a decoder. The field of coding theory deals with developing and analyzing codes and decoding algorithms. Although coding theory is fairly abstract and generally involves a great deal of math, our knowledge of factor graphs will allow us to derive decoding algorithms without delving too deep. As we will see, using factor graphs, decoding becomes a fairly straightforward matter. In contrast to conventional decoding algorithms, our notation will be the same for all types of codes, which makes it easier to understand and interpret the algorithms.

In this chapter, we will deal with four types of error-correcting codes: repeat–accumulate (RA) codes, low-density parity-check (LDPC) codes, convolutional codes, and turbo codes. Repeat–accumulate codes were introduced in 1998 as a type of toy code. Later they turned out to have a great deal of practical importance [81]. We then move on to the LDPC codes, which were invented by Gallager in 1963 [50], and re-introduced in the early 1990s by MacKay [82]. Both types of codes can easily be cast into factor graphs; these factor graphs turn out to have cycles, leading to iterative decoding algorithms. Convolutional codes, on the other hand, are based on state-space models and thus lead to cycle-free factor graphs [83]. Finally, we will consider turbo codes, which were introduced in 1993 by Berrou *et al.* [3], consisting of the concatenation of two convolutional encoders separated by an interleaver. The RA, LDPC, and turbo codes all have the following in common: they are decoded using iterative decoding algorithms, and they contain a pseudo-random component (such as an interleaver). Standard reference works on coding include [84, 85].

This chapter is organized as follows.

- In **Section 8.2** we will describe the main goals of this chapter.
- **Section 8.3** gives a brief overview of block codes and their relation to factor graphs.
- Four types of codes will described in detail. We start with the simplest type in **Section 8.4**: RA codes.
- This is followed by LDPC codes in **Section 8.5**.

- Convolutional codes and two flavors of turbo codes will be the topics of **Sections 8.6** and **8.7**.
- A performance illustration of a turbo code will be given in **Section 8.8**.

8.2   Goals

We have two goals in this chapter: first of all, we will replace the node $p\,(\mathbf{C}\,|\mathbf{B}, \mathcal{M})$ in Fig. 8.1 by a more detailed factor graph depending on the particular type of error-correcting code. Secondly, we will show how the SPA can be executed on the resulting factor graphs. For every type of code, we will draw the factor graph, locate the basic building blocks in the graph, and show how to implement the SPA on these blocks. We will end by deriving the decoding algorithms in the probability domain, the log-domain, and the LLR domain. We remind the reader that these message types were discussed in Section 5.3.3.

To help the reader understand the decoding algorithms, one can imagine the coded bits being mapped to BPSK symbols, such that $a_k = 2c_k - 1$, $k = 1, \ldots, N_c$, and that $y_k = a_k + n_k$, where $n_k \sim \mathcal{N}\left(0, \sigma^2\right)$. In that case, $p(\mathbf{A}|\mathbf{C}, \mathcal{M}) = \prod_{k=1}^{N_c} \boxminus(A_k, 2C_k - 1)$ and $p(\mathbf{Y} = \mathbf{y}|\mathbf{A}, \mathcal{M}) \propto \prod_{k=1}^{N_c} \exp[-(y_k - A_k)^2/(2\sigma^2)]$. Hence, the upward messages on the C_k-edges in Fig. 8.1 are given by $\mu_{C_k \to \mathrm{dec}}(c_k) \propto \exp[-(y_k - 2c_k + 1)^2/(2\sigma^2)]$. Here, the subscript "dec" stands for the decoding node. From the previous chapter we know that the downward messages on the B_k-edges are given by $\mu_{B_k \to \mathrm{dec}}(b_k) = p_{B_k}(b_k)$, the a-priori probabilities.

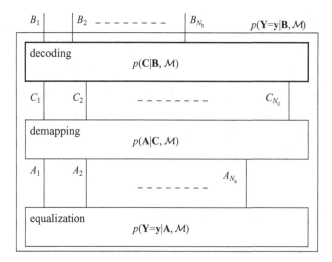

Figure 8.1. A factor graph of $p(\mathbf{Y} = \mathbf{y}|\mathbf{B}, \mathcal{M})$. The node is opened to reveal its structure. The node in bold is the topic of this chapter.

Table 8.1. Addition and multiplication in the binary field

b_1	b_2	$b_1 + b_2$	$b_1 \times b_2$
0	0	0	0
0	1	1	0
1	0	1	0
1	1	0	1

8.3   Block codes

Before we delve into specific types of error-correcting codes, let us first review some basic terminology, show how codes can be related to factor graphs, and introduce the concept of puncturing.

8.3.1   Basic concepts

We will limit ourselves to binary codes, where the components of **b** and **c** belong to $\mathbb{B}$. The binary field $\mathbb{B}$ contains two elements, commonly denoted 0 and 1. The field is endowed with two operators: addition ($+$) and multiplication ($\times$), defined in Table 8.1. Multiplication is distributive over addition.

A code is defined as a set of codewords $\mathcal{C} \subset \mathbb{B}^{N_c}$, and the field of coding theory is mainly concerned with searching and analyzing these sets. For practical purposes, we also need a function that maps an information word **b** onto a codeword $\mathbf{c} \in \mathcal{C}$: $\mathbf{c} = f_c(\mathbf{b})$. This is a process known as encoding. The encoding process is reversible, so, for every codeword $\mathbf{c} \in \mathcal{C}$, we can write $\mathbf{b} = f_c^{-1}(\mathbf{c})$ for some $\mathbf{b} \in \mathbb{B}^{N_b}$.

DEFINITION 8.1 (Block code).   *An (N_c, N_b) block code is a set $\mathcal{C} \subset \mathbb{B}^{N_c}$ of 2^{N_b} distinct elements, with $N_c \geq N_b$. The elements in $\mathcal{C}$ are known as the codewords. The ratio N_b/N_c is called the rate of the code.*

Many state-of-the-art error-correcting codes make use of interleavers, so let us define the concept of interleaving.

DEFINITION 8.2 (Interleaver).   *An interleaving function of size N is any bijective function from $\pi : \{1, \ldots, N\} \to \{1, \ldots, N\}$. Given an array **x** of length N, $\mathbf{x} = [x_1, \ldots, x_N]^\mathsf{T} \in \mathcal{X}^N$, an interleaver is a function $f_\pi : \mathcal{X}^N \to \mathcal{X}^N$, where $\tilde{\mathbf{x}} = f_\pi(\mathbf{x})$ with*

$$\tilde{x}_k = x_{\pi(k)} \tag{8.1}$$

for $k \in \{1, \ldots, N\}$. We abuse the notation and write $\tilde{\mathbf{x}} = \pi(\mathbf{x})$ instead of $\tilde{\mathbf{x}} = f_\pi(\mathbf{x})$.

Interleaving functions are usually chosen to exhibit particular properties related to randomness and spreading [86]. For the purpose of this book, think of an interleaver as a function that spreads bits around in some random fashion.

8.3.2 Two types of codes

We will consider two types of block codes: linear block codes and trellis block codes. The latter are based on state-space models.

8.3.2.1 Linear block codes

DEFINITION 8.3 (Linear block code). *An* (N_c, N_b) *linear block code is a set* $\mathcal{C} \subset \mathbb{B}^{N_c}$ *of* 2^{N_b} *distinct elements that form a linear subspace of* $\mathbb{B}^{N_c}$ *of dimension* N_b.

Consider a linear block code $\mathcal{C}$. Since the codewords form a linear subspace, there exists a set of N_b basis vectors in $\mathbb{B}^{N_c}$. Let us write these column-vectors as $\mathbf{g}_1, \mathbf{g}_2, \ldots, \mathbf{g}_{N_b}$. Note that this basis is not unique. Any linear combination of the N_b basis vectors yields an element in $\mathcal{C}$, so for every N_b binary elements $b_1, b_2, \ldots, b_{N_b}$

$$\mathbf{c} = \sum_{k=1}^{N_b} b_k \mathbf{g}_k \in \mathcal{C}. \tag{8.2}$$

If we write $\mathbf{b} = \begin{bmatrix} b_1 b_2 \ldots b_{N_b} \end{bmatrix}^{\mathsf{T}}$, then (8.2) can be re-written as

$$\mathbf{c} = \mathbf{G}\mathbf{b}, \tag{8.3}$$

where $\mathbf{G}$ is an $N_c \times N_b$ binary matrix with its kth column equal to $\mathbf{g}_k$. We call $\mathbf{G}$ the *generator matrix*. We see that $\mathbf{G}$ maps the information sequence $\mathbf{b}$ onto the codeword $\mathbf{c}$. Hence, it is a way to encode information. Usually the generator matrix is reduced by suitable operations to a systematic form

$$\mathbf{G}_s = \begin{bmatrix} \mathbf{I}_{N_b} \\ \mathbf{P} \end{bmatrix}, \tag{8.4}$$

where $\mathbf{P}$ is an $(N_c - N_b) \times N_b$ *parity matrix*. Note that $\mathbf{G}_s$ and $\mathbf{G}$ generate exactly the same code $\mathcal{C}$; they merely correspond to different sets of basis vectors.

For every linear block code, we can introduce a *parity-check matrix* $\mathbf{H}$, an $N_c \times (N_c - N_b)$ binary matrix whose columns generate the null space of $\mathbf{G}$. One can then show that $\mathbf{c} \in \mathcal{C} \iff \mathbf{H}^{\mathsf{T}}\mathbf{c} = \mathbf{0}$. For a systematic code with generator matrix $\mathbf{G}_s$, a corresponding parity-check matrix $\mathbf{H}_s$ can be constructed as

$$\mathbf{H}_s^{\mathsf{T}} = \begin{bmatrix} \mathbf{P} & \mathbf{I}_{N_c - N_b} \end{bmatrix}. \tag{8.5}$$

Note that $\mathbf{H}_s^{\mathsf{T}}\mathbf{G}_s = \mathbf{0}$, and also $\mathbf{H}^{\mathsf{T}}\mathbf{G} = \mathbf{H}_s^{\mathsf{T}}\mathbf{G} = \mathbf{H}^{\mathsf{T}}\mathbf{G}_s = \mathbf{0}$.

8.3.2.2 Trellis block codes

Trellis block codes are based on state space systems with a state space $\mathcal{S}$. Although some trellis block codes can be seen as linear block codes, this connection is not important for our purposes. Trellis block codes start with a description of the encoder (as opposed to a description of the code space).

At time instant $k-1$, the encoder is in a certain state s_{k-1}. Given an input of $N_{in} \geq 1$ information bits, the encoder transitions into a state s_k and generates an output. Let us break up $\mathbf{b}$ into segments $\mathbf{b}_1, \mathbf{b}_2, \ldots, \mathbf{b}_{N_b/N_{in}}$, each of length N_{in} (assuming that $N_b/N_{in} \in \mathbb{N}$). The state at time 0 is given by $s_{start} \in \mathcal{S}$, which is known both to the transmitter (encoder) and to the receiver (decoder). At every time instant k, the state s_k is related to s_{k-1} by

$$s_k = f_s(s_{k-1}, \mathbf{b}_k). \tag{8.6}$$

The output at time k is a string of $N_{out} > N_{in}$ bits $\mathbf{c}_k$ given by

$$\mathbf{c}_k = f_o(s_{k-1}, \mathbf{b}_k). \tag{8.7}$$

The functions $f_s(\cdot)$ and $f_o(\cdot)$ are deterministic functions from $\mathcal{S} \times \mathbb{B}^{N_{in}}$ to $\mathcal{S}$ and to $\mathbb{B}^{N_{out}}$, respectively. A systematic code an be obtained by selecting $f_o(\cdot)$ such that the first N_{in} bits of $\mathbf{c}_k$ are equal to $\mathbf{b}_k$. The codeword $\mathbf{c}$ is given by the concatenation of $\mathbf{c}_1, \mathbf{c}_2, \ldots, \mathbf{c}_{N_b/N_{in}}$, so $N_c = N_{out} N_b/N_{in}$. Hence, the rate of this code is $N_b/N_c = N_{in}/N_{out}$. Since the receiver has no knowledge of the final state $s_{N_b/N_{in}}$, such a code is said to be *unterminated*.

Termination

A trellis block code can be *terminated* by appending a number of termination segments to $\mathbf{b}$ such that the encoder ends up in a predetermined state, say $s_{end} \in \mathcal{S}$. We assume that any state can be reached from any other state in at most L time instants. This is generally always true by virtue of the design of the function $f_s(\cdot)$. After processing the N_b/N_{in} segments from $\mathbf{b}$, the encoder is in a certain state $s_{N_b/N_{in}}$, which is known to the transmitter. The transmitter then selects L termination segments $\mathbf{t}_{N_b/N_{in}+1}, \ldots, \mathbf{t}_{N_b/N_{in}+L}$, each of length N_{in}. Then, for $k = N_b/N_{in} + 1, \ldots, N_b/N_{in} + L$:

$$s_k = f_s(s_{k-1}, \mathbf{t}_k), \tag{8.8}$$

$$\mathbf{c}_k = f_o(s_{k-1}, \mathbf{t}_k). \tag{8.9}$$

By proper selection of $\mathbf{t}_{N_b/N_{in}+1}, \ldots, \mathbf{t}_{N_b/N_{in}+L}$, we can ensure that $s_{N_b/N_{in}+L} = s_{end}$. The rate of this terminated code is

$$\frac{N_b}{N_c} = \frac{N_{in}}{N_{out}} \frac{1}{1 + LN_{in}/N_b}. \tag{8.10}$$

In other words, termination results in a rate loss.

8.3.3   Codes and factor graphs

8.3.3.1   Introduction

Our goal is to re-write the function $p(\mathbf{C}|\mathbf{B}, \mathcal{M})$ to reveal the structure of the decoding node in the factor graph in Fig. 8.1. There are two common ways to achieve this: a *generative* approach or a *descriptive* approach.

- In a generative approach, we factorize $f_c(\mathbf{b})$, possibly introducing new variables. Since

$$p(\mathbf{C} = \mathbf{c}|\mathbf{B} = \mathbf{b}, \mathcal{M}) = \boxminus(\mathbf{c}, f_c(\mathbf{b})) \tag{8.11}$$

we can use the factorization of $f_c(\mathbf{b})$ to open the node $p(\mathbf{C}|\mathbf{B}, \mathcal{M})$.
- In a descriptive approach, we assume that we have available a description of the code-space $\mathcal{C}$. In other words, we require a factorization of $\mathbb{I}\{\mathbf{c} \in \mathcal{C}\}$. When $f_c^{-1}(\mathbf{c})$ is very simple (for instance $f_c^{-1}(\mathbf{c}) = \mathbf{c}_{1:N_b}$ for a systematic code), then we can write

$$p(\mathbf{C} = \mathbf{c}|\mathbf{B} = \mathbf{b}, \mathcal{M}) = \mathbb{I}\left\{\mathbf{c} \in \mathcal{C} \text{ and } \mathbf{b} = f_c^{-1}(\mathbf{c})\right\} \tag{8.12}$$

$$= \mathbb{I}\{\mathbf{c} \in \mathcal{C}\} \boxminus\left(\mathbf{b}, f_c^{-1}(\mathbf{c})\right). \tag{8.13}$$

As we will see, RA codes, convolutional codes and turbo codes are based on the generative approach, whereas LDPC codes are based on the descriptive approach. Before we tackle these advanced codes, let us first see how the concepts from linear block codes and trellis block codes can be used in factorizing the function $p(\mathbf{C}|\mathbf{B}, \mathcal{M})$.

8.3.3.2   Linear block codes

For linear block codes we know that $f_c(\mathbf{b}) = \mathbf{Gb}$, so

$$p(\mathbf{C} = \mathbf{c}|\mathbf{B} = \mathbf{b}, \mathcal{M}) = \boxminus(\mathbf{c}, \mathbf{Gb}). \tag{8.14}$$

On the other hand, since $\mathbf{c} \in \mathcal{C} \iff \mathbf{H}^{\mathrm{T}}\mathbf{c} = \mathbf{0}$, we also have

$$p(\mathbf{C} = \mathbf{c}|\mathbf{B} = \mathbf{b}, \mathcal{M}) = \boxminus(\mathbf{H}^{\mathrm{T}}\mathbf{c}, \mathbf{0})\boxminus(\mathbf{b}, f_c^{-1}(\mathbf{c})). \tag{8.15}$$

Note that when the code is systematic $\boxminus(\mathbf{b}, f_c^{-1}(\mathbf{c})) = \prod_{l=1}^{N_b} \boxminus(b_l, c_l)$.

8.3.3.3   Trellis block codes

For trellis block codes, the node $p(\mathbf{C}|\mathbf{B}, \mathcal{M})$ can be replaced by a factorization of $p(\mathbf{C}, \mathbf{S}|\mathbf{B}, \mathcal{M})$ (for unterminated codes) or of $p(\mathbf{C}, \mathbf{S}, \mathbf{T}|\mathbf{B}, \mathcal{M})$ (for terminated codes), with

$$p(\mathbf{C} = \mathbf{c}, \mathbf{S} = \mathbf{s}|\mathbf{B} = \mathbf{b}, \mathcal{M}) = \boxminus(s_0, s_{\text{start}}) \prod_{k=1}^{N_b/N_{\text{in}}} \boxminus(s_k, f_s(s_{k-1}, \mathbf{b}_k))$$

$$\times \boxminus(\mathbf{c}_k, f_o(s_{k-1}, \mathbf{b}_k)) \tag{8.16}$$

and

$$p(\mathbf{C} = \mathbf{c}, \mathbf{S} = \mathbf{s}, \mathbf{T} = \mathbf{t} | \mathbf{B} = \mathbf{b}, \mathcal{M})$$

$$= \boxed{=}(s_0, s_{\text{start}}) \prod_{k=1}^{N_b/N_{\text{in}}} \boxed{=}(s_k, f_s(s_{k-1}, \mathbf{b}_k)) \boxed{=}(\mathbf{c}_k, f_o(s_{k-1}, \mathbf{b}_k)) \tag{8.17}$$

$$\times \underbrace{\prod_{l=N_b/N_{\text{in}}+1}^{N_b/N_{\text{in}}+L} \boxed{=}(s_l, f_s(s_{l-1}, \mathbf{t}_l)) \boxed{=}(\mathbf{c}_l, f_o(s_{l-1}, \mathbf{t}_l)) \boxed{=}(s_{N_b/N_{\text{in}}+L}, s_{\text{end}})}_{\text{termination part}}.$$

8.3.4 Puncturing

In some cases the intrinsic rate of the code may be very small. For practical reasons, the transmitter may choose to *puncture* a number of the coded bits so as to reduce N_c. Puncturing a coded bit means that we do not transmit that particular bit. For instance, by puncturing every third coded bit in a rate $1/3$ code, we obtain a rate $1/2$ code. At the receiver side, a punctured coded bit C_k translates into a half-edge in the factor graph, with $\mu_{C_k \to \text{dec}}(C_k)$ set to a uniform distribution. An example is shown in Fig. 8.2.

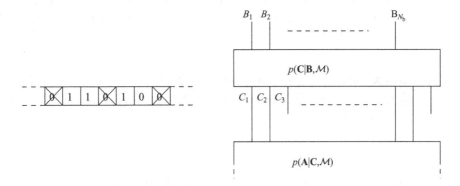

Figure 8.2. An example of puncturing: every third bit in the coded sequence on the left is punctured (not transmitted). At the receiver, the corresponding variables become half-edges in the factor graph.

8.4 Repeat–accumulate codes

8.4.1 Description

Repeat–accumulate (RA) codes are a very simple but powerful type of error-correcting code. An RA codeword is created as follows: every bit in $\mathbf{b}$ is copied N_c/N_b times,[1]

---
[1] We choose $N_c/N_b \in \mathbb{N}$.

resulting in a sequence $\mathbf{d}$ of length N_c. This sequence is passed to an interleaver, which essentially shuffles the bits around so that they look more random. We will denote the interleaver function by $\pi(\cdot)$ and set $\mathbf{e} = \pi(\mathbf{d})$. Finally, the coded sequence $\mathbf{c}$ is passed through an accumulator:

$$c_1 = e_1, \tag{8.18}$$

$$c_2 = c_1 + e_2,$$

$$c_3 = c_2 + e_3,$$

$$\dots$$

$$c_{N_c} = c_{N_c-1} + e_{N_c}.$$

It is easily verified that we can express the relationship between $\mathbf{b}$ and $\mathbf{c}$ in a generator-matrix representation as

$$\mathbf{c} = \mathbf{G}\mathbf{b} \tag{8.19}$$

$$= \mathbf{G}_{\mathrm{acc}}\mathbf{G}_\pi \mathbf{G}_{\mathrm{rep}}\mathbf{b}, \tag{8.20}$$

where $\mathbf{G}_{\mathrm{rep}}$ is an $N_c \times N_b$ repeater matrix with N_c/N_b consecutive 1s per column and one 1 per row, the matrix $\mathbf{G}_\pi$ is an $N_c \times N_c$ interleaver matrix (more commonly referred to as a permutation matrix), with exactly one 1 per row and per column, and $\mathbf{G}_{\mathrm{acc}}$ is an $N_c \times N_c$ accumulator matrix. An accumulator matrix is defined as follows: introduce $\mathbf{1}_k$ as a row-vector of k 1s, and $\mathbf{0}_k$ as a row-vector of k 0s, then the kth row ($1 \leq k \leq N_c$) is given by $\left[\mathbf{1}_k \mathbf{0}_{N_c-k}\right]$.

Example 8.4. *Let us look at an example, for $N_b = 3$ and $N_c/N_b = 2$. When $\mathbf{b} = [b_1 b_2 b_3]^{\mathrm{T}}$, then $\mathbf{d} = [b_1 b_1 b_2 b_2 b_3 b_3]^{\mathrm{T}}$ can be formed by $\mathbf{d} = \mathbf{G}_{\mathrm{rep}}\mathbf{b}$, where the repeater matrix is given by*

$$\mathbf{G}_{\mathrm{rep}} = \begin{bmatrix} 1 & 0 & 0 \\ 1 & 0 & 0 \\ 0 & 1 & 0 \\ 0 & 1 & 0 \\ 0 & 0 & 1 \\ 0 & 0 & 1 \end{bmatrix}.$$

Suppose that our interleaver operates as follows: $\mathbf{e} = \pi(\mathbf{d})$ with $e_1 = d_1$, $e_2 = d_3$, $e_3 = d_6$, $e_4 = d_4$, $e_5 = d_2$, and $e_6 = d_5$. We can write $\mathbf{e} = \mathbf{G}_\pi \mathbf{d}$, where

$$\mathbf{G}_\pi = \begin{bmatrix} 1 & 0 & 0 & 0 & 0 & 0 \\ 0 & 0 & 1 & 0 & 0 & 0 \\ 0 & 0 & 0 & 0 & 0 & 1 \\ 0 & 0 & 0 & 1 & 0 & 0 \\ 0 & 1 & 0 & 0 & 0 & 0 \\ 0 & 0 & 0 & 0 & 1 & 0 \end{bmatrix}.$$

Finally, the matrix $\mathbf{G}_{\mathrm{acc}}$ *for* $N_c = 6$ *is given by*

$$
\mathbf{G}_{\mathrm{acc}} = \begin{bmatrix}
1 & 0 & 0 & 0 & 0 & 0 \\
1 & 1 & 0 & 0 & 0 & 0 \\
1 & 1 & 1 & 0 & 0 & 0 \\
1 & 1 & 1 & 1 & 0 & 0 \\
1 & 1 & 1 & 1 & 1 & 0 \\
1 & 1 & 1 & 1 & 1 & 1
\end{bmatrix}.
$$

8.4.2 Factor graphs

With the generative approach described in Section 8.3.3, we know that we can write

$$
p(\mathbf{C} = \mathbf{c}|\mathbf{B} = \mathbf{b}, \mathcal{M}) = \boxminus(\mathbf{c}, \mathbf{Gb}). \tag{8.21}
$$

We can now open the node $p(\mathbf{C}|\mathbf{B}, \mathcal{M})$ by replacing it by a factorization of $p(\mathbf{C}, \mathbf{D}, \mathbf{E}|\mathbf{B}, \mathcal{M})$:

$$
p(\mathbf{C} = \mathbf{c}, \mathbf{D} = \mathbf{d}, \mathbf{E} = \mathbf{e}|\mathbf{B} = \mathbf{b}, \mathcal{M}) = \boxminus(\mathbf{c}, \mathbf{G}_{\mathrm{acc}}\mathbf{e}) \boxminus (\mathbf{e}, \mathbf{G}_\pi \mathbf{d}) \boxminus (\mathbf{d}, \mathbf{G}_{\mathrm{rep}}\mathbf{b}). \tag{8.22}
$$

Let us introduce an additional function $\boxplus : \mathbb{B}^D \to \mathbb{B}$ for any $D \in \mathbb{N}$ as

$$
\boxplus(x_1, x_2, \ldots, x_D) = \begin{cases} 1 & \sum_{k=1}^{D} x_k = 0, \\ 0 & \text{else,} \end{cases} \tag{8.23}
$$

where the summation takes place in the binary domain (see Table 8.1). The $\boxplus$-operator allows us to express the three factors in (8.22) as follows:

$$
\boxminus(\mathbf{c}, \mathbf{G}_{\mathrm{acc}}\mathbf{e}) = \boxminus(c_1, e_1) \prod_{k=2}^{N_c} \boxplus(c_k, c_{k-1}, e_k), \tag{8.24}
$$

$$
\boxminus(\mathbf{e}, \mathbf{G}_\pi \mathbf{d}) = \prod_{k=1}^{N_c} \boxminus(e_k, d_{\pi(k)}), \tag{8.25}
$$

and finally

$$
\boxminus(\mathbf{d}, \mathbf{G}_{\mathrm{rep}}\mathbf{b}) = \prod_{l=1}^{N_b} \prod_{n=0}^{N_c/N_b - 1} \boxminus(b_l, d_{lN_c/N_b - n}). \tag{8.26}
$$

The factor graph for our example RA code is shown in Fig. 8.3.

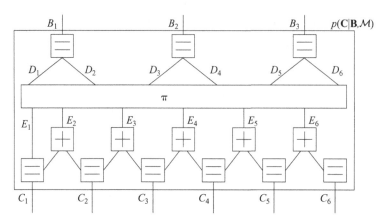

Figure 8.3. Opening the decoding node for an RA code. The node marked π is a shorthand for $\prod_{k=1}^{N_c} \boxminus(E_k, D_{\pi(k)})$.

8.4.3 Building blocks

From the factor graph in Fig. 8.3, we discern two types of nodes: equality nodes and check nodes (marked $\boxplus$). We will now show how messages are computed in these nodes. Note that each check node is connected to exactly three edges.

8.4.3.1 Equality nodes
Probability domain
Given an equality node $\boxminus(X_1, X_2, \ldots, X_D)$ with incoming messages $\mathbf{p}_{X_k \to \boxminus}$, we compute an outgoing message $\mathbf{p}_{\boxminus \to X_l}$ as

$$\mathbf{P}_{\boxminus \to X_l}(x_l) \propto \sum_{\sim\{x_l\}} \boxminus(x_1, x_2, \ldots, x_D) \prod_{k \neq l} \mathbf{p}_{X_k \to \boxminus}(x_k) \tag{8.27}$$

$$= \prod_{k \neq l} \mathbf{p}_{X_k \to \boxminus}(x_l). \tag{8.28}$$

We remind the reader that the probability vectors need to be normalized so that $\sum_{x_l \in \mathbb{B}} \mathbf{P}_{\boxminus \to X_l}(x_l) = 1$.

The log domain
Usually, we prefer log-domain messages, with $\mathbf{L}_{X_k \to \boxminus}(x_k) = \log \mathbf{p}_{X_k \to \boxminus}(x_k)$. In that case, we know from Section 5.3.3 that

$$\mathbf{L}_{\boxminus \to X_l}(x_l) = \sum_{k \neq l} \mathbf{L}_{X_k \to \boxminus}(x_l). \tag{8.29}$$

The LLR domain
Since we can add any real number to a log-domain message, let us always make sure that the first entry in any log-domain vector (say, $\mathbf{L}$) is zero. This can always be achieved

by subtracting $\mathbf{L}(0)$ from the message $\mathbf{L}$. Then messages can be represented by a single number $\lambda_{X_k \to \boxminus} = \mathbf{L}_{X_k \to \boxminus}(1) - \mathbf{L}_{X_k \to \boxminus}(0) = \mathbf{L}_{X_k \to \boxminus}(1)$. In that case (8.29) becomes

$$\lambda_{\boxminus \to X_l} = \sum_{k \neq l} \lambda_{X_k \to \boxminus}. \tag{8.30}$$

8.4.3.2 Check nodes

The probability domain

Given a check node $\boxplus(X_1, X_2, X_3)$ with incoming messages $\mathbf{p}_{X_k \to \boxplus}$, we compute an outgoing message $\mathbf{p}_{\boxplus \to X_l}$. Since the function $\boxplus$ is symmetric in each of its arguments, let us focus on $\mathbf{p}_{\boxplus \to X_1}$:

$$\mathbf{p}_{\boxplus \to X_1}(x_1) \propto \sum_{x_2, x_3} \boxplus(x_1, x_2, x_3) \prod_{k \neq 1} \mathbf{p}_{X_k \to \boxplus}(x_k) \tag{8.31}$$

$$= \sum_{x_3 \in \mathbb{B}} \mathbf{p}_{X_3 \to \boxplus}(x_3) \, \mathbf{p}_{X_2 \to \boxplus}(x_1 + x_3) \tag{8.32}$$

since, for a fixed x_1 and x_3, $\boxplus(x_1, x_2, x_3) \neq 0$ only when $x_2 = x_1 + x_3$.

The log domain

Transforming (8.32) into the log domain yields

$$\mathbf{L}_{\boxplus \to X_1}(x_1) = \mathbb{M}_{\sim\{x_1\}}\left(\log \boxminus(x_1 + x_2 + x_3, 0) + \sum_{k \neq 1} \mathbf{L}_{X_k \to \boxplus}(x_k) \right) \tag{8.33}$$

$$= \mathbb{M}_{x_3}\left(\mathbf{L}_{X_3 \to \boxplus}(x_3) + \mathbf{L}_{X_2 \to \boxplus}(x_1 + x_3) \right) \tag{8.34}$$

$$= \mathbb{M}\left(\mathbf{L}_{X_3 \to \boxplus}(0) + \mathbf{L}_{X_2 \to \boxplus}(x_1), \mathbf{L}_{X_3 \to \boxplus}(1) + \mathbf{L}_{X_2 \to \boxplus}(x_1 + 1) \right). \tag{8.35}$$

The LLR domain

As for the equality nodes, we can modify the log-domain messages such that the first component $\mathbf{L}(0)$ in any message $\mathbf{L}$ is always zero. Let us introduce $\lambda_{X_k \to \boxplus} = \mathbf{L}_{X_k \to \boxplus}(1) - \mathbf{L}_{X_k \to \boxplus}(0) = \mathbf{L}_{X_k \to \boxplus}(1)$. Let us compute $\mathbf{L}_{\boxplus \to X_1}(1)$ and $\mathbf{L}_{\boxplus \to X_1}(0)$ using (8.35):

$$\mathbf{L}_{\boxplus \to X_1}(1) = \mathbb{M}\left(\mathbf{L}_{X_3 \to \boxplus}(0) + \mathbf{L}_{X_2 \to \boxplus}(1), \mathbf{L}_{X_3 \to \boxplus}(1) + \mathbf{L}_{X_2 \to \boxplus}(0) \right) \tag{8.36}$$

$$= \mathbb{M}\left(\lambda_{X_2 \to \boxplus}, \lambda_{X_3 \to \boxplus} \right) \tag{8.37}$$

and

$$\mathbf{L}_{\boxplus \to X_1}(0) = \mathbb{M}\left(\mathbf{L}_{X_3 \to \boxplus}(0) + \mathbf{L}_{X_2 \to \boxplus}(0), \mathbf{L}_{X_3 \to \boxplus}(1) + \mathbf{L}_{X_2 \to \boxplus}(1) \right) \tag{8.38}$$

$$= \mathbb{M}\left(0, \lambda_{X_3 \to \boxplus} + \lambda_{X_2 \to \boxplus} \right) \tag{8.39}$$

The final outgoing message becomes

$$\lambda_{\boxplus \to X_1} = \mathbf{L}_{\boxplus \to X_1}(1) - \mathbf{L}_{\boxplus \to X_1}(0) \qquad (8.40)$$

$$= \mathrm{M}\left(\lambda_{X_2 \to \boxplus}, \lambda_{X_3 \to \boxplus}\right) - \mathrm{M}\left(0, \lambda_{X_3 \to \boxplus} + \lambda_{X_2 \to \boxplus}\right). \qquad (8.41)$$

It will turn out to be convenient to introduce an additional function $f_\boxplus : \mathbb{R} \times \mathbb{R} \to \mathbb{R}$ with

$$f_\boxplus (x, y) = \mathrm{M}(x, y) - \mathrm{M}(0, x + y) \qquad (8.42)$$

so that $\lambda_{\boxplus \to X_1} = f_\boxplus \left(\lambda_{X_2 \to \boxplus}, \lambda_{X_3 \to \boxplus}\right)$.

8.4.4   Decoding repeat–accumulate codes

Since the factor graph of the RA code contains cycles, we must decide on a particular scheduling strategy. We will work in the LLR domain and consider the most common scheduling.

1. **Initialization** The downward messages over the edges E_k, $k = 1, \ldots, N_c$ are initialized with uniform distributions (or, equivalently, uniform log-domain messages, or zero LLRs). This is depicted in Fig. 8.4.
2. **Forward–Backward step** A forward–backward-type algorithm is executed on the lower part of the graph. The forward and backward phases are shown in Fig 8.5s. and 8.6, respectively, and described in Algorithm 8.1 (see Fig. 8.7 for a description of the variables).
3. **Upward messages** Upward messages on the E_k-edges are computed (see Fig. 8.8). These messages are then de-interleaved to obtain messages over the D_k-edges. This step is described in Algorithm 8.2.
4. **Downward messages** Downward messages on the D_k-edges are computed, as shown in Fig. 8.9. These messages are then de-interleaved to obtain messages over the E_k-edges. This step is described in Algorithm 8.3. Go back to step 2 (forward–backward step).
5. **Termination** After a number of iterations, the decoding algorithm is halted, and outgoing messages $\lambda_{\boxminus \to B_k}$ and $\lambda_{\boxminus \to C_k}$ are computed.

The entire algorithm is shown in Algorithm 8.4.

8.5   Low-density parity-check codes

8.5.1   Description

Low-density parity-check (LDPC) codes are a class of codes whose parity-check matrices are sparse in the sense that they contain very few ones per column. They are generally created starting from a sparse parity-check matrix $\mathbf{H}$, from which we then derive a

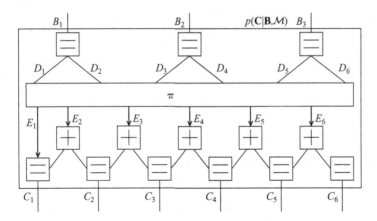

Figure 8.4. Repeat–accumulate codes: initialization. Bold arrows indicate that uniform pmfs are passed over the corresponding edges.

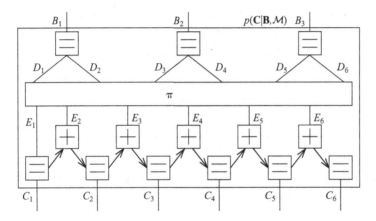

Figure 8.5. Repeat–accumulate codes: forward phase. Bold arrows indicate that messages are passed over the corresponding edges.

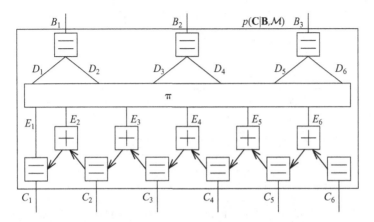

Figure 8.6. Repeat–accumulate codes: backward phase. Bold arrows indicate that messages are passed over the corresponding edges.

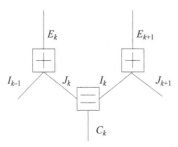

Figure 8.7. Repeat–accumulate codes: a detailed view.

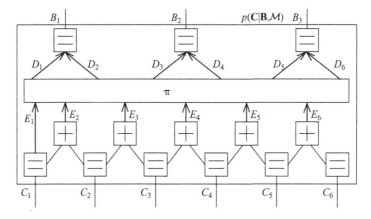

Figure 8.8. Repeat–accumulate codes: upward messages. Bold arrows indicate that messages are passed over the corresponding edges.

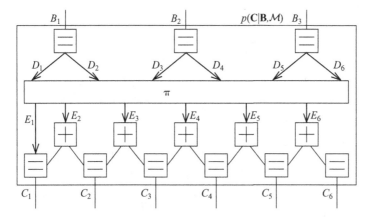

Figure 8.9. Repeat–accumulate codes: downward messages. Bold arrows indicate that messages are passed over the corresponding edges.

Algorithm 8.1 Decoding RA codes: forward–backward phase

1: take care of special cases:

$$\lambda_{\boxminus \to I_1} = \lambda_{E_1 \to \boxminus} + \lambda_{C_1 \to \boxminus}$$
$$\lambda_{J_2 \to \boxminus} = f_{\boxplus}\left(\lambda_{E_2 \to \boxplus}, \lambda_{I_1 \to \boxplus}\right)$$
$$\lambda_{\boxminus \to J_{N_c}} = \lambda_{C_{N_c} \to \boxminus}$$
$$\lambda_{\boxplus \to I_{N_c-1}} = f_{\boxplus}\left(\lambda_{E_{N_c} \to \boxplus}, \lambda_{J_{N_c} \to \boxplus}\right)$$

2: **for** $k = 2$ to $N_c - 1$ **do**

3:     compute forward messages:

$$\lambda_{\boxminus \to I_k} = \lambda_{J_k \to \boxminus} + \lambda_{C_k \to \boxminus}$$
$$\lambda_{J_{k+1} \to \boxminus} = f_{\boxplus}\left(\lambda_{E_{k+1} \to \boxplus}, \lambda_{I_k \to \boxplus}\right)$$

4:     compute backward message:

set $l = N_c - k + 1$

$$\lambda_{\boxminus \to J_l} = \lambda_{C_l \to \boxminus} + \lambda_{I_l \to \boxminus}$$
$$\lambda_{\boxplus \to I_{l-1}} = f_{\boxplus}\left(\lambda_{E_l \to \boxplus}, \lambda_{J_l \to \boxplus}\right)$$

5: **end for**

Algorithm 8.2 Decoding RA codes: upward messages

1: $\lambda_{E_1 \to \pi} = \lambda_{C_1 \to \boxminus} + \lambda_{I_1 \to \boxminus}$

2: **for** $k = 2$ to N_c **do**

3:     compute upward message:

$$\lambda_{E_k \to \pi} = f_{\boxplus}\left(\lambda_{I_{k-1} \to \boxplus}, \lambda_{J_k \to \boxplus}\right)$$

4: **end for**

5: **for** $k = 1$ to N_c **do**

6:     deinterleave message:

$$\lambda_{D_{\pi^{-1}(k)} \to \boxminus} = \lambda_{E_k \to \pi}$$

7: **end for**

Algorithm 8.3 Decoding RA codes: downward messages

1: **for** $k = 1$ to N_b **do**

2:     **for** $n = 0$ to $N_c/N_b - 1$ **do**

3:         compute downward message, with $K = N_c/N_b$

$$\lambda_{\boxminus \to D_{kK-n}} = \lambda_{B_k \to \boxminus} + \sum_{m=0, m \neq n}^{K-1} \lambda_{D_{kK-m}}$$

4:         interleave message

$$\lambda_{\pi \to E_{\pi(kK-n)}} = \lambda_{\boxminus \to D_{kK-n}}$$

5:     **end for**

6: **end for**

Algorithm 8.4 Decoding RA codes: complete decoding algorithm

1: *input*: $\lambda_{B_k \to \boxminus}$, $k = 1, \dots, N_b$
2: *input*: $\lambda_{C_k \to \boxminus}$, $k = 1, \dots, N_c$
3: initialization:

$\qquad \lambda_{E_1 \to \boxminus} = 0$
$\qquad \lambda_{E_k \to \boxplus} = 0, k > 1.$

4: **for** iter $= 1$ to N_{iter} **do**
5: $\qquad$ compute forward and backward messages using Algorithm 8.1
6: $\qquad$ compute upward messages using Algorithm 8.2
7: $\qquad$ compute downward messages using Algorithm 8.3
8: **end for**
9: **for** $k = 1$ to N_b **do**
10: $\qquad \lambda_{\boxminus \to B_k} = \sum_{n=0}^{K-1} \lambda_{D_{kK-n} \to \boxminus}$ with $K = N_c/N_b$
11: **end for**
12: $\lambda_{\boxminus \to C_1} = \lambda_{E_1 \to \boxminus} + \lambda_{I_1 \to \boxminus}$
13: **for** $k = 2$ to N_c **do**
14: $\qquad \lambda_{\boxminus \to C_k} = \lambda_{I_k \to \boxminus} + \lambda_{J_k \to \boxminus}$
15: **end for**

---

systematic generator matrix $\mathbf{G}$ [46, 87]. In this section we will cover how LDPC codes are created, what their factor graphs look like, and how they are decoded using the SPA. As with RA codes, we will work with a running example. We start from a parity-check matrix $\mathbf{H}$ given by

$$\mathbf{H}^{\mathrm{T}} = \begin{bmatrix} 1 & 0 & 0 & 1 & 0 & 1 & 0 & 1 \\ 0 & 1 & 1 & 1 & 0 & 0 & 1 & 0 \\ 1 & 1 & 1 & 0 & 1 & 0 & 0 & 1 \\ 0 & 0 & 1 & 0 & 1 & 1 & 1 & 1 \end{bmatrix}. \tag{8.43}$$

Since $\mathbf{H}$ is an 8×4 matrix, it is the parity-check matrix of a ($N_c = 8, N_b = 4$) code. Let us find a systematic generator matrix for this code. We first transform (through row operations) $\mathbf{H}^{\mathrm{T}}$ into the form $[\mathbf{P} \quad \mathbf{I}_{N_c-N_b}]$ as

$$\mathbf{H}_s^{\mathrm{T}} = \begin{bmatrix} 1 & 1 & 0 & 0 & 1 & 0 & 0 & 0 \\ 1 & 0 & 1 & 1 & 0 & 1 & 0 & 0 \\ 0 & 1 & 1 & 1 & 0 & 0 & 1 & 0 \\ 0 & 0 & 1 & 0 & 0 & 0 & 0 & 1 \end{bmatrix} \tag{8.44}$$

$$= [\mathbf{P} \quad \mathbf{I}_4]. \tag{8.45}$$

Note that $\mathbf{H}_s$ is a parity-check matrix for the same code C as $\mathbf{H}$. We find that the systematic generator matrix is given by

$$\mathbf{G}_s = \begin{bmatrix} \mathbf{I}_4 \\ \mathbf{P} \end{bmatrix} \tag{8.46}$$

$$= \begin{bmatrix} 1 & 0 & 0 & 0 \\ 0 & 1 & 0 & 0 \\ 0 & 0 & 1 & 0 \\ 0 & 0 & 0 & 1 \\ 1 & 1 & 0 & 0 \\ 1 & 0 & 1 & 1 \\ 0 & 1 & 1 & 1 \\ 0 & 0 & 1 & 0 \end{bmatrix}. \tag{8.47}$$

Clearly, $\mathbf{H}_s^{\mathsf{T}}\mathbf{G}_s = \mathbf{H}^{\mathsf{T}}\mathbf{G}_s = \mathbf{0}$.

8.5.2   Factor graphs

We can now apply (8.15) to our systematic code, leading to

$$p(\mathbf{C} = \mathbf{c}|\mathbf{B} = \mathbf{b}, \mathcal{M}) = \prod_{k=1}^{N_c - N_b} \boxminus(\mathbf{h}_k^{\mathsf{T}}\mathbf{c}, 0) \prod_{l=1}^{N_b} \boxminus(b_l, c_l), \tag{8.48}$$

where $\mathbf{h}_k^{\mathsf{T}}$ is the kth row of $\mathbf{H}^{\mathsf{T}}$. For our example, the factor graph is shown in Fig. 8.10. As for RA codes, the check nodes (marked $\boxplus$) are defined as follows, for $x_k \in \mathbb{B}$:

$$\boxplus(x_1, x_2, \ldots, x_D) = \begin{cases} 1 & \sum_{k=1}^{D} x_k = 0 \\ 0 & \text{else.} \end{cases} \tag{8.49}$$

For instance, consider the first check (the first row in $\mathbf{H}^{\mathsf{T}}$):

$$\boxminus\left(\mathbf{h}_1^{\mathsf{T}}\mathbf{c}, 0\right) = \boxminus(c_1 + c_4 + c_6 + c_8, 0) = \boxplus(c_1, c_4, c_6, c_8).$$

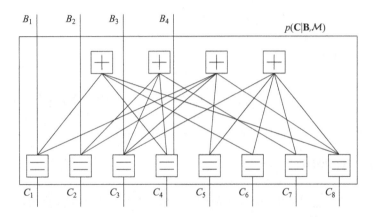

Figure 8.10. Opening the decoding node.

8.5.3   Building blocks

Looking at the factor graph from Fig. 8.10, we discern two types of nodes: equality nodes and check nodes (marked $\boxplus$). The equality nodes were discussed in Section 8.4.3 for RA codes. Note that check nodes can now be of any degree, as opposed to their being of degree three for RA codes. As we will see, the computational complexity is exponential in the degree D of the check node. This can be avoided by opening the check node, leading to a complexity linear in D.

8.5.3.1   Check nodes

Given an equality node $\boxplus (X_1, X_2, \ldots, X_D)$ with incoming messages $\mathbf{p}_{X_k \to \boxplus}$, we compute an outgoing message $\mathbf{p}_{\boxplus \to X_1}$ as

$$\mathbf{p}_{\boxplus \to X_1}(x_1) \propto \sum_{\sim\{x_1\}} \boxminus(x_1 + x_2 + \cdots + x_D, 0) \prod_{k \neq 1} \mathbf{p}_{X_k \to \boxplus}(x_k) \tag{8.50}$$

$$= \sum_{x_2, x_3, \ldots, x_D} \boxminus(x_1 + x_2 + \cdots + x_D, 0) \prod_{k \neq 1} \mathbf{p}_{X_k \to \boxplus}(x_k) \tag{8.51}$$

$$= \sum_{x_3, \ldots, x_D} \mathbf{p}_{X_2 \to \boxplus}(x_1 + x_3 + \cdots + x_D) \prod_{k \neq 1,2} \mathbf{p}_{X_k \to \boxplus}(x_k). \tag{8.52}$$

The summation goes over 2^{D-2} terms. Computing all messages $\mathbf{p}_{\boxplus \to X_k}$ for $k = 1, \ldots, D$ requires on the order of 2^D computations.

8.5.3.2   Opening the check node

The computational complexity can be reduced by opening the node $\boxplus (X_1, X_2, \ldots, X_D)$ and considering only $\boxplus$-nodes of three variables. We can introduce binary variables $U_2, \ldots, U_{D-2}$ such that

$$\overbrace{\underbrace{X_1 + \underbrace{X_2}_{U_2} + \overbrace{X_3}^{U_3} + X_4 + \cdots + X_{D-2}}_{U_4} + X_{D-1} + X_D}^{U_{D-2}} = 0$$

and a function

$$f(X_1, \ldots, X_D, U_2, \ldots, U_{D-2}) = \boxplus (X_1, X_2, U_2) \boxplus (U_{D-2}, X_{D-1}, X_D) \prod_{k=3}^{D-2} \boxplus (U_{k-1}, X_k, U_k).$$

$$\tag{8.53}$$

It is easily verified that

$$\sum_{u_2, \ldots, u_{D-1}} f(x_1, x_2, \ldots, x_D, u_2, \ldots, u_{D-2}) = \boxplus (x_1, x_2, \ldots, x_D). \tag{8.54}$$

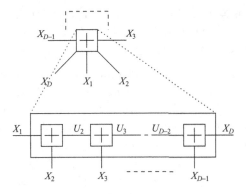

Figure 8.11. Opening a check node.

This implies (see Section 4.6.2) that we can replace $\boxplus(X_1, X_2, \ldots, X_D)$ by the factorization (8.53). The corresponding factor graph is shown in Fig. 8.11. We see that only check nodes of degree three remain. Such nodes were discussed in Section 8.4.3 for RA codes. Let us focus on LLR-domain messages. Re-using the function $f_\boxplus(\cdot)$ (see (8.42)) from RA codes, the new algorithm for computing the messages in a node $\boxplus(X_1, X_2, \ldots, X_D)$ is then given in Algorithm 8.5, where we again refer to Fig. 8.11, with $\lambda_{U_k}^{(\rightarrow)}$ representing a message from left to right on edge U_k in Fig. 8.11, and $\lambda_{U_k}^{(\leftarrow)}$ representing a message from right to left on edge U_k in Fig. 8.11. It is easy to see that the complexity is now linear in D.

Algorithm 8.5 LDPC code: SPA in check nodes

1: $\lambda_{U_2}^{(\rightarrow)} = f_\boxplus\left(\lambda_{X_1 \rightarrow \boxplus}, \lambda_{X_2 \rightarrow \boxplus}\right)$

2: $\lambda_{U_{D-2}}^{(\leftarrow)} = f_\boxplus\left(\lambda_{X_D \rightarrow \boxplus}, \lambda_{X_{D-1} \rightarrow \boxplus}\right)$

3: **for** $k = 3$ to $D - 2$ **do**

4: $\quad \lambda_{U_k}^{(\rightarrow)} = f_\boxplus\left(\lambda_{U_{k-1}}^{(\rightarrow)}, \lambda_{X_k \rightarrow \boxplus}\right)$

5: $\quad \lambda_{U_{D-k}}^{(\leftarrow)} = f_\boxplus\left(\lambda_{U_{D-k+1}}^{(\leftarrow)}, \lambda_{X_{D-k+1} \rightarrow \boxplus}\right)$

6: **end for**

7: $\lambda_{\boxplus \rightarrow X_1} = f_\boxplus\left(\lambda_{U_2}^{(\leftarrow)}, \lambda_{X_2 \rightarrow \boxplus}\right)$, $\lambda_{\boxplus \rightarrow X_2} = f_\boxplus\left(\lambda_{U_2}^{(\leftarrow)}, \lambda_{X_1 \rightarrow \boxplus}\right)$

8: $\lambda_{\boxplus \rightarrow X_D} = f_\boxplus\left(\lambda_{U_{D-2}}^{(\rightarrow)}, \lambda_{X_{D-1} \rightarrow \boxplus}\right)$, $\lambda_{\boxplus \rightarrow X_{D-1}} = f_\boxplus\left(\lambda_{U_{D-2}}^{(\rightarrow)}, \lambda_{X_D \rightarrow \boxplus}\right)$

9: **for** $k = 3$ to $D - 2$ **do**

10: $\quad \lambda_{\boxplus \rightarrow X_k} = f_\boxplus\left(\lambda_{U_{k-1}}^{(\rightarrow)}, \lambda_{U_k}^{(\leftarrow)}\right)$

11: **end for**

8.5.4 Decoding low-density parity-check codes

We will again work in the LLR domain. Remember that $\mathbf{H}^{\mathrm{T}}\mathbf{c} = \mathbf{0}$ for any codeword $\mathbf{c}$. Let us introduce the following notation.

- The check nodes (marked as $\boxplus$ in the factor graph) are numbered $1, 2, \ldots, N_c - N_b$. The nth check node corresponds to the nth row in $\mathbf{H}^{\mathrm{T}}$.

- The equality nodes (marked as ⊟ in the factor graph) are numbered $1, 2, \ldots, N_c$. The kth equality node corresponds to the kth column in $\mathbf{H}^T$ (to C_k).
- The message from the kth equality node to the nth check node is denoted $\lambda_{\boxminus k \to \boxplus n}$.

Since the matrix $\mathbf{H}^T$ is sparse, we need to keep track only of the locations of the non-zero entries. Let us assume that some suitable data structure is available, so that we can introduce functions $\psi_r(\cdot)$ and $\psi_c(\cdot)$ such that $\psi_r(n)$ returns the indices of the non-zero elements in nth row of $\mathbf{H}^T$ and $\psi_c(k)$ returns the indices of the kth column of $\mathbf{H}^T$. For instance, in our example with $\mathbf{H}^T$ from (8.43), $\psi_r(3) = \{1, 2, 3, 5, 8\}$ and $\psi_c(4) = \{1, 2\}$. Since the factor graph of an LDPC code has cycles, we again need to decide on a scheduling strategy. We will discuss the most common strategy.

- **Initialization.** The decoding algorithm is initialized by setting the messages $\lambda_{\boxplus n \to \boxminus k} = 0, \forall k \in \{1, \ldots, N_c\}, n \in \psi_c(k)$ (see Fig. 8.12).
- **Upward messages** $\lambda_{\boxminus k \to \boxplus n}$ are computed for all k and all $n \in \psi_c(k)$ (see Fig. 8.13).
- **Downward messages** $\lambda_{\boxplus n \to \boxminus k}$ are computed for all n and all $k \in \psi_r(n)$ (see Fig. 8.14).

The decoding algorithm iterates between the upward and downward phases, as described in Algorithm 8.6. In LDPC codes, we commonly check after every iteration whether the hard decisions on the coded bits form a codeword. The hard decisions are determined by the mode of the beliefs (the approximation of the a-posteriori distribution at that iteration):

$$\hat{c}_k = \arg \max_{c_k \in \mathbb{B}} p(C_k = c_k | \mathbf{Y} = \mathbf{y}, \mathcal{M}) \tag{8.55}$$

$$= \begin{cases} 0 & \lambda_{\boxminus \to C_k} + \lambda_{C_k \to \boxminus} < 0 \\ 1 & \lambda_{\boxminus \to C_k} + \lambda_{C_k \to \boxminus} > 0. \end{cases} \tag{8.56}$$

We then check whether $\mathbf{H}^T \hat{c} = \mathbf{0}$. If so, the SPA on the factor graph of $p(\mathbf{B}, \mathbf{Y} = \mathbf{y} | \mathcal{M})$ can be halted.

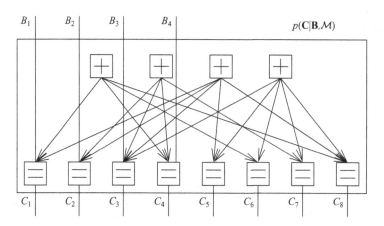

Figure 8.12. LDPC codes: initialization. Bold arrows indicate that uniform pmfs are passed over the corresponding edges.

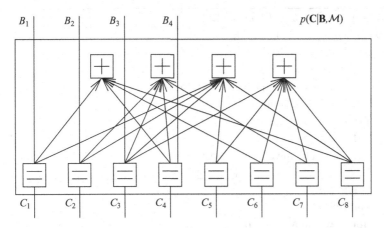

Figure 8.13. LDPC codes: upward phase. Bold arrows indicate that messages are passed over the corresponding edges.

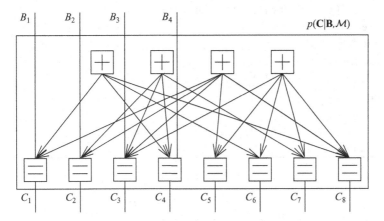

Figure 8.14. LDPC codes: downward phase. Bold arrows indicate that messages are passed over the corresponding edges.

8.6   Convolutional codes

8.6.1   Description

In convolutional codes, a codeword is obtained by passing a binary information sequence through a finite-length shift register. A typical convolutional encoder is depicted in Fig. 8.15. Note that $N_{in} = 1$ and $N_{out} = 2$. We will consider only codes with $N_{in} = 1$. The encoder consists of a sequence of L memory blocks (registers), and binary adders. In our example, $L = 3$. Observe that the code is systematic, since $c_k^{(1)} = b_k$.

Convolutional codes are usually described by feedforward and feedback polynomials, reflecting the relation between the outputs and the values in the registers. In our example, the feedback polynomial is $g_{FB}(D) = 1 + D^2 + D^3$ (since we feed back the output of the

---

Algorithm 8.6 Decoding LDPC codes

1: *input:* $\lambda_{B_k \to \boxminus}, k = 1, \ldots, N_{\mathrm{b}}$

2: *input:* $\lambda_{C_k \to \boxminus}, k = 1, \ldots, N_{\mathrm{c}}$

3: initialization:

$$\lambda_{\boxplus_n \to \boxminus_k} = 0, \forall k \in \{1, 2, \ldots, N_{\mathrm{c}}\}, \forall n \in \psi_{\mathrm{c}}(k)$$

4: **for** iter $= 1$ to N_{iter} **do**

5:    **for** $k = 1$ to N_{c} **do**

6:      **for** $n \in \psi_{\mathrm{c}}(k)$ **do**

7:        compute upward messages:

$$\lambda_{\boxminus_k \to \boxplus_n} = \lambda_{C_k \to \boxminus} + \mathbb{I}\{k \leq N_{\mathrm{b}}\}\lambda_{B_k \to \boxminus} + \sum_{m \in \psi_{\mathrm{c}}(k) \setminus \{n\}} \lambda_{\boxplus_m \to \boxminus_k}$$

8:      **end for**

9:    **end for**

10:    **for** $n = 1$ to $N_{\mathrm{c}} - N_{\mathrm{b}}$ **do**

11:      compute downward messages $\lambda_{\boxplus_n \to \boxminus_k}$ using Algorithm 8.5 for all $k \in \psi_{\mathrm{r}}(n)$.

12:    **end for**

13:    **if** $\mathbf{H}^{\mathrm{T}}\hat{\mathbf{c}} = \mathbf{0}$ **then**

14:      STOP iterations {this step is optional}

15:    **end if**

16: **end for**

17: **for** $k = 1$ to N_{b} **do**

18:    $\lambda_{\boxminus \to B_k} = \lambda_{C_k \to \boxminus} + \sum_{m \in \psi_{\mathrm{c}}(k)} \lambda_{\boxplus_m \to \boxminus_k}$

19: **end for**

20: **for** $k = 1$ to N_{c} **do**

21:    $\lambda_{\boxminus \to C_k} = \mathbb{I}\{k \leq N_{\mathrm{b}}\}\lambda_{B_k \to \boxminus} + \sum_{m \in \psi_{\mathrm{c}}(k)} \lambda_{\boxplus_m \to \boxminus_k}$

22: **end for**

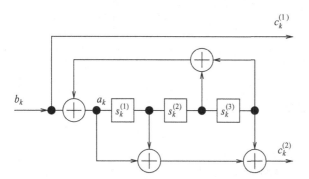

Figure 8.15. A recursive systematic convolutional encoder with $N_{\mathrm{in}} = 1$ and $N_{\mathrm{out}} = 2$.

second and third register), while the feedforward polynomial is $g_{\mathrm{FF}}(D) = 1 + D + D^3$ (since we feed forward the output of the first and third register). When the feedback polynomial is trivial (i.e., $g_{\mathrm{FB}}(D) = 1$), we say that the code is non-recursive. In our example from Fig. 8.15, we have a recursive systematic convolutional code.

The bit values in the registers are known as the *state*. Since there are L registers, the state space $\mathcal{S}$ contains 2^L elements (all binary sequences of length L). The state s_k and the output c_k depend on the input b_k and the previous state s_{k-1}. Hence, we have a state-space system, as described in Section 8.3.2.2, where the functions $f_s(\cdot)$ and $f_o(\cdot)$ depend on the feedback and feedforward polynomials. In our example, introducing $a_k = b_k + s_{k-1}^{(2)} + s_{k-1}^{(3)}$, we have

$$s_k = f_s(s_{k-1}, b_k) = \left[a_k \; s_{k-1}^{(1)} \; s_{k-1}^{(2)} \right], \tag{8.57}$$

$$c_k = f_o(s_{k-1}, b_k) = \left[c_k^{(1)} \; c_k^{(2)} \right], \tag{8.58}$$

where $c_k^{(1)} = b_k$ and $c_k^{(2)} = a_k + s_{k-1}^{(1)} + s_{k-1}^{(3)}$.

Termination
Termination of a convolutional code can be achieved by adding L termination input bits $\mathbf{t} = \left[t_{N_b+1}, \ldots, t_{N_b+L} \right]$. Suppose that we wish to terminate our code so that the final state s_{N_b+L} is the all-zero state. From the state s_{N_b}, we input $t_{N_b+1}, \ldots, t_{N_b+L}$, so that, when we input t_{N_b+k}, we make sure that $s_{N_b+k}^{(k)}$ becomes zero. In that way, we can insure that s_{N_b+L} is the all-zero state. In the remainder of this chapter we will deal only with terminated convolutional codes. Unterminated codes can be obtained by puncturing the last LN_{out} coded bits.

Example 8.2. *Let us consider a simple example in which initially the register values are all zero:* $s_{start} = \left[s_0^{(1)} \; s_0^{(2)} \; s_0^{(3)} \right] = [0 \; 0 \; 0]$ *and* $\mathbf{b} = [0 \; 1 \; 1 \; 0]$.

• **Time 1:** $b_1 = 0$ *is the input. Then* $a_1 = b_1 + s_0^{(2)} + s_0^{(3)} = 0$. *Now,* $c_1^{(1)} = b_1 = 0$ *and* $c_1^{(2)} = a_1 + s_0^{(1)} + s_0^{(3)} = 0$. *Finally, the state is changed from* $s_0 = [0 \; 0 \; 0]$ *to* $s_1 = [a_1 \; 0 \; 0] = [0 \; 0 \; 0]$.

• **Time 2:** $b_2 = 1$ *is the input. Then* $a_2 = b_2 + s_1^{(2)} + s_1^{(3)} = 1$. *Now,* $c_2^{(1)} = b_2 = 1$ *and* $c_2^{(2)} = a_2 + s_1^{(1)} + s_1^{(3)} = 1$. *Finally, the state is changed from* $s_1 = [0 \; 0 \; 0]$ *to* $s_2 = [a_2 \; 0 \; 0] = [1 \; 0 \; 0]$.

• **Time 3:** $b_3 = 1$ *is the input. Then* $a_3 = b_3 + s_2^{(2)} + s_2^{(3)} = 1$. *Now,* $c_3^{(1)} = b_3 = 1$ *and* $c_3^{(2)} = a_3 + s_2^{(1)} + s_2^{(3)} = 0$. *Finally, the state is changed from* $s_2 = [1 \; 0 \; 0]$ *to* $s_3 = [a_3 \; 1 \; 0] = [1 \; 1 \; 0]$.

• **Time 4:** $b_4 = 0$ *is the input. Then* $a_4 = b_4 + s_3^{(2)} + s_3^{(3)} = 1$. *Now,* $c_4^{(1)} = b_4 = 0$ *and* $c_4^{(2)} = a_4 + s_3^{(1)} + s_3^{(3)} = 0$. *Finally, the state is changed from* $s_3 = [1 \; 1 \; 0]$ *to* $s_4 = [a_4 \; 1 \; 1] = [1 \; 1 \; 1]$.

Starting from the state $s_0 = [0 \; 0 \; 0]$, *we have encoded* $\mathbf{b} = [0 \; 1 \; 1 \; 0]$ *into* $\mathbf{c} = [0 \; 0 \; 1 \; 1 \; 1 \; 0 \; 0 \; 0]$. *The final state is* $s_4 = [1 \; 1 \; 1]$. *The code is unterminated. The reader can verify that we can terminate the code into* $s_{end} = s_{start}$ *by providing the encoder with* $\mathbf{t} = [0 \; 0 \; 1]$. *The corresponding output is* $[0 \; 0 \; 0 \; 1 \; 1 \; 1]$ *and is concatenated to* **c**. *Hence, for a terminated code,*

$$[0 \; 0 \; 1 \; 1 \; 1 \; 0 \; 0 \; 0 \; 0 \; 0 \; 0 \; 1 \; 1 \; 1] = f_c([0 \; 1 \; 1 \; 0]).$$

Puncturing the last $LN_{out} = 6$ *coded bits yields the sequence* $[0 \; 0 \; 1 \; 1 \; 1 \; 0 \; 0 \; 0]$ *of the unterminated code.*

8.6.2   Factor graphs

Following Section 8.3.3, we open the node $p(\mathbf{C}|\mathbf{B}, \mathcal{M})$ and replace it by the factorization of $p(\mathbf{C}, \mathbf{S}, \mathbf{T}|\mathbf{B}, \mathcal{M})$, where

$$p(\mathbf{C} = \mathbf{c}, \mathbf{S} = \mathbf{s}, \mathbf{T} = \mathbf{t}|\mathbf{B} = \mathbf{b}, \mathcal{M})$$

$$= \boxminus(s_0, s_{start}) \prod_{k=1}^{N_b} \boxminus(s_k, f_s(s_{k-1}, b_k)) \boxminus(\mathbf{c}_k, f_o(s_{k-1}, b_k)) \qquad (8.59)$$

$$\times \underbrace{\prod_{l=N_b+1}^{N_b+L} \boxminus(s_l, f_s(s_{l-1}, t_l)) \boxminus(\mathbf{c}_l, f_o(s_{l-1}, t_l)) \boxminus(s_{N_b+L}, s_{end})}_{\text{termination part}}.$$

The corresponding factor graph is shown in Fig. 8.16 for our example convolutional code. Note that for a terminated code $N_c = N_{out}(N_b + L)$, whereas for an unterminated code the bits $c_{N_{out}N_b+1}$ until $c_{N_{out}(N_b+L)}$ are punctured so that $N_c = N_{out}N_b$.

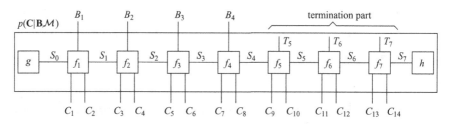

Figure 8.16. A factor graph for a convolutional code with $N_{in} = 1$ and $N_{out} = 2$, with $g(s_0) = \boxminus(s_0, s_{start})$, $h(s_{N_b+L}) = \boxminus(s_{N_b+L}, s_{end})$ and $f_k(s_{k-1}, s_k, b_k, c_{2k-1}, c_{2k}) = \boxminus(s_k, f_s(s_{k-1}, b_k)) \boxminus([c_{2k-1}, c_{2k}], f_o(s_{k-1}, b_k))$.

8.6.3   Building blocks

In the factor graph from Fig. 8.16 we see two types of nodes: nodes of degree five (in general of degree $N_{in} + N_{out} + 2$) marked f_k and nodes of degree one (g and h). Note that messages over the edges B_k, C_k, and T_k are functions of binary variables, whereas messages over the edges S_k are functions of variables defined over $\mathcal{S}$. Remember that $|\mathcal{S}| = 2^L$, where L is the number of one-bit registers. Since the receiver does not know the termination bits, we always have $\mu_{T_k \to f_k}(t_k) = 1/2$, for all k, $t_k \in \mathbb{B}$. When the code is unterminated, the bits $c_{N_{out}N_b+1}$ until $c_{N_{out}(N_b+L)}$ are punctured so that $\mu_{C_{2k} \to f_k}(c_{2k}) = 1/2 = \mu_{C_{2k-1} \to f_k}(c_{2k-1})$, $k > N_b$.

8.6.3.1   Nodes of degree one
The probability domain

Clearly $\mathbf{p}_{g \to S_0}$ is a vector of size 2^L, with $\mathbf{p}_{g \to S_0}(s) = 1$ when $s = s_{start}$ and zero otherwise. Similarly, $\mathbf{p}_{h \to S_{N_b+L}}(s) = 1$ when $s = s_{end}$ and zero otherwise. Note that for an unterminated code $\mathbf{p}_{S_{N_b} \to f_{N_b}}(s) = 1/2^L$ for all $s \in S$. In other words, the termination part in Fig. 8.16 can be omitted for an unterminated convolutional code.

The log domain

In the log domain, we just take the logarithm of the messages in the probability domain. For instance, $\mathbf{L}_{g \to S_0}(s) = 0$ when $s = s_{start}$ and $-\infty$ otherwise.

8.6.3.2   Nodes of degree five
We will consider the scenario where $N_{in} = 1$ and $N_{out} = 2$. Generalizations are straightforward.

The probability domain

Let us consider time instant k, with

$$f_k(s_{k-1}, s_k, b_k, c_{2k-1}, c_{2k}) = \boxminus(s_k, f_s(s_{k-1}, b_k)) \boxminus ([c_{2k-1} \quad c_{2k}], f_o(s_{k-1}, b_k)). \tag{8.60}$$

For a terminated code and $k > N_b$, we replace b_k by t_k. Assuming that we have available all incoming messages, we would like to compute all outgoing messages: $\mathbf{p}_{f_k \to B_k}$, $\mathbf{p}_{f_k \to C_{2k-1}}$, $\mathbf{p}_{f_k \to C_{2k}}$, $\mathbf{p}_{f_k \to S_k}$, and $\mathbf{p}_{f_k \to S_{k-1}}$. Since the input b_k and the current state s_{k-1} uniquely determine the output c_k as well as the next state s_k, the most efficient way to compute all outgoing messages is as shown in Algorithm 8.7, provided that we initialize $\mathbf{p}_{f_k \to B_k}$, $\mathbf{p}_{f_k \to C_{2k-1}}$, $\mathbf{p}_{f_k \to C_{2k}}$, $\mathbf{p}_{f_k \to S_k}$, and $\mathbf{p}_{f_k \to S_{k-1}}$ to all-zero vectors.

The log domain

We simply take the logarithm of the messages in the probability domain, replacing $\times$ by $+$ and $+$ by $\mathbb{M}(\cdot)$. The resulting algorithm is shown in Algorithm 8.8, provided that we initialize $\mathbf{L}_{f_k \to B_k}$, $\mathbf{L}_{f_k \to C_{2k-1}}$, $\mathbf{L}_{f_k \to C_{2k}}$, $\mathbf{L}_{f_k \to S_k}$, and $\mathbf{L}_{f_k \to S_{k-1}}$ to all-minus-infinity vectors.

The LLR domain

For reasons of computational complexity, the messages of binary variables are usually represented as LLRs. The algorithm remains essentially unmodified from the log-domain algorithm, except that the incoming messages (LLRs) are converted into log-domain vectors. Conversion from LLR to log-domain vectors is commonly achieved in one of

---

Algorithm 8.7 Convolutional code: building block

1: **for** $s_{k-1} \in \mathcal{S}$ **do**
2:   **for** $b_k \in \{0,1\}$ **do**
3:     $s_k = f_s(s_{k-1}, b_k)$
4:     $[c_{2k-1} \quad c_{2k}] = f_0(s_{k-1}, b_k)$
5:     let $p_{B_k} = \mathbf{p}_{B_k \to f_k}(b_k)$, $p_{C_{2k-1}} = \mathbf{p}_{C_{2k-1} \to f_k}(c_{2k-1})$, $p_{C_{2k}} = \mathbf{p}_{C_{2k} \to f_k}(c_{2k})$, $p_{S_{k-1}} = \mathbf{p}_{S_{k-1} \to f_k}(s_{k-1})$, $p_{S_k} = \mathbf{p}_{S_k \to f_k}(s_k)$.
6:     $\mathbf{p}_{f_k \to B_k}(b_k) = \mathbf{p}_{f_k \to B_k}(b_k) + p_{C_{2k-1}} p_{C_{2k}} p_{S_k} p_{S_{k-1}}$
7:     $\mathbf{p}_{f_k \to C_{2k-1}}(c_{2k-1}) = \mathbf{p}_{f_k \to C_{2k-1}}(c_{2k-1}) + p_{B_k} p_{C_{2k}} p_{S_k} p_{S_{k-1}}$
8:     $\mathbf{p}_{f_k \to C_{2k}}(c_{2k}) = \mathbf{p}_{f_k \to C_{2k}}(c_{2k}) + p_{C_{2k-1}} p_{B_k} p_{S_k} p_{S_{k-1}}$
9:     $\mathbf{p}_{f_k \to S_k}(s_k) = \mathbf{p}_{f_k \to S_k}(s_k) + p_{C_{2k-1}} p_{C_{2k}} p_{B_k} p_{S_{k-1}}$
10:    $\mathbf{p}_{f_k \to S_{k-1}}(s_{k-1}) = \mathbf{p}_{f_k \to S_{k-1}}(s_{k-1}) + p_{C_{2k-1}} p_{C_{2k}} p_{S_k} p_{B_k}$
11:   **end for**
12: **end for**
13: normalize $\mathbf{p}_{f_k \to B_k}$, $\mathbf{p}_{f_k \to C_{2k}}$, $\mathbf{p}_{f_k \to C_{2k-1}}$, $\mathbf{p}_{f_k \to S_k}$, and $\mathbf{p}_{f_k \to S_{k-1}}$

Algorithm 8.8 Convolutional code: building block in the log domain

1: **for** $s_{k-1} \in \mathcal{S}$ **do**
2:   **for** $b_k \in \{0,1\}$ **do**
3:     $s_k = f_s(s_{k-1}, b_k)$
4:     $[c_{2k-1} \quad c_{2k}] = f_0(s_{k-1}, b_k)$
5:     let $L_{B_k} = \mathbf{L}_{B_k \to f_k}(b_k)$, $L_{C_{2k-1}} = \mathbf{L}_{C_{2k-1} \to f_k}(c_{2k-1})$, $L_{C_{2k}} = \mathbf{L}_{C_{2k} \to f_k}(c_{2k})$, $L_{S_{k-1}} = \mathbf{L}_{S_{k-1} \to f_k}(s_{k-1})$, $L_{S_k} = \mathbf{L}_{S_k \to f_k}(s_k)$.
6:     $\mathbf{L}_{f_k \to B_k}(b_k) = \mathbb{M}\left(\mathbf{L}_{f_k \to B_k}(b_k), L_{C_{2k-1}} + L_{C_{2k}} + L_{S_k} + L_{S_{k-1}}\right)$
7:     $\mathbf{L}_{f_k \to C_{2k-1}}(c_{2k-1}) = \mathbb{M}\left(\mathbf{L}_{f_k \to C_{2k-1}}(c_{2k-1}), L_{B_k} + L_{C_{2k}} + L_{S_k} + L_{S_{k-1}}\right)$
8:     $\mathbf{L}_{f_k \to C_{2k}}(c_{2k}) = \mathbb{M}\left(\mathbf{L}_{f_k \to C_{2k}}(c_{2k}), L_{C_{2k-1}} + L_{B_k} + L_{S_k} + L_{S_{k-1}}\right)$
9:     $\mathbf{L}_{f_k \to S_k}(s_k) = \mathbb{M}\left(\mathbf{L}_{f_k \to S_k}(s_k), L_{C_{2k-1}} + L_{C_{2k}} + L_{B_k} + L_{S_{k-1}}\right)$
10:    $\mathbf{L}_{f_k \to S_{k-1}}(s_{k-1}) = \mathbb{M}\left(\mathbf{L}_{f_k \to S_{k-1}}(s_{k-1}), L_{C_{2k-1}} + L_{C_{2k}} + L_{S_k} + L_{B_k}\right)$
11:   **end for**
12: **end for**

---

the following two ways (we focus on $\lambda_{B_k \to f_k}$; the same technique is applied to $\lambda_{C_{2k-1} \to f_k}$ and $\lambda_{C_{2k} \to f_k}$):

- given $\lambda_{B_k \to f_k}$, set $\mathbf{L}_{B_k \to f_k}(1) = \lambda_{B_k \to f_k}$ and $\mathbf{L}_{B_k \to f}(0) = 0$;
- given $\lambda_{B_k \to f_k}$, set $\mathbf{L}_{B_k \to f_k}(1) = \lambda_{B_k \to f_k}/2$ and $\mathbf{L}_{B_k \to f_k}(0) = -\lambda_{B_k \to f_k}/2$. This allows us to write $\mathbf{L}_{B_k \to f_k}(b_k) = (2b_k - 1)\lambda_{B_k \to f_k}/2$.

At the end of the algorithm, the outgoing log-domain vectors are converted back into LLRs, for instance $\lambda_{f_k \to B_k} = \mathbf{L}_{f_k \to B_k}(1) - \mathbf{L}_{f_k \to B_k}(0)$.

8.6.4 Decoding convolutional codes

Since a convolutional code is a state-space model, decoding is based on the forward–backward algorithm. This algorithm was originally derived in [88] and is known as the BCJR algorithm (after the four authors). The entire algorithm is given in Algorithm 8.9. Observe that the complexity scales as $\mathcal{O}\left(N_b 2^L\right)$ and that, for an unterminated code, lines 11–13 can be omitted.

Algorithm 8.9 Decoding of a convolutional code

1: *input*: $\lambda_{B_k \to \text{dec}}$, $k = 1, \ldots, N_b$
2: *input*: $\lambda_{C_k \to \text{dec}}$, $k = 1, \ldots, N_c$
3: initialization:

   $L_{g \to S_0}(s_0) = 0$ when $s_0 = s_{\text{start}}$ and $-\infty$ otherwise,

   $L_{h \to S_{N_b+L}}(s) = 0$ when $s = s_{\text{end}}$ and $-\infty$ otherwise

   $L_{T_k \to f_k}(t_k) = 0$, $t_k \in \mathbb{B}$, $\forall k > N_b$

   $L_{C_{2k} \to f_k}(c) = 0 = L_{C_{2k-1} \to f_k}(c)$, $c \in \mathbb{B}$, $k > N_{\text{out}} N_b$ for an unterminated code

   set $L_{f_k \to B_k}$, $L_{f_k \to T_k}$, $L_{f_k \to C_{2k}}$, $L_{f_k \to C_{2k-1}}$, $L_{f_k \to S_k}$, and $L_{f_k \to S_{k-1}}$ to all-minus-infinity vectors, $\forall k$.

4: **for** $k = 1$ to $N_b + L - 1$ **do**
5:     compute forward message $L_{f_k \to S_k}$ using Algorithm 8.8 (without lines 6, 7, 8, and 10)
6:     compute backward message $L_{S_{N_b+L-k} \to f_{N_b+L-k}}$ using Algorithm 8.8 (without lines 6–9)
7: **end for**
8: **for** $k = 1$ to N_b **do**
9:     compute messages $\lambda_{f_k \to B_k}$, $\lambda_{f_k \to C_{2k}}$, and $\lambda_{f_k \to C_{2k-1}}$ using Algorithm 8.8 (without lines 9 and 10)
10: **end for**
11: **for** $k = N_b + 1$ to $N_b + L$ **do**
12:    compute messages $\lambda_{f_k \to C_{2k}}$ and $\lambda_{f_k \to C_{2k-1}}$ using Algorithm 8.8 (without lines 9 and 10)
13: **end for**

8.6.5 Sequence detection

Although we have considered only the standard SPA for convolutional codes, it is clear that, when the factor graph of $p(\mathbf{B}, \mathbf{Y} = \mathbf{y}|\mathcal{M})$ contains no cycles, we can also apply the max–sum algorithm (Viterbi algorithm) to the factorization of $\log p(\mathbf{B}, \mathbf{Y} = \mathbf{y}|\mathcal{M})$. This leads to the most likely sequence

$$\hat{\mathbf{b}} = \arg\max_{\mathbf{b}} \log p(\mathbf{B} = \mathbf{b}|\mathbf{Y} = \mathbf{y}, \mathcal{M}). \qquad (8.61)$$

In view of our observations from Section 5.3.3.3, we can find the most likely sequence by implementing the SPA in the log domain and replacing $\mathbb{M}(\cdot)$ by the max operation.

Note that for RA codes and LDPC codes (and also for turbo codes, as we will see shortly), the factor graph of $p(\mathbf{B}, \mathbf{Y} = \mathbf{y}|\mathcal{M})$ necessarily contains cycles, so optimal sequence detection is not possible.

8.7   Turbo codes

8.7.1   Description

We will consider the two most important types of turbo codes: parallel concatenation of convolutional codes (PCCC) and serial concatenation of convolutional codes (SCCC). The former were the original turbo codes, as described by Berrou *et al.* in 1993 [3]. The latter were introduced a few years later by Benedetto *et al.* [89]. As the names suggest, turbo codes consist of the concatenation of two convolutional encoders, separated by an interleaver.

PCCC

In PCCC, we encode our information sequence $\mathbf{b}$ using a $(N_{\text{in}} = 1, N_{\text{out}} = 2)$ systematic convolutional encoder $(CC^{(A)})$. This leads to a codeword $\mathbf{c}^{(A)} = f_c^{(A)}(\mathbf{b})$. We will write the coded bits in a funny way,

$$
\mathbf{c}^{(A)} = \left[d_1, c_3, \ldots, d_{N_b}, c_{3N_b}, \underbrace{c_{3N_b+1}, c_{3N_b+2}, \ldots, c_{3N_b+2L^{(A)}}}_{\text{termination part}} \right] \qquad (8.62)
$$

such that $d_k = b_k$ for $k = 1, \ldots, N_b$ (since the code is systematic). The last $2L^{(A)}$ bits correspond to the termination. As always, puncturing these last $2L^{(A)}$ bits leads to an unterminated code. We then interleave the information bits, $\mathbf{e} = \pi(\mathbf{d}) = \pi(\mathbf{b})$, and encode this interleaved sequence using a $(N_{\text{in}} = 1, N_{\text{out}} = 2)$ systematic convolutional encoder, leading to $\mathbf{c}^{(B)} = f_c^{(B)}(\mathbf{e}) = f_c^{(B)}(\pi(\mathbf{b}))$. We will write the coded bits as follows:

$$
\mathbf{c}^{(B)} = \left[c_1, c_2, c_4, c_5, \ldots, c_{3N_b-2}, c_{3N_b-1}, \underbrace{c_{3N_b+2L^{(A)}+1}, \ldots, c_{3N_b+2L^{(A)}+2L^{(B)}}}_{\text{termination part}} \right]. \qquad (8.63)
$$

Since the code is systematic, we know that $c_{3k-2} = e_k = b_{\pi(k)}$, for $k = 1, \ldots, N_b$. We can again choose to puncture the last $2L^{(B)}$ bits from $\mathbf{c}^{(B)}$ to obtain an unterminated code. The final codeword is given by concatenating all the coded bits, and dropping the bits d_k:

$$
\mathbf{c} = f_c(\mathbf{b}) \qquad (8.64)
$$

$$
= \left[c_1, c_2, c_3, \ldots, c_{3N_b+2L^{(A)}+2L^{(B)}} \right], \qquad (8.65)
$$

where we set $L^{(A)}$ ($L^{(B)}$) to zero when $CC^{(A)}$ ($CC^{(B)}$) is unterminated. Generally $f_c^{(B)}(\cdot) = f_c^{(A)}(\cdot)$. The rate of this code is given by

$$r = \frac{N_b}{N_c} \tag{8.66}$$

$$= \frac{1}{3} \frac{1}{1 + 2(L^{(A)} + L^{(B)})/N_b}, \tag{8.67}$$

where $L^{(A)}$ and $L^{(B)}$ can be zero depending on whether or not encoder A or encoder B is terminated. Observe that termination leads to a small rate loss.

SCCC

In SCCC, we again encode our information sequence $\mathbf{b}$ using a ($N_{in} = 1, N_{out} = 2$) systematic convolutional encoder. This leads to a codeword $\mathbf{d} = f_c^{(A)}(\mathbf{b})$ of length $N^{(A)} = 2N_b + 2L^{(A)}$, where $L^{(A)} = 0$ when the encoder is unterminated. In contrast to PCCC, we now interleave the *entire* sequence $\mathbf{d}$ leading to $\mathbf{e} = \pi(\mathbf{d})$. We encode $\mathbf{e}$ to $\mathbf{c} = f^{(B)}(\mathbf{e})$ using a ($N_{in} = 1, N_{out} = 2$) systematic convolutional encoder. The codeword has length $N_c = 2N^{(A)} + 2L^{(B)}$, where $L^{(B)} = 0$ when the second encoder is unterminated. Hence, the overall rate is

$$r = \frac{N_b}{N_c} \tag{8.68}$$

$$= \frac{1}{4} \frac{1}{1 + 2(2L^{(A)} + L^{(B)})/N_b}. \tag{8.69}$$

8.7.2 Factor graphs

PCCC

We replace the node $p(\mathbf{C}|\mathbf{B}, \mathcal{M})$ by the following factorization:

$$p\Big(\mathbf{C} = \mathbf{c}, \mathbf{D} = \mathbf{d}, \mathbf{E} = \mathbf{e}, \mathbf{C}^{(A)} = \mathbf{c}^{(A)}, \mathbf{C}^{(B)} = \mathbf{c}^{(B)} | \mathbf{B} = \mathbf{b}, \mathcal{M}\Big)$$

$$= \boxdot\Big(\mathbf{c}^{(A)}, f_c^{(A)}(\mathbf{b})\Big)\boxdot\Big(\mathbf{c}^{(B)}, f_c^{(B)}(\mathbf{e})\Big)$$

$$\times \prod_{k=1}^{N_b} \boxdot\Big(d_k, c_{2k-1}^{(A)}\Big)\boxdot(\mathbf{e}, \pi(\mathbf{d}))$$

$$\times \prod_{k=1}^{N_b} \boxdot\Big(c_{3k-2}, c_{2k-1}^{(B)}\Big)\boxdot\Big(c_{3k-1}, c_{2k}^{(B)}\Big)\boxdot\Big(c_{3k}, c_{2k}^{(A)}\Big)$$

$$\times \underbrace{\prod_{k=1}^{2L^{(A)}} \boxdot\Big(c_{3N_b+k}, c_{2N_b+k}^{(A)}\Big) \prod_{k=1}^{2L^{(B)}} \boxdot\Big(c_{3N_b+2L^{(A)}+k}, c_{2N_b+k}^{(B)}\Big)}_{\text{termination part}} \tag{8.70}$$

with the factor graph shown in Fig. 8.17.

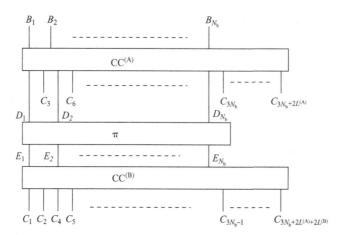

Figure 8.17. A factor graph for a PCCC turbo code. The variables $C_k^{(A)}$ and $C_k^{(B)}$ are omitted for notational convenience. Both codes are assumed to be terminated. The factor graph for unterminated codes can be obtained by puncturing the coded bits corresponding to termination.

SCCC
We replace the node $p(\mathbf{C}|\mathbf{B}, \mathcal{M})$ by the following factorization

$$p(\mathbf{C} = \mathbf{c}, \mathbf{D} = \mathbf{d}, \mathbf{E} = \mathbf{e}|\mathbf{B} = \mathbf{b}, \mathcal{M}) = \boxminus\left(\mathbf{d}, f_c^{(A)}(\mathbf{b})\right) \boxminus(\mathbf{e}, \pi(\mathbf{d})) \boxminus\left(\mathbf{c}, f_c^{(B)}(\mathbf{e})\right).$$

$$(8.71)$$

The corresponding factor graph is shown in Fig. 8.18.

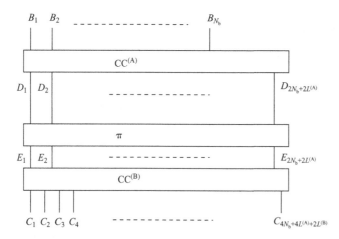

Figure 8.18. A factor graph for an SCCC turbo code. Both codes are assumed to be terminated. The factor graph for unterminated codes can be obtained by puncturing the coded bits corresponding to termination.

8.7.3   Decoding turbo codes

Since the factor graphs in Figs. 8.17 and 8.18 have cycles, the SPA becomes iterative. We choose the following scheduling.

1. Initialize the downward messages on the E_k-edges with uniform distributions: $\mathbf{L}_{E_k \to CC^{(B)}} = \mathbf{0}$. See Fig. 8.19.
2. Decode $CC^{(B)}$ using Algorithm 8.9. This results in upward messages $\lambda_{CC^{(B)} \to E_k}$.
3. de-interleave the upward messages on the E_k-edges. See Fig. 8.20. This results in upward messages $\lambda_{D_k \to CC^{(A)}}$.
4. Decode $CC^{(A)}$ using Algorithm 8.9. This results in downward messages $\lambda_{CC^{(A)} \to D_k}$.

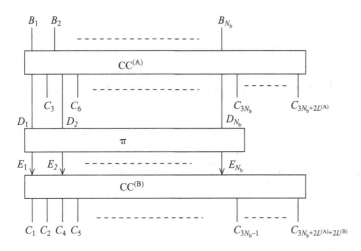

Figure 8.19. Decoding a turbo code: initialization.

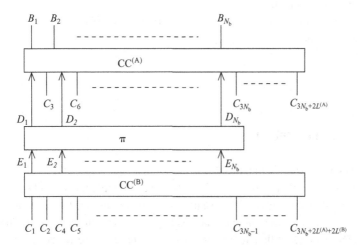

Figure 8.20. Decoding a turbo code: upward messages after decoding code $CC^{(B)}$.

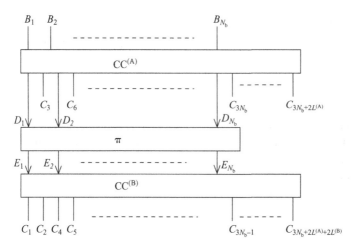

Figure 8.21. Decoding a turbo code: downward messages after decoding code $CC^{(A)}$.

5. Interleave the downward messages on the D_k-edges. See Fig. 8.21. This results in downward messages $\lambda_{E_k \to CC^{(B)}}$. Go to step 2.

The final algorithm for decoding a turbo code is described in Algorithm 8.10.

Algorithm 8.10 Decoding of a PCCC/SCCC turbo code

1: *input*: $\lambda_{B_k \to \text{dec}}$, $k = 1, \ldots, N_b$
2: *input*: $\lambda_{C_k \to \text{dec}}$, $k = 1, \ldots, N_c$
3: initialization
$$\lambda_{E_k \to CC^{(B)}} = 0, \forall k$$
4: **for** iter $= 1$ to N_{iter} **do**
5:     decode convolutional code B using Algorithm 8.9. This yields $\lambda_{CC^{(B)} \to E_k}$, $\forall k$
6:     de-interleave the messages: $\lambda_{\pi \to D_k} = \lambda_{CC^{(B)} \to E_{\pi(k)}}$, $\forall k$
7:     decode convolutional code A using Algorithm 8.9. This yields $\lambda_{D_k \to \pi}$, $\forall k$
8:     interleave the messages: $\lambda_{E_k \to CC^{(B)}} = \lambda_{D_{\pi^{-1}(k)} \to \pi}$, $\forall k$
9: **end for**

8.8   Performance illustration

Let us consider an example of a PCCC turbo code with $g_{\text{FF}}(D) = 1 + D^4$ and $g_{\text{FB}}(D) = 1 + D + D^2 + D^3 + D^4$. The first encoder is terminated, while the second encoder is not. We set $N_b = 330$, so that $N_c = 998$. We use BPSK modulation. The interleaver between the encoders is pseudo-random and changes from codeword to codeword. The channel is an AWGN channel so that $y_k = \sqrt{E_s} a_k + n_k$, where n_k are iid complex Gaussian, zero mean and $\mathbb{E}\{|N_k|^2\} = N_0$. We evaluate the performance in terms of bit error rate

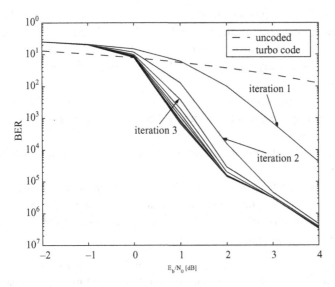

Figure 8.22. Turbo code: BER versus SNR performance. Observe the gain in BER with every iteration. After four iterations, the gain becomes negligible. As a reference, we include the performance of an uncoded system (where $E_s = E_b$).

(BER) versus signal-to-noise ratio (SNR). The SNR is expressed as E_b/N_0 (in decibels[2]), with $E_b = E_s/(R \log_2 |\Omega|)$, where R denotes the code rate. The resulting performance is shown in Fig. 8.22. As the iterations proceed, the BER drops by several orders of magnitude. We discern three SNR regions for the turbo code: below 0 dB, the BER is fairly high. Then, above 0 dB the BER suddenly drops. The SNR value at which the BER suddenly drops is known as the *pinch-off point* (0 dB in our case). Above 0 dB the turbo code enters the so-called *waterfall region*, where the BER drops dramatically with increasing number of iterations. Above 2 dB, the BER becomes flatter. The high-SNR region where the BER flattens out is known as the *error-floor region*. The BER in this region is mainly dependent on the Hamming distance of the turbo code. As a reference we include the BER for an uncoded system (with $R = 1$). The performance gain of the turbo code is quite impressive.

8.9   Main points

In this chapter we have focused on opening the node representing the function $p(\mathbf{C}|\mathbf{B}, \mathcal{M})$ and describing the SPA on the resulting factor graph, assuming that messages from the C_k-edges and the B_k-edges are available. We covered four types of error-correcting codes.

---

[2] We can express an SNR in decibels (dB) as $10 \log_{10}$ SNR.

- *RA codes*: the factor graph consists of equality nodes and check nodes of degree three, as well as an interleaver. The graph contains cycles, so the SPA is iterative.
- *LDPC codes*: these codes are based on sparse parity-check matrices. The factor graph consists again of equality nodes and check nodes. The graph contains cycles, so the SPA is iterative.
- *Convolutional codes*: these codes are based on state-space models. The factor graph is a variation of those presented in Chapter 6. The SPA is not iterative. When the factor graph of the factorization of $p(\mathbf{B}, \mathbf{Y} = \mathbf{y}|\mathcal{M})$ contains no cycles, we can apply the max–sum algorithm (the Viterbi algorithm) to log $p(\mathbf{B}, \mathbf{Y} = \mathbf{y}|\mathcal{M})$ to obtain the most likely sequence.
- *Turbo codes*: by concatenating two convolutional encoders separated by an interleaver we obtain a turbo code. The factor graph consists of linking together the two factor graphs of the convolutional codes. This creates cycles, resulting in an iterative SPA. We covered PCCC and SCCC turbo codes.

The decoder requires messages $\mu_{B_k \to \text{dec}}(B_k)$, $\forall k$ and $\mu_{C_k \to \text{dec}}(C_k)$, $\forall k$. The messages $\mu_{B_k \to \text{dec}}(B_k)$ are the a-priori distributions of the information bits, and are assumed to be known by the receiver. The messages $\mu_{C_k \to \text{dec}}(C_k)$ are obtained by executing the SPA on the demapping node, representing the distribution $p(\mathbf{A}|\mathbf{C}, \mathcal{M})$. How these messages are to be computed will be the topic of the next chapter.

9 Demapping

9.1 Introduction

In the previous chapter we have seen how messages are computed in the decoding node in the factor graph of the distribution $p(\mathbf{B}, \mathbf{Y} = \mathbf{y}|\mathcal{M})$ (see Fig. 9.1). Now we will deal with the second node, the demapping node, representing the distribution $p(\mathbf{A}|\mathbf{C}, \mathcal{M})$, where $\mathbf{C}$ is a sequence of N_c coded bits, and $\mathbf{A}$ is a sequence of N_s coded symbols, each belonging to some constellation Ω (for instance, 16-QAM). By mapping the coded bits onto a signaling constellation, we can tune the spectral efficiency of the system: the more bits we map onto any constellation point, the fewer complex symbols we need to transmit.

While there exists a wide variety of mapping schemes, we will focus on two popular instances: bit-interleaved coded modulation (BICM) and trellis-coded modulation (TCM). The former was first introduced by Zehavi [90] and later analyzed in detail by Caire *et al.* [91]. It was only with [92–94] that it was realized that employing BICM at the transmitter naturally leads to an iterative receiver. Trellis-coded modulation, on the

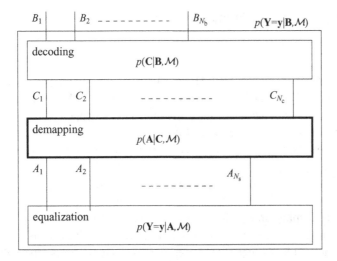

Figure 9.1. A factor graph of $p(\mathbf{Y} = \mathbf{y}|\mathbf{B}, \mathcal{M})$. The node is opened to reveal its structure. The node in bold is the topic of this chapter.

other hand, was proposed by Ungerboeck in [95] as a way to combine (convolutional) coding and mapping to obtain a receiver that could perform optimal sequence detection.

This chapter is organized as follows.

- In **Section 9.2** we will describe the main goal of this chapter.
- **Section 9.3** deals with BICM, while TCM is covered in **Section 9.4**. For both modulation schemes, we will describe the corresponding demapping algorithm in the probability domain, the log domain, and the LLR domain.
- A performance illustration of BICM will be given in **Section 9.5**.

9.2   Goals

We have two goals in this chapter: first of all, we will replace the node $p(\mathbf{A}|\mathbf{C}, \mathcal{M})$ in Fig. 9.1 by a more detailed factor graph depending on the particular mapping scheme. Secondly, we will show how the SPA can be executed on the resulting factor graphs. We will derive the demapping algorithms in the probability domain, the log domain, and the LLR domain.

The node $p(\mathbf{A}|\mathbf{C}, \mathcal{M})$ has incoming messages from the decoder block $\mu_{C_k \to \text{dem}}(C_k)$ (which is just another notation for $\mu_{\text{dec} \to C_k}(C_k)$), as well as the equalization block $\mu_{A_k \to \text{dem}}(A_k)$. Here "dem" stands for the demapper node. When the messages $\mu_{C_k \to \text{dem}}(C_k)$ are not available because of cycles in the factor graph of $p(\mathbf{A}|\mathbf{C}, \mathcal{M})p(\mathbf{C}|\mathbf{B}, \mathcal{M})$, they are assumed to be uniform distributions over $\mathbb{B}$. To help the reader understand the demapping algorithms, let us assume that $y_k = a_k + n_k$, where $n_k \sim \mathcal{N}^{\mathbb{C}}(0, \sigma^2)$, $p(\mathbf{Y} = \mathbf{y}|\mathbf{A} = \mathbf{a}, \mathcal{M}) \propto \prod_{k=1}^{N_c} \exp(-|y_k - a_k|^2/\sigma^2)$. This means that $\mu_{A_k \to \text{dem}}(a_k) \propto \exp(-|y_k - a_k|^2/\sigma^2)$.

9.3   Bit-interleaved coded modulation

9.3.1   Principles

Bit-interleaved coded modulation (BICM) operates as follows: the sequence of N_c coded bits is interleaved,[1] resulting in a sequence $\mathbf{d} = \pi(\mathbf{c})$. The bits in $\mathbf{d}$ are then grouped into consecutive blocks of length $m = \log_2|\Omega|$. The kth block is mapped onto a constellation point $a_k \in \Omega$, leading to a sequence $\mathbf{a}$ of $N_s = N_c/m$ coded symbols. The mapping $\mathbb{B}^m \to \Omega$ is achieved using a mapping function $\phi(\cdot)$, such that, for $k = 1, \ldots, N_c/m$,

$$a_k = \phi\big(d_{(k-1)m+1}, d_{(k-1)m+2}, \ldots, d_{km}\big). \tag{9.1}$$

An example of a mapping strategy for a 16-QAM constellation is shown in Fig. 9.2. The interleaver ensures that when a symbol a_k is greatly affected either by the channel or by noise, this will not result in a sequence of m erroneous bits in $\mathbf{c}$. This is especially critical

---

[1] The interleaver for BICM is not the same interleaver as in RA or turbo codes. It is a separate interleaver.

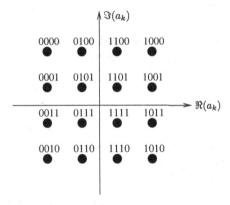

Figure 9.2. Gray mapping for 16-QAM.

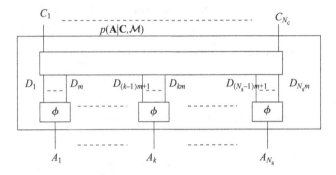

Figure 9.3. A factor graph for BICM. The nodes ϕ are shorthand for the mapping constraint: $\phi\big(\mathbf{d}_{(k-1)m+1:km}\big)$. The node marked π represents the function $\boxminus(\mathbf{d}, \pi(\mathbf{c}))$.

when combined with certain types of error-correcting codes (such as convolutional codes), which are notoriously bad at coping with burst errors.

9.3.2   Factor graphs

We replace the node $p(\mathbf{A}|\mathbf{C}, \mathcal{M})$ by a factorization of $p(\mathbf{A}, \mathbf{D}|\mathbf{C}, \mathcal{M})$ with

$$p(\mathbf{A} = \mathbf{a}, \mathbf{D} = \mathbf{d}|\mathbf{C} = \mathbf{c}, \mathcal{M}) = \prod_{k=1}^{N_s} \boxminus(a_k, \phi(\mathbf{d}_{(k-1)m+1:km})) \prod_{n=1}^{N_c} \boxminus(d_k, c_{\pi(k)}), \quad (9.2)$$

where we have abbreviated $d_{(k-1)m+1}, d_{(k-1)m+2}, \ldots, d_{km}$ by $\mathbf{d}_{(k-1)m+1:km}$. The corresponding factor graph is shown in Fig. 9.3.

9.3.3   Building blocks

Since the factor graph in Fig. 9.3 consists of N_s disjoint nodes, we have to deal with only a single building node, say $\boxminus(a_k, \phi(\mathbf{d}_{(k-1)m+1:km}))$. Let us see how to implement the SPA on this node.

The probability domain
We are given the incoming messages $\mathbf{p}_{D_{(k-1)m+n}\to\phi}$, $n \in \{1, \dots, m\}$ and $\mathbf{p}_{A_k\to\phi}$. Note that $\mathbf{p}_{D_{(k-1)m+n}\to\phi}$ is a vector of size 2, while $\mathbf{p}_{A_k\to\phi}$ is a vector of size $|\Omega| = 2^m$. Let us compute an outgoing message on edge $D_{(k-1)m+n}$:

$$\mathbf{p}_{\phi\to D_{(k-1)m+n}}(d) \propto \sum_{\tilde{\mathbf{d}}\in\mathbb{B}^m:\tilde{d}_n=d} \mathbf{p}_{A_k\to\phi}\left(\phi\left(\tilde{\mathbf{d}}\right)\right) \prod_{l\neq n} \mathbf{p}_{D_{(k-1)m+l}\to\phi}\left(\tilde{d}_l\right). \tag{9.3}$$

This computation requires summation over all binary sequences of length m, with the nth entry fixed to $d \in \mathbb{B}$. An outgoing message on the edge A_k is simply

$$\mathbf{p}_{\phi\to A_k}(a) \propto \prod_{n=1}^{m} \mathbf{p}_{D_{(k-1)m+n}\to\phi}\left(\left[\phi^{-1}(a)\right]_n\right). \tag{9.4}$$

The log domain
On replacing $\times$ by $+$ and $+$ by $\mathbb{M}$, we immediately find that

$$\mathbf{L}_{\phi\to D_{(k-1)m+n}}(d) = \mathbb{M}_{\tilde{\mathbf{d}}\in\mathbb{B}^m:\tilde{d}_n=d}\left(\mathbf{L}_{A_k\to\phi}\left(\phi\left(\tilde{\mathbf{d}}\right)\right) + \sum_{l\neq n}\mathbf{L}_{D_{(k-1)m+l}\to\phi}\left(\tilde{d}_l\right)\right) \tag{9.5}$$

and

$$\mathbf{L}_{\phi\to A_k}(a) = \sum_{n=1}^{m}\mathbf{L}_{D_{(k-1)m+n}\to\phi}\left(\left[\phi^{-1}(a)\right]_n\right). \tag{9.6}$$

The LLR domain
Given messages (scalars) $\lambda_{D_{(k-1)m+n}\to\phi}$, $n \in \{1, \dots, m\}$ and (vectors) $\mathbf{L}_{A_k\to\phi}$, we find that

$$\lambda_{\phi\to D_{(k-1)m+n}} = \mathbb{M}_{\tilde{\mathbf{d}}\in\mathbb{B}^m:\tilde{d}_n=1}\left(\mathbf{L}_{A_k\to\phi}\left(\phi\left(\tilde{\mathbf{d}}\right)\right) + \sum_{l\neq n}\boxminus\left(\tilde{d}_l, 1\right)\lambda_{D_{(k-1)m+l}\to\phi}\right)$$

$$- \mathbb{M}_{\tilde{\mathbf{d}}\in\mathbb{B}^m:\tilde{d}_n=0}\left(\mathbf{L}_{A_k\to\phi}\left(\phi\left(\tilde{\mathbf{d}}\right)\right) + \sum_{l\neq n}\boxminus\left(\tilde{d}_l, 1\right)\lambda_{D_{(k-1)m+l}\to\phi}\right) \tag{9.7}$$

and

$$\mathbf{L}_{\phi \to A_k}(a) = \sum_{n=1}^{m} \boxminus\left(\left[\phi^{-1}(a)\right]_n, 1\right)\lambda_{D_{(k-1)m+n} \to \phi}.$$ (9.8)

9.3.4 Demapping algorithm

The complete demapping algorithm for BICM is given in Algorithm 9.1.

Algorithm 9.1 Demapping for BICM

1: *input:* $\lambda_{C_k \to \mathrm{dem}}, k = 1, \ldots, N_c$
2: *input:* $\mathbf{L}_{A_k \to \mathrm{dem}}, k = 1, \ldots, N_s$
3: interleave the messages
    $\lambda_{D_{\pi(k)} \to \phi} = \lambda_{C_k \to \mathrm{dem}}, \forall k$
4: **for** $k = 1$ to N_s **do**
5:      **for** $a \in \Omega$ **do**
6:          $\mathbf{L}_{\phi \to A_k}(a) = \sum_{n=1}^{m} \boxminus([\phi^{-1}(a)]_n, 1)\lambda_{D_{(k-1)m+n} \to \phi}, \forall a \in \Omega$
7:      **end for**
8:      **for** $n = 1$ to m **do**
9:          compute $\lambda_{\phi \to D_{(k-1)m+n}}$ using (9.7)
10:     **end for**
11: **end for**
12: de-interleave the messages
    $\lambda_{C_k \to \mathrm{dem}} = \lambda_{D_{\pi(k)} \to \phi}, \forall k$

9.3.5 Interaction with decoding node

Using BICM in combination with an error-correcting code always leads to cycles between the demapping node and the decoding node, even when the decoding node has no cycles by itself. This cyclic dependency can be seen from Eq. (9.7): an upward message on the edge $D_{(k-1)m+n}$ depends on the downward messages on the edges $D_{(k-1)m+l}$, $l \neq n$ (except in the case of BPSK modulation, where $m = 1$). Hence, given messages from the equalization block, $\mu_{A_k \to \phi}(a_k)$, $\forall k$, we can iterate between the demapping node and the decoding node. This is known as bit-interleaved coded modulation with iterative decoding (BICM-ID). Initially, the messages from the decoder to the demapper are set to uniform distributions. When the decoding node also contains cycles, we end up with a doubly iterative system. For instance, in an LDPC code with BICM, we can choose the number of decoding iterations we perform with each BICM-ID iteration. Different scheduling strategies may lead to differences in BER performance.

9.4   Trellis-coded modulation

9.4.1   Description

Trellis-coded modulation (TCM) is a technique to combine convolutional coding and mapping in such a way that the factor graph of the decoder and demapper has no cycles. We start with a convolutional code with $N_{\text{in}} = 1$. Convolutional codes were covered in Section 8.6. At every time k, the encoder is in a state $\mathbf{s}_{k-1}$. When we input b_k, the encoder goes into state

$$\mathbf{s}_k = f_{\text{s}}(\mathbf{s}_{k-1}, b_k) \tag{9.9}$$

and outputs a sequence of N_{out} bits $\mathbf{c}_k$, with

$$\mathbf{c}_k = f_{\text{o}}(\mathbf{s}_{k-1}, b_k). \tag{9.10}$$

Instead of interleaving the coded bits as in BICM, we map $\mathbf{c}_k$ onto a coded symbol $a_k \in \Omega$, where $\log_2 |\Omega| = N_{\text{out}}$:

$$a_k = \phi(\mathbf{c}_k) \tag{9.11}$$

for $k = 1, \ldots, N_{\text{s}}$.

9.4.2   Factor graphs

We find that

$$p(\mathbf{A} = \mathbf{a} | \mathbf{C} = \mathbf{c}, \mathcal{M}) = \prod_{k=1}^{N_{\text{s}}} \boxed{=}(a_k, \phi(\mathbf{c}_k)). \tag{9.12}$$

The node of $p(\mathbf{C}|\mathbf{B}, \mathcal{M})$ can be replaced by the following factorization (see Eq. (8.63)):

$$p(\mathbf{C} = \mathbf{c}, \mathbf{S} = \mathbf{s}, \mathbf{T} = \mathbf{t} | \mathbf{B} = \mathbf{b}, \mathcal{M})$$

$$= \boxed{=}(s_0, s_{\text{start}}) \prod_{k=1}^{N_{\text{b}}} \boxed{=}(s_k, f_{\text{s}}(s_{k-1}, b_k)) \boxed{=}(\mathbf{c}_k, f_{\text{o}}(s_{k-1}, b_k))$$

$$\times \underbrace{\prod_{l=N_{\text{b}}+1}^{N_{\text{b}}+L} \boxed{=}(s_l, f_{\text{s}}(s_{l-1}, t_l)) \boxed{=}(\mathbf{c}_l, f_{\text{o}}(s_{l-1}, t_l)) \boxed{=}(s_{N_{\text{b}}+L}, s_{\text{end}})}_{\text{termination part}}. \tag{9.13}$$

The factor graph of the decoding and demapping node is shown in Fig. 9.4. Note the following.

- This factor graph has no cycles. This is because we consider bit-sequences $\mathbf{c}_k$ (rather than individual bits) as variables.

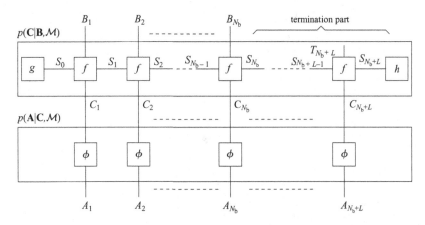

Figure 9.4. A factor graph for TCM for $N_{in} = 1$, with $g(s_0) = \boxed{=}(s_0, s_{start})$, $h(s_{N_b+L}) = \boxed{=}(s_{N_b+L}, s_{end})$ and $f(s_{k-1}, s_k, b_k, \mathbf{c}_k) = \boxed{=}(s_k, f_s(s_{k-1}, b_k))$ $\boxed{=}(\mathbf{c}_k, f_0(s_{k-1}, b_k))$, where $\mathbf{c}_k$ is a binary sequence of length N_{out} and the nodes ϕ are shorthand for the mapping constraint $\boxed{=}(a_k, \phi(\mathbf{c}_k))$.

- The variables $\mathbf{c}_k$ are elements of $\mathbb{B}^{N_{out}}$, $N_{out} = \log_2 |\Omega|$.
- $N_s = N_b + L$ when $N_{in} = 1$.
- When some coded bit-sequences $\mathbf{c}_k$ are punctured, the messages $\mu_{\phi \to C_k}(\mathbf{c}_k)$ become uniform distributions (for instance in an unterminated code).

9.4.3   Demapping algorithms

We will focus exclusively on the mapper nodes from Fig. 9.4 (marked ϕ). The SPA on the convolutional code is very similar to what we discussed in Section 8.6. The only difference is that now we treat the bit-sequences $\mathbf{c}_k$ as variables over $\mathbb{B}^{N_{out}}$, rather than N_{out} binary variables. This results in messages $\mathbf{p}_{\phi \to C_k}$ and $\mathbf{p}_{C_k \to \phi}$, which are vectors of size $2^{N_{out}}$.

Given $\mathbf{p}_{A_k \to \phi}$ (a vector of length $2^{N_{out}} = |\Omega|$), we find that $\mathbf{p}_{\phi \to C_k}$ (another vector of length $2^{N_{out}} = |\Omega|$) is given by

$$\mathbf{p}_{\phi \to C_k}(\mathbf{c}_k) = \mathbf{p}_{A_k \to \phi}(\phi^{-1}(\mathbf{c}_k)). \tag{9.14}$$

Similarly,

$$\mathbf{p}_{\phi \to A_k}(a_k) = \mathbf{p}_{C_k \to \phi}(\phi(a_k)). \tag{9.15}$$

Transformation to the log domain is trivial.

9.4.4   Sequence detection

Although we have considered only the standard SPA for TCM, it is clear that we can also apply the max–sum/Viterbi algorithm when the factor graph of $\log p(\mathbf{B}, \mathbf{Y} = \mathbf{y}|\mathcal{M})$

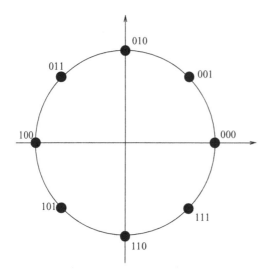

Figure 9.5. Set-partitioning mapping for 8-PSK.

is cycle-free. This leads to the most likely sequence

$$\hat{\mathbf{b}} = \arg \max_{\mathbf{b}} \log p(\mathbf{B} = \mathbf{b}|\mathbf{Y} = \mathbf{y}, \mathcal{M}). \qquad (9.16)$$

In view of our observations from Section 5.3.3.3, we can find the most likely sequence by implementing the SPA in the log domain and replacing $\mathbb{M}(\cdot)$ by the max operation.

9.5   Performance illustration

Let us evaluate the performance of BICM with a terminated systematic convolutional code with generators $g_{\mathrm{FF}}(D) = 1+D^2$ and $g_{\mathrm{FB}}(D) = 1+D+D^2$, and coded sequences of length $N_c = 240$. The modulation format is 8-PSK with set-partitioning mapping (see Fig. 9.5). The bit-interleaver is pseudo-random and changes from codeword to codeword. The channel is an AWGN channel so that $y_k = \sqrt{E_s}a_k + n_k$, where n_k are iid complex Gaussian, zero mean and $\mathbb{E}\{|N_k|^2\} = N_0$. We evaluate the performance in terms of bit error rate (BER) versus signal-to-noise ratio (SNR). The SNR is expressed as E_b/N_0 (in decibels[2]), with $E_b = E_s/(R \log_2 |\Omega|)$, where R denotes the code rate. The resulting performance is shown in Fig. 9.6. The BER drops by several orders of magnitude with increasing number of iterations between demapper and decoder.

9.6   Main points

We have described two important modulation schemes, leading to elegant factor-graph representations.

---

[2] We can express an SNR in decibels (dB) as $10 \log_{10} \mathrm{SNR}$.

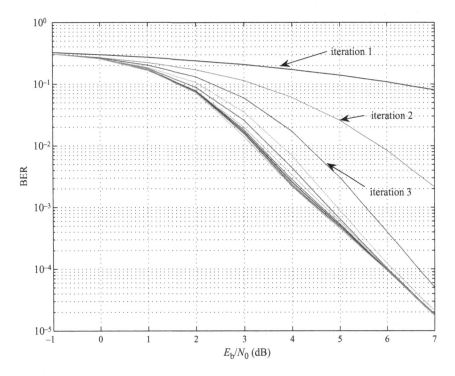

Figure 9.6. The BER versus SNR performance for BICM. Observe the reduction in BER with every iteration. After four iterations, the reduction becomes negligible.

- *Bit-interleaved coded modulation (BICM)*: the factor graph consists of an interleaver node and a set of disjoint nodes, each node relating a group of coded bits to a single coded symbol. In combination with an error-correcting code, the factor graph of $p(\mathbf{A}|\mathbf{C}, \mathcal{M})p(\mathbf{C}|\mathbf{B}, \mathcal{M})$ always contains cycles, resulting in an iterative demapping algorithm. This scheme can be combined with any decoding block from Chapter 8.
- *Trellis-coded modulation (TCM)*: trellis-coded modulation combines a convolutional code with mapping in such a way that the factor graph of $p(\mathbf{A}|\mathbf{C}, \mathcal{M})p(\mathbf{C}|\mathbf{B}, \mathcal{M})$ contains no cycles. When combined with a transmission scheme such that the factor graph of $p(\mathbf{B}, \mathbf{Y} = \mathbf{y}|\mathcal{M})$ is cycle-free, we can also implement the max–sum algorithm on $\log p(\mathbf{B}, \mathbf{Y} = \mathbf{y}|\mathcal{M})$ leading to optimal sequence detection.

The demapping node requires messages $\mu_{A_k \to \text{dem}}(A_k)$, $\forall k$. These are obtained by executing the SPA on the equalization node, representing the distribution $p(\mathbf{Y} = \mathbf{y}|\mathbf{A}, \mathcal{M})$. How these messages can be computed will be the topic of the next chapter.

10 Equalization–general formulation

10.1 Introduction

Apart from decoding and demapping, a factor-graph-based receiver also requires an equalizer. The equalizer performs the SPA on the equalization node $p(\mathbf{Y} = \mathbf{y}|\mathbf{A}, \mathcal{M})$, seen in Fig. 10.1, where $\mathbf{a}$ is the sequence of unknown coded data symbols and $\mathbf{y}$ is the known observation. The observation is obtained by suitable processing of the received waveform. This processing depends highly on the specific communication scenario (number of users, number of antennas, etc.) and will be dealt with in Chapters 11–13. Here we work in a more abstract setting. This chapter builds on the works [96–100].

The observation $\mathbf{y}$ is generally a non-deterministic function of the data symbols $\mathbf{a}$. As we have seen in Chapter 2, the signal at the receiver is corrupted by noise, has been passed through a physical channel, and may depend on symbols other than the N_s data symbols we are interested in. This gives rise to the following very general relation between the observation $\mathbf{y}$ and the coded symbols $\mathbf{a}$:

$$\mathbf{y} = \mathbf{h}(\mathbf{a}, \tilde{\mathbf{a}}) + \mathbf{n}, \qquad (10.1)$$

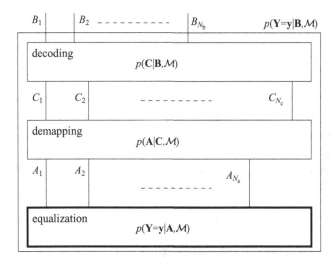

Figure 10.1. A factor graph of $p(\mathbf{Y} = \mathbf{y}|\mathbf{B}, \mathcal{M})$. The node is opened to reveal its structure. The node in bold is the topic of this chapter.

where **n** is a complex Gaussian noise vector, $\tilde{\mathbf{a}}$ represents symbols that affect the observation **y** but are not part of **a** and $\mathbf{h}(\cdot)$ is the transformation encapsulating the physical channels as well as any processing at the transmitter and the receiver. Since such a model will recur throughout the rest of this book, we will devote some effort to seeing how the equalization process is executed. As in the previous two chapters, we assume that we have messages $\mu_{A_k \to \text{eq}}(A_k)$ available, for all k, and are interested in computing $\mu_{\text{eq} \to A_k}(A_k)$. Here "eq" stands for equalizer.

This chapter is organized as follows.

• We will start by outlining the equalization problem in **Section 10.2**.
• In **Section 10.3** we will describe a collection of general-purpose equalization techniques, including a sliding-window equalizer, a Gaussian equalizer, and a Monte Carlo equalizer.
• In **Section 10.4** we give a brief discussion related to how equalization interacts with the demapping and decoding.
• A performance illustration will be given in **Section 10.5**.

10.2   Problem description

We consider the following problem: we have a complex-valued sequence $\mathbf{a} = [a_1, \ldots, a_{N_s}]^{\mathsf{T}}$, with each element belonging to a finite set $\Omega \subset \mathbb{C}$. We observe a complex vector $\mathbf{y} = [y_1, \ldots, y_{N_o}]^{\mathsf{T}}$, which is related to **a** as

$$\mathbf{y} = \mathbf{h}(\mathbf{a}, \tilde{\mathbf{a}}) + \mathbf{n}, \tag{10.2}$$

where $\mathbf{h} : \Omega^{N_s} \times \Omega^{N_d} \to \mathbb{C}^{N_o}$ is a *known* function, $\tilde{\mathbf{a}}$ are N_d other symbols that affect **y**, and **n** is an $N_o \times 1$ vector, independent of **a**, with $\mathbf{n} \sim p_{\mathbf{N}}(\mathbf{n})$. The constellation Ω has the properties $\sum_{a \in \Omega} a = 0$ and $\sum_{a \in \Omega} |a|^2 = |\Omega|$. The model (10.2) immediately tells us that the node $p(\mathbf{Y} = \mathbf{y} | \mathbf{A}, \mathcal{M})$ can be opened and replaced by a factorization of

$$p(\mathbf{Y} = \mathbf{y}, \tilde{\mathbf{A}} = \tilde{\mathbf{a}} | \mathbf{A} = \mathbf{a}, \mathcal{M}) = p_{\mathbf{N}}(\mathbf{y} - \mathbf{h}(\mathbf{a}, \tilde{\mathbf{a}})) p(\tilde{\mathbf{A}} = \tilde{\mathbf{a}}). \tag{10.3}$$

The symbols in $\tilde{\mathbf{a}}$ are either unknown or known. We can replace $p(\tilde{\mathbf{A}} = \tilde{\mathbf{a}})$ by $\prod_{k=1}^{N_d} p(\tilde{A}_k = \tilde{a}_k)$, where $p(\tilde{A}_k = \tilde{a}_k)$ is either a uniform distribution (for unknown symbols) or a discrete Dirac distribution (for known symbols). For notational convenience, we will combine **a** and $\tilde{\mathbf{a}}$ into a single vector denoted by **a** and introduce $\mu_{A_k \to \text{eq}}(A_k)$ as either uniform or discrete Dirac for those a_k not belonging to the coded data symbols of interest. This allows us to write

$$\mathbf{y} = \mathbf{h}(\mathbf{a}) + \mathbf{n}, \tag{10.4}$$

where **a** should be understood as a vector of length $N_s + N_d$, including the coded symbols of interest as well as additional symbols. We will generally make no distinction between **a** and $[\mathbf{a}, \tilde{\mathbf{a}}]$, or between N_s and $N_s + N_d$. An example should make things clearer.

Example 10.1. *Consider the situation in which* **a** *is part of a much longer data stream, and that, for* $n \in \mathbb{Z}$,

$$y_n = \sum_{l=0}^{L-1} h_l a_{n-l} + n_n,$$

where the noise samples n_n *are iid zero-mean complex Gaussian with* $\mathbb{E}\{|N_n|^2\} = 2\sigma^2$. *Suppose that* $\mathbf{y} = [y_1, \dots, y_{N_s}]^{\mathrm{T}}$. *In that case* $\mathbf{a} = [a_1, \dots, a_{N_s}]^{\mathrm{T}}$ *and* $\tilde{\mathbf{a}} = [a_{-L+1}, \dots, a_0]^{\mathrm{T}}$ *and we can write*

$$\mathbf{y} = \mathbf{H} \begin{bmatrix} \tilde{\mathbf{a}} \\ \mathbf{a} \end{bmatrix} + \mathbf{n},$$

where **H** *is a suitable Toeplitz matrix. In some cases, the symbols in* $\tilde{\mathbf{a}}$ *are known, for instance when they belong to a block that the receiver has already correctly decoded. In other cases, the symbols are unknown. As we mentioned, we shall write*

$$\mathbf{y} = \mathbf{H}\mathbf{a} + \mathbf{n},$$

where **a** *should now be understood as being the vector of data symbols of interest as well as any known or unknown symbols.*

10.3 Equalization methods

10.3.1 Overview

In the following sections we will describe a set of techniques that solve the equalization problem in a very general setting. We distinguish between *exact* techniques and *approximate* techniques, depending on whether or not the messages $\mu_{\mathrm{eq}\to A_k}(A_k)$ to the demapper are computed exactly. Often different methods can be combined.

Exact techniques
- **SPA equalizer:** straightforward application of the SPA will give rise to the messages $\mu_{\mathrm{eq}\to A_k}(A_k)$ to the demapper. The computational complexity is exponential in the length of **a**.
- **Structured equalizer:** opening the node $p(\mathbf{Y} = \mathbf{y}|\mathbf{A}, \mathcal{M})$ and exploiting the underlying structure results in the messages $\mu_{\mathrm{eq}\to A_k}(A_k)$ at a much smaller computational cost.
- **State-space-models equalizer:** in some cases, exploiting structure naturally leads to a state-space model (SSM). Computation of the messages $\mu_{\mathrm{eq}\to A_k}(A_k)$ can then be performed in a time linear in the length of **a**, using the forward–backward algorithm.

Approximate techniques

- **Sliding-window equalizer:** rather than considering the entire observation $\mathbf{y}$, we look at a suitable window $\mathbf{y}_k$ and compute $\mu_{\mathrm{eq}\rightarrow A_k}(A_k)$ on the basis of $p(\mathbf{Y}_k = \mathbf{y}_k | \mathbf{A}_k = \mathbf{a}_k, \mathcal{M})$, where $\mathbf{a}_k$ is a window around a_k. A sliding-window equalizer artificially creates structure in the factor graph, and thus leads to a lower complexity.
- **Monte Carlo equalizer:** Monte Carlo sampling methods approximate $\mu_{\mathrm{eq}\rightarrow A_k}(A_k)$ through sampling. The complexity scales linearly with the number of samples. Monte Carlo equalization can be combined with a structured equalizer or sliding-window equalizer.
- **Gaussian/MMSE equalizer:** when we temporarily work under the assumption that a_k is Gaussian, computing the message $\mu_{\mathrm{eq}\rightarrow A_k}(A_k)$ is straightforward as long as $\mathbf{y}$ is a linear function of $\mathbf{a}$. The complexity of the resulting algorithm heavily depends on the structure of $\mathbf{H}$. Gaussian equalizers also go by the name of *MMSE equalizers*. MMSE equalization can be combined with a structured equalizer or a sliding-window equalizer.

We will derive only algorithms in the probability domain. The messages obtained can be transformed to the log domain by taking the natural logarithm.

10.3.2   Sum–product-algorithm equalizers

Our goal is to compute the messages $\mu_{\mathrm{eq}\rightarrow A_k}(A_k)$, on the basis of incoming messages from the demodulator $\mu_{A_l\rightarrow \mathrm{eq}}(A_l)$, $l \neq k$. Using the SPA, we find that the message is given by

$$\mu_{\mathrm{eq}\rightarrow A_k}(a_k) \propto \sum_{\sim\{a_k\}} p(\mathbf{Y} = \mathbf{y} | \mathbf{A} = \mathbf{a}, \mathcal{M}) \prod_{l\neq k} \mu_{A_l\rightarrow \mathrm{eq}}(a_l). \tag{10.5}$$

Complexity

For any $a_k \in \Omega$, the complexity of (10.5) scales as $\mathcal{O}(|\Omega|^{N_\mathrm{s}-1})$, so the total message $\mu_{\mathrm{eq}\rightarrow A_k}(A_k)$ requires a complexity of $\mathcal{O}(|\Omega|^{N_\mathrm{s}})$. This is the complexity per symbol. Since there are N_s symbols in $\mathbf{a}$, this leads to an overall complexity of $\mathcal{O}(N_\mathrm{s}|\Omega|^{N_\mathrm{s}})$.

Interpretation of messages

As we have seen in Section 5.3.2, since the incoming messages $\mu_{A_l\rightarrow \mathrm{eq}}(A_l)$ are normalized, we can *interpret* them as (artificial) a priori distributions of the coded symbols. We can then introduce an *artificial* joint a-posteriori distribution $p(\mathbf{A}|\mathbf{Y} = \mathbf{y}, \mathcal{M})$, where the symbols A_l are independent with a-priori probabilities $p(A_l = a_l) = \mu_{A_l\rightarrow \mathrm{eq}}(a_l)$. The marginal with respect to A_k of this artificial distribution is given by (by definition)

$$p(A_k = a | \mathbf{Y} = \mathbf{y}, \mathcal{M}) = \sum_{\mathbf{a}:a_k=a} p(\mathbf{A} = \mathbf{a} | \mathbf{Y} = \mathbf{y}, \mathcal{M}) \tag{10.6}$$

$$\propto \sum_{\mathbf{a}:a_k=a} p(\mathbf{Y} = \mathbf{y} | \mathbf{A} = \mathbf{a}, \mathcal{M}) p(\mathbf{A} = \mathbf{a}) \tag{10.7}$$

$$= \sum_{\mathbf{a}:a_k=a} p(\mathbf{Y} = \mathbf{y} | \mathbf{A} = \mathbf{a}, \mathcal{M}) \prod_l p(A_l = a_l) \tag{10.8}$$

$$\propto \mu_{A_k \to eq}(a) \mu_{eq \to A_k}(a), \tag{10.9}$$

where we have used (10.5) in the last transition. Note that, when the factor graph is cycle-free, relation (10.9) is the actual a-posteriori distribution of A_k. When the factor graph has cycles, relation (10.9) is the current[1] approximation of the actual a-posteriori distribution of A_k (the belief). In many technical papers, the message $\mu_{eq \to A_k}(A_k)$ is known as the *extrinsic probability* of A_k, so that (10.9) can be interpreted as

a-posteriori probability $\propto$ a-priori probability $\times$ extrinsic probability.

This relationship will be useful in deriving equalizers. In particular, for the MC and Gaussian equalizers, we can determine the message $\mu_{eq \to A_k}(A_k)$ by first computing the (artificial) marginal a-posteriori distribution $p(A_k | \mathbf{Y} = \mathbf{y}, \mathcal{M})$, interpreting the various coded symbols as independent with a-priori distributions given by $\mu_{A_l \to eq}(A_l)$, $\forall l$. Once $p(A_k | \mathbf{Y} = \mathbf{y}, \mathcal{M})$ has been found, we divide by $\mu_{A_k \to eq}(A_k)$, normalize, and obtain the message $\mu_{eq \to A_k}(A_k)$.

10.3.3 Structured equalizers

In many cases the node $p(\mathbf{Y} = \mathbf{y} | \mathbf{A}, \mathcal{M})$ can be opened to reveal the underlying structure. A common situation occurs when $p(\mathbf{Y} = \mathbf{y} | \mathbf{A}, \mathcal{M})$ can be factored as a function of the symbols A_k or small blocks of symbols. In other words, there exists a factorization

$$p(\mathbf{Y} = \mathbf{y} | \mathbf{A}, \mathcal{M}) \propto \prod_{n=1}^{N_p} f_n(\mathbf{A}_{\mathcal{S}_n}, \mathbf{y}), \tag{10.10}$$

where $\mathbf{A}_{\mathcal{S}_n}$ denotes the elements of $\mathbf{A}$ whose indices fall in some set $\mathcal{S}_n$. The sets $\mathcal{S}_n$ are disjoint. Computing an outgoing message $\mu_{eq \to A_k}(A_k)$, where $k \in \mathcal{S}_m$, for some unique

---

[1] As the iterations proceed, this approximation changes (because the messages $\mu_{A_l \to eq}(A_l)$ change), and, we hope, improves.

m then gives

$$\mu_{\text{eq}\to A_k}(a_k) \propto \sum_{\sim\{a_k\}} p(\mathbf{Y} = \mathbf{y}|\mathbf{A} = \mathbf{a}, \mathcal{M}) \prod_{l\neq k} \mu_{A_l\to\text{eq}}(a_l) \qquad (10.11)$$

$$\propto \sum_{\sim\{a_k\}} f_m(\mathbf{A}_{\mathcal{S}_m} = \mathbf{a}_{\mathcal{S}_m}, \mathbf{y}) \prod_{l\in\mathcal{S}_m, l\neq k} \mu_{A_l\to\text{eq}}(a_l), \qquad (10.12)$$

which reduces the complexity from $\mathcal{O}(|\Omega|^{N_s})$ to $\mathcal{O}(|\Omega|^{|\mathcal{S}_m|})$ per data symbol. In other cases, a more complicated factorization of $p(\mathbf{Y} = \mathbf{y}|\mathbf{A}, \mathcal{M})$ may be required. An example will be covered in the next section on SSM equalizers.

Example 10.2. *Consider the observation*

$$\mathbf{y} = \mathbf{Ha} + \mathbf{n},$$

where $\mathbf{y}$ *is an* $N_s \times 1$ *vector,* $\mathbf{n} \sim \mathcal{N}_{\mathbf{n}}^{\mathbb{C}}(\mathbf{0}, 2\sigma^2\mathbf{I}_{N_s})$, *and*

$$\mathbf{H} = \begin{bmatrix} h_1 & 0 & \cdots & 0 \\ 0 & h_2 & \cdots & 0 \\ \vdots & \vdots & \ddots & \vdots \\ 0 & 0 & \cdots & h_{N_s} \end{bmatrix}.$$

We can then factorize $p(\mathbf{Y} = \mathbf{y}|\mathbf{A}, \mathcal{M})$ *as follows:*

$$p(\mathbf{Y} = \mathbf{y}|\mathbf{A} = \mathbf{a}, \mathcal{M}) \propto \exp\left(-\frac{1}{2\sigma^2}\|\mathbf{y} - \mathbf{Ha}\|^2\right)$$

$$= \prod_{n=1}^{N_s} \exp\left(-\frac{1}{2\sigma^2}|y_n - h_n a_n|^2\right).$$

The message $\mu_{\text{eq}\to A_k}(A_k)$ *is then given by*

$$\mu_{\text{eq}\to A_k}(a_k) \propto \sum_{\sim\{a_k\}} p(\mathbf{Y} = \mathbf{y}|\mathbf{A} = \mathbf{a}, \mathcal{M}) \prod_{l=1, l\neq k}^{N_s} \mu_{A_l\to\text{eq}}(a_l)$$

$$\propto \exp\left(-\frac{1}{2\sigma^2}|y_k - h_k a_k|^2\right).$$

Complexity

The overall complexity depends highly on the structure of the function $\mathbf{h}(\cdot)$ and can range from $\mathcal{O}(N_s|\Omega|^{N_s})$ to $\mathcal{O}(N_s)$.

10.3.4 State-space-model equalizers

An important special case of structure is that of a state-space model (SSM). The observation $\mathbf{y}$ and the data symbols $\mathbf{a}$ are related by[2]

$$y_k = h_k(a_k, a_{k-1}, \ldots, a_{k-L+1}) + n_k, \tag{10.13}$$

where the noise samples n_k are iid complex Gaussian with $\mathbb{E}\{|N_k|^2\} = 2\sigma^2$. We refer to L as the *channel length*.

Assume that we have observations $\mathbf{y} = [y_1, y_2, \ldots, y_{N_s+L-1}]^{\mathrm{T}}$, where the data symbols a_k for $k < 1$ and for $k > N_s$ are unknown. We can transform (10.13) into a transition-emitting state-space model with state at time $k-1$ given by $\mathbf{s}_{k-1} = [a_{k-1}, \ldots, a_{k-L+1}] \in \Omega^{L-1}$ and the following update equations:

$$\mathbf{s}_k = f_s(\mathbf{s}_{k-1}, a_k) \tag{10.14}$$

$$= [a_k \quad [\mathbf{s}_{k-1}]_{1:L-2}], \tag{10.15}$$

where $[\mathbf{s}_{k-1}]_{1:L-2}$ denotes the vector of length $L-2$ containing the first until the second-last entry in $\mathbf{s}_{k-1}$ and

$$y_k = h_k(a_k, [\mathbf{s}_{k-1}]_1, [\mathbf{s}_{k-1}]_2, \ldots, [\mathbf{s}_{k-1}]_{L-2}) + n_k, \tag{10.16}$$

where $[\mathbf{s}_{k-1}]_i$ is the ith entry in $\mathbf{s}_{k-1} = [[\mathbf{s}_{k-1}]_1, \ldots, [\mathbf{s}_{k-1}]_{L-1}]$.

The SSM allows us to replace the node $p(\mathbf{Y} = \mathbf{y}|\mathbf{A}, \mathcal{M})$ by a factorization of $p(\mathbf{Y} = \mathbf{y}, \mathbf{S}|\mathbf{A}, \mathcal{M})$:

$$p(\mathbf{Y} = \mathbf{y}, \mathbf{S} = \mathbf{s}|\mathbf{A} = \mathbf{a}, \mathcal{M})$$

$$= p(\mathbf{Y} = \mathbf{y}|\mathbf{S} = \mathbf{s}, \mathbf{A} = \mathbf{a}, \mathcal{M})p(\mathbf{S} = \mathbf{s}|\mathbf{A} = \mathbf{a}, \mathcal{M}) \tag{10.17}$$

$$\propto \prod_{k=1}^{N_s+L-1} \boxminus(\mathbf{s}_k, f_s(\mathbf{s}_{k-1}, a_k))p(Y_k = y_k|\mathbf{S}_{k-1} = \mathbf{s}_{k-1}, A_k = a_k, \mathcal{M}). \tag{10.18}$$

The corresponding factor graph is shown in Fig. 10.2. Given messages $\mu_{A_l \to \text{eq}}(A_l)$, we can execute the forward–backward algorithm from Section 6.3 on this graph and obtain $\mu_{\text{eq} \to A_k}(A_k)$. Observe that the messages $\mu_{\mathbf{S}_0 \to f}(\mathbf{s}_0)$ and $\mu_{\mathbf{S}_{N_s+L} \to f}(\mathbf{s}_{N_s+L})$ as well as $\mu_{A_k \to \text{eq}}(A_k)$ for $k > N_s$ are all uniform distributions over their corresponding domains.

Complexity
The overall complexity of the SPA on this factor graph scales as $\mathcal{O}(N_s|\Omega|^L)$.

---

[2] In particular, we have $y_k = \sum_{l=0}^{L-1} h_l a_{k-l} + n_k$.

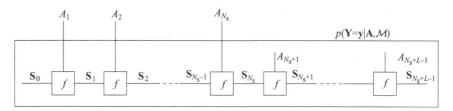

Figure 10.2. A factor graph of $p(\mathbf{Y} = \mathbf{y}|\mathbf{A}, \mathcal{M})$. The node is opened to reveal its structure, where $f(s_{k-1}, s_k, a_k) = \boxminus(s_k, f_s(s_{k-1}, a_k))p(Y_k = y_k|S_{k-1} = s_{k-1}, A_k = a_k, \mathcal{M})$.

10.3.5   Sliding-window equalizers

In many equalization problems, the structured approach may still be too complex for a practical implementation. In such cases, applying a sliding window is a useful technique. We focus again on the model

$$y_k = h_k(a_k, a_{k-1}, \ldots, a_{k-L+1}) + n_k, \tag{10.19}$$

where we now allow the total noise vector to be $\mathbf{n} \sim \mathcal{N}_{\mathbf{n}}^{\mathbb{C}}(\mathbf{0}, \mathbf{\Sigma})$ (i.e., not necessarily iid). Suppose that we are interested in computing the message $\mu_{\text{eq}\to A_k}(A_k)$ to the demapper. This message is given by

$$\mu_{\text{eq}\to A_k}(a_k) \propto \sum_{\sim\{a_k\}} p(\mathbf{Y} = \mathbf{y}|\mathbf{A} = \mathbf{a}, \mathcal{M}) \prod_{l\neq k} \mu_{A_l\to\text{eq}}(a_l). \tag{10.20}$$

We now select a suitable segment or window $\mathbf{y}_k$ in $\mathbf{y}$ of length W and approximate (10.20) as

$$\mu_{\text{eq}\to A_k}(a_k) \propto \sum_{\sim\{a_k\}} p(\mathbf{Y}_k = \mathbf{y}_k|\mathbf{A} = \mathbf{a}, \mathcal{M}) \prod_{l\neq k} \mu_{A_l\to\text{eq}}(a_l). \tag{10.21}$$

The window is chosen such that it contains most of the observations which depend on a_k. Because of the model (10.19), $p(\mathbf{Y}_k = \mathbf{y}_k|\mathbf{A} = \mathbf{a}, \mathcal{M}) = p(\mathbf{Y}_k = \mathbf{y}_k|\mathbf{A}_k = \mathbf{a}_k, \mathcal{M})$ where $\mathbf{a}_k$ is a window in $\mathbf{a}$ of length $W + L - 1$, allowing us to write

$$\mu_{\text{eq}\to A_k}(a_k) \propto \sum_{\mathbf{a}_k : a_k} p(\mathbf{Y}_k = \mathbf{y}_k|\mathbf{A}_k = \mathbf{a}_k, \mathcal{M}) \prod_{l\neq k} \mu_{A_l\to\text{eq}}(a_l). \tag{10.22}$$

It is up to the designer of the receiver to select suitable windows that result in (i) good performance and (ii) a significant reduction in computational complexity.

Complexity

The overall complexity of the SPA on this factor graph scales as $\mathcal{O}(N_s|\Omega|^{W+L-1})$. This complexity can be further reduced by combining the sliding-window equalizer with an MC or a Gaussian/MMSE equalizer.

Example 10.3 (Sliding window equalizer for linear models). *We will work with the observation model*

$$y_k = \alpha a_k + \beta a_{k-1} + n_k,$$

where $\alpha = 0.01$ *and* $\beta = 1$ *and the noise samples* n_k *are iid zero-mean complex Gaussian with* $\mathbb{E}\{|N_k|^2\} = 2\sigma^2$. *This corresponds to a channel of length* $L = 2$. *We observe* $\mathbf{y} = [y_1 \ \ y_2 \ \ y_3 \ \ y_4 \ \ y_5]^{\mathrm{T}}$, *which can be written as*

$$\mathbf{y} = \mathbf{H}\mathbf{a} + \mathbf{n}$$

with

$$\begin{bmatrix} y_1 \\ y_2 \\ y_3 \\ y_4 \\ y_5 \end{bmatrix} = \begin{bmatrix} \beta & \alpha & 0 & 0 & 0 & 0 \\ 0 & \beta & \alpha & 0 & 0 & 0 \\ 0 & 0 & \beta & \alpha & 0 & 0 \\ 0 & 0 & 0 & \beta & \alpha & 0 \\ 0 & 0 & 0 & 0 & \beta & \alpha \end{bmatrix} \begin{bmatrix} a_0 \\ a_1 \\ a_2 \\ a_3 \\ a_4 \\ a_5 \end{bmatrix} + \begin{bmatrix} n_1 \\ n_2 \\ n_3 \\ n_4 \\ n_5 \end{bmatrix}.$$

Note that $\mathbf{H}$ *is a Toeplitz matrix. Suppose that our* $N_{\mathrm{s}} = 4$ *coded symbols of interest are* $[a_1 \ \ a_2 \ \ a_3 \ \ a_4]^{\mathrm{T}}$ *and we wish to compute* $\mu_{\mathrm{eq} \to A_k}(A_k)$, *for* $k = 1, \ldots, 4$. *The symbols* a_0 *and* a_5 *unknown, so we set* $\mu_{A_k \to \mathrm{eq}}(A_k)$ *to uniform distributions for* $k = 0$ *and* $k = 5$. *Note that executing the SPA leads to an overall complexity of* $\mathcal{O}(N_{\mathrm{s}}|\Omega|^{N_{\mathrm{s}}+L})$, *whereas transformation into an SSM will result in a complexity* $O(N_{\mathrm{s}}|\Omega|^L)$.
To approximate $\mu_{\mathrm{eq} \to A_k}(A_k)$, $k = 1, \ldots, 4$, *we can select different windows, for instance the following.*
(1) $\mathbf{y}_k = y_k$, *leading to*

$$\mathbf{y}_k = \begin{bmatrix} \beta & \alpha \end{bmatrix} \begin{bmatrix} a_{k-1} \\ a_k \end{bmatrix} + n_k,$$

so that

$$\mathbf{a}_k = \begin{bmatrix} a_{k-1} \\ a_k \end{bmatrix}.$$

The SPA rule becomes

$$\mu_{\mathrm{eq} \to A_k}(a_k) \propto \sum_{a_{k-1} \in \Omega} p(\mathbf{Y}_k = \mathbf{y}_k | \mathbf{A}_k = \mathbf{a}_k, \mathcal{M}) \mu_{A_{k-1} \to \mathrm{eq}}(a_{k-1}).$$

(2) $\mathbf{y}_k = y_{k+1}$, *leading to*

$$\mathbf{y}_k = \begin{bmatrix} \beta & \alpha \end{bmatrix} \begin{bmatrix} a_k \\ a_{k+1} \end{bmatrix} + n_{k+1},$$

so that

$$\mathbf{a}_k = \left[\begin{array}{c} a_k \\ a_{k+1} \end{array} \right].$$

The SPA rule becomes

$$\mu_{\mathrm{eq} \to A_k}(a_k) \propto \sum_{a_{k+1} \in \Omega} p(\mathbf{Y}_k = \mathbf{y}_k | \mathbf{A}_k = \mathbf{a}_k, \mathcal{M}) \mu_{A_{k+1} \to \mathrm{eq}}(a_{k+1}).$$

(3)

$$\mathbf{y}_k = \left[\begin{array}{c} y_k \\ y_{k+1} \end{array} \right],$$

leading to

$$\mathbf{y}_k = \left[\begin{array}{ccc} \beta & \alpha & 0 \\ 0 & \beta & \alpha \end{array} \right] \left[\begin{array}{c} a_{k-1} \\ a_k \\ a_{k+1} \end{array} \right] + \left[\begin{array}{c} n_k \\ n_{k+1} \end{array} \right],$$

so that

$$\mathbf{a}_k = \left[\begin{array}{c} a_{k-1} \\ a_k \\ a_{k+1} \end{array} \right].$$

The SPA rule becomes

$$\mu_{\mathrm{eq} \to A_k}(a_k) \propto \sum_{(a_{k-1}, a_{k+1}) \in \Omega^2} p(\mathbf{Y}_k = \mathbf{y}_k | \mathbf{A}_k = \mathbf{a}_k, \mathcal{M})$$

$$\times \mu_{A_{k+1} \to \mathrm{eq}}(a_{k+1}) \mu_{A_{k-1} \to \mathrm{eq}}(a_{k-1}).$$

The overall computational complexity of SPA using the first two windows is the same, namely $\mathcal{O}(N_{\mathrm{s}}|\Omega|^L)$ (or $\mathcal{O}(|\Omega|^L)$ per data symbol). However, since $|\alpha| \ll |\beta|$, it is reasonable to expect that $\mathbf{y}_k = y_{k+1}$ is a more suitable window for approximating $\mu_{\mathrm{eq} \to A_k}(A_k)$. The third window results in an overall complexity $\mathcal{O}(N_{\mathrm{s}}|\Omega|^{L+1})$.

10.3.6   Monte Carlo equalizers

While the structured, SSM, and sliding-window equalizers work well in most circumstances, their computational complexity may still be prohibitively large. We call on Monte Carlo (MC) techniques to reduce the computation requirements. Let us introduce $\mathbf{A}_{\bar{k}} = \left[A_1, \ldots, A_{k-1}, A_{k+1}, \ldots, A_{N_s}\right]^{\mathrm{T}}$. Then the message from equalizer to demapper can be written as

$$\mu_{\mathrm{eq}\to A_k}(a_k) \propto \sum_{\sim\{a_k\}} p(\mathbf{Y} = \mathbf{y}|\mathbf{A} = \mathbf{a}, \mathcal{M}) \prod_{l=1, l\neq k}^{N_s} p(A_l = a_l) \qquad (10.23)$$

$$= \sum_{\mathbf{a}_{\bar{k}} \in \Omega^{N_s-1}} p(\mathbf{Y} = \mathbf{y}|\mathbf{A}_{\bar{k}} = \mathbf{a}_{\bar{k}}, A_k = a_k, \mathcal{M}) \prod_{l=1, l\neq k}^{N_s} p(A_l = a_l). \qquad (10.24)$$

We know from Section 10.3.2 that we can interpret $\mu_{\mathrm{eq}\to A_k}(a_k)$ as follows:

$$\mu_{\mathrm{eq}\to A_k}(a_k) \propto \frac{p(A_k = a_k|\mathbf{Y} = \mathbf{y}, \mathcal{M})}{\mu_{A_k\to\mathrm{eq}}(a_k)}, \qquad (10.25)$$

where $p(A_k|\mathbf{Y} = \mathbf{y}, \mathcal{M})$ is the marginal of an artificial joint a-posteriori distribution $p(\mathbf{A}|\mathbf{Y} = \mathbf{y}, \mathcal{M})$. We will describe three common MC equalization techniques, which approximate the distribution $p(A_k|\mathbf{Y} = \mathbf{y}, \mathcal{M})$ by means of sampling methods. Each of these techniques requires sampling from a specific target distribution. In Section 10.3.6.2, we will then show how this sampling can be performed. The methods require sampling from a distribution $q_{\mathbf{A}_{\bar{k}}}(\cdot)$ or $q_{\mathbf{A}}(\cdot)$ (we remind the reader that, when we have samples from a distribution $q_{\mathbf{A}}(\cdot)$, samples from $q_{\mathbf{A}_{\bar{k}}}(\cdot)$ can be obtained by dropping the kth component of every sample). According to (10.25), we can divide the approximate a-posteriori distribution $p(A_k|\mathbf{Y} = \mathbf{y}, \mathcal{M})$ by $\mu_{A_k\to\mathrm{eq}}(A_k)$ to obtain the message $\mu_{\mathrm{eq}\to A_k}(A_k)$.

Monte Carlo equalizers can be combined with sliding-window equalizers by replacing $p(\mathbf{Y} = \mathbf{y}|\mathbf{A} = \mathbf{a}, \mathcal{M})$ by $p(\mathbf{Y}_k = \mathbf{y}_k|\mathbf{A}_k = \mathbf{a}_k, \mathcal{M})$ in (10.23), where $\mathbf{y}_k$ is a windowed observation and $\mathbf{a}_k$ the corresponding window of data symbols. Monte Carlo equalizers can also be combined with structured equalizers by applying the MC techniques on a single factor.

10.3.6.1 Three equalization methods
Distinct samples from $p(\mathbf{A}_{\bar{k}}|\mathbf{Y} = \mathbf{y}, \mathcal{M})$

The idea behind this method is that only a few terms in (10.23) actually contribute to the summation. Most terms will be very close to zero. When we limit the summation to the most relevant terms, the result should be very close to the correct value.

Given the list $\mathcal{L}_{\bar{k}}$ of samples from $p(\mathbf{A}_{\bar{k}}|\mathbf{Y} = \mathbf{y}, \mathcal{M})$, we first remove all duplicates. We end up with a new list $\mathcal{L}_{\bar{k}}^{\mathrm{d}}$ of $L_{\mathrm{d}} \leq L$ distinct samples. We now perform the following approximation:

$$\mu_{\text{eq}\to A_k}(a_k) \propto \sum_{\mathbf{a}_{\bar{k}}\in\Omega^{N_s-1}} p(\mathbf{Y}=\mathbf{y}|A_{\bar{k}}=\mathbf{a}_{\bar{k}}, A_k=a_k, \mathcal{M}) \prod_{l=1, l\neq k}^{N_s} p(A_l=a_l) \quad (10.26)$$

$$\approx \sum_{\mathbf{a}_{\bar{k}}\in\mathcal{L}_k^d} p(\mathbf{Y}=\mathbf{y}|A_{\bar{k}}=\mathbf{a}_{\bar{k}}, A_k=a_k, \mathcal{M}) \prod_{l=1, l\neq k}^{N_s} p(A_l=a_l). \quad (10.27)$$

When $L_{\text{d}} \ll |\Omega|^{N_s-1}$, this results in a significant decrease in computational complexity.

Unweighted sampling

A second approach is first to determine an approximation of $p(A_k = a_k|\mathbf{Y}=\mathbf{y}, \mathcal{M})$, and then divide by $p(A_k = a_k)$ to obtain $\mu_{\text{eq}\to A_k}(a_k)$, as described in (10.25). Now, suppose that we have a list $\mathcal{L}$ of samples from the joint distribution $p(\mathbf{A}|\mathbf{Y}=\mathbf{y}, \mathcal{M})$. These samples form a particle representation $\mathcal{R}_L(p(\mathbf{A}|\mathbf{Y}=\mathbf{y}, \mathcal{M})) = \{1/L, \mathbf{a}^{(n)}\}_{n=1}^L$. It then follows that $\{1/L, a_k^{(n)}\}_{n=1}^L$ is a particle representation of the marginal $p(A_k|\mathbf{Y}=\mathbf{y}, \mathcal{M})$. In other words, we can approximate

$$p(A_k = a_k|\mathbf{Y}=\mathbf{y}, \mathcal{M}) \approx \sum_{n=1}^L \frac{1}{L} \boxminus\left(a_k, a_k^{(n)}\right). \quad (10.28)$$

Dividing by $p(A_k = a_k)$, followed by normalization, leads to $\mu_{\text{eq}\to A_k}(a_k)$.

Importance sampling

This technique is similar to the previous one, in that we again try to approximate $p(A_k|\mathbf{Y}=\mathbf{y}, \mathcal{M})$. Suppose that we have a list of L samples from a distribution $q_{A_{\bar{k}}}(\mathbf{a}_{\bar{k}})$. Then

$$p(A_k = a_k|\mathbf{Y}=\mathbf{y}, \mathcal{M}) = \sum_{\mathbf{a}_{\bar{k}}\in\Omega^{N_s-1}} p\left(A_k = a_k, A_{\bar{k}} = \mathbf{a}_{\bar{k}}|\mathbf{Y}=\mathbf{y}, \mathcal{M}\right)$$

$$= \sum_{\mathbf{a}_{\bar{k}}\in\Omega^{N_s-1}} p\left(A_k = a_k|A_{\bar{k}} = \mathbf{a}_{\bar{k}}, \mathbf{Y}=\mathbf{y}, \mathcal{M}\right) p(A_{\bar{k}} = \mathbf{a}_{\bar{k}}|\mathbf{Y}=\mathbf{y}, \mathcal{M}).$$

Given L samples from $q_{A_{\bar{k}}}(\mathbf{a}_{\bar{k}})$, we have the following sample representation: $\mathcal{R}_L(p(A_{\bar{k}}|\mathbf{Y}=\mathbf{y}, \mathcal{M})) = \{(w^{(n)}, \mathbf{a}_{\bar{k}}^{(n)})\}_{n=1}^L$, where we set

$$w^{(n)} \propto \frac{p\left(A_{\bar{k}} = \mathbf{a}_{\bar{k}}^{(n)}|\mathbf{Y}=\mathbf{y}, \mathcal{M}\right)}{q_{A_{\bar{k}}}\left(\mathbf{a}_{\bar{k}}^{(n)}\right)}. \quad (10.29)$$

This enables us to approximate $p(A_k = a_k|\mathbf{Y} = \mathbf{y}, \mathcal{M})$ by

$$p(A_k = a_k|\mathbf{Y} = \mathbf{y}, \mathcal{M}) \approx \sum_{n=1}^{L} w^{(n)} p\left(A_k = a_k \middle| \mathbf{A}_{\bar{k}} = \mathbf{a}_{\bar{k}}^{(n)}, \mathbf{Y} = \mathbf{y}, \mathcal{M}\right). \qquad (10.30)$$

As we will see shortly, the factor $p(A_k = a_k|\mathbf{A}_{\bar{k}} = \mathbf{a}_{\bar{k}}^{(n)}, \mathbf{Y} = \mathbf{y}, \mathcal{M})$ is very easy to compute. The weights are normalized such that $\sum_{n=1}^{L} w^{(n)} = 1$. In the special case $q_{\mathbf{A}_{\bar{k}}}(\mathbf{a}_{\bar{k}}) = p(\mathbf{A}_{\bar{k}} = \mathbf{a}_{\bar{k}}|\mathbf{Y} = \mathbf{y}, \mathcal{M})$, $w_n = 1/L$. Dividing by $p(A_k = a_k)$, followed by normalization, leads to $\mu_{\text{eq} \to A_k}(a_k)$.

10.3.6.2 Sampling

The above methods require sampling from a distribution $q_{\mathbf{A}_{\bar{k}}}(\mathbf{a}_{\bar{k}})$, with as a special case $p(\mathbf{A}_{\bar{k}}|\mathbf{Y} = \mathbf{y}, \mathcal{M})$. We will focus only on the special case, since this is most interesting. It is clear that when we can draw samples from $p(\mathbf{A}|\mathbf{Y} = \mathbf{y}, \mathcal{M})$, samples from $p(\mathbf{A}_{\bar{k}}|\mathbf{Y} = \mathbf{y}, \mathcal{M})$ are obtained by simply dropping the kth component. So, how do we draw samples from $p(\mathbf{A}|\mathbf{Y} = \mathbf{y}, \mathcal{M})$? We resort to a Gibbs sampler as described in Algorithm 10.1. Note that sampling from conditional distributions $p(A_l|\mathbf{A}_{\bar{l}} = \mathbf{A}_{\bar{l}}, \mathbf{Y} = \mathbf{y}, \mathcal{M})$ can be achieved as follows. We know that

$$p(A_l = a_l|\mathbf{A}_{\bar{l}} = \mathbf{a}_{\bar{l}}, \mathbf{Y} = \mathbf{y}, \mathcal{M}) \propto p(A_l = a_l) p(\mathbf{Y} = \mathbf{y}|\mathbf{A} = \mathbf{a}, \mathcal{M}). \qquad (10.31)$$

Since A_l is defined over a small set (Ω), we can determine, for a fixed $\mathbf{a}_{\bar{k}}$, $p(A_l = a_l) p(\mathbf{Y} = \mathbf{y}|\mathbf{A} = \mathbf{a}, \mathcal{M})$ for every value $a_l \in \Omega$. We then determine the normalization constant. This creates the pmf $p(A_l|\mathbf{A}_{\bar{l}} = \mathbf{a}_{\bar{l}}, \mathbf{Y} = \mathbf{y}, \mathcal{M})$, from which we can draw a sample. Note that $p(A_l|\mathbf{A}_{\bar{l}} = \mathbf{a}_{\bar{l}}, \mathbf{Y} = \mathbf{y}, \mathcal{M})$ is also required in the computation (10.30).

Algorithm 10.1 Gibbs sampler for $p(\mathbf{A}|\mathbf{Y} = \mathbf{y}, \mathcal{M})$

1: initialization: choose an initial state $\mathbf{a}^{(-N_{\text{burn}}-1)}$

2: **for** $i = -N_{\text{burn}}$ to L **do**

3:      **for** $l = 1$ to N_s **do**

4:          determine $p\left(A_l \middle| \mathbf{A}_{1:l-1} = \mathbf{a}_{1:l-1}^{(i)}, \mathbf{A}_{l+1:N_s} = \mathbf{a}_{l+1:N_s}^{(i-1)}, \mathbf{Y} = \mathbf{y}, \mathcal{M}\right)$

5:          draw $a_l^{(i)} \sim p\left(A_l \middle| \mathbf{A}_{1:l-1} = \mathbf{a}_{1:l-1}^{(i)}, \mathbf{A}_{l+1:N_s} = \mathbf{a}_{l+1:N_s}^{(i-1)}, \mathbf{Y} = \mathbf{y}, \mathcal{M}\right)$

6:      **end for**

7: **end for**

10.3.7   Gaussian/MMSE equalizers

10.3.7.1   Model

In this section, we will use a specialization of the model (10.2), namely a linear model with Gaussian noise, so that

$$y = Ha + n, \tag{10.32}$$

where H is a complex $N_o \times N_s$ matrix, and $n \sim \mathcal{N}_n^{\mathbb{C}}(0, \Sigma)$. The MMSE equalizers operate by temporarily assuming a to be a Gaussian vector. We will describe only the most common approaches to MMSE equalization. The MMSE equalizers can be combined with sliding-window equalizers by replacing $y = Ha + n$ by $y_k = H_k a_k + n_k$ in (10.23), where y_k is a windowed observation, and H_k, a_k, and n_k are the corresponding data symbols, sub-matrix of the matrix H, and noise vector. The MMSE equalizers can also be combined with structured equalizers by applying the MMSE techniques per factor.

10.3.7.2   Step 1: the Gaussian assumption

Although a_k belongs to a finite set Ω, let us temporarily assume that a has the following a-priori distribution: $a \sim \mathcal{N}_a^{\mathbb{C}}(m_a, \Sigma_a)$, where $m_a = [m_{a,1}, \ldots, m_{a,N_s}]^T$ and $\Sigma_a = \mathrm{diag}\left(\sigma_{a,1}^2, \ldots, \sigma_{a,N_s}^2\right)$, an $N_s \times N_s$ matrix, with

$$m_{a,l} = \sum_{a_l \in \Omega} a_l p(A_l = a_l), \tag{10.33}$$

$$\sigma_{a,l}^2 = \sum_{a_l \in \Omega} |a_l - m_{a,l}|^2 p(A_l = a_l). \tag{10.34}$$

We will use the subscripts "a," "p" and "e" for a priori, a posteriori, and extrinsic, respectively. Now Y and A are jointly Gaussian, so A has the following a-posteriori distribution:

$$p(A = a|Y = y, \mathcal{M}) = \mathcal{N}_a^{\mathbb{C}}(m_p, \Sigma_p), \tag{10.35}$$

where

$$\Sigma_p = \left(\Sigma_a^{-1} + H^H \Sigma^{-1} H\right)^{-1}$$

$$= \left(I_{N_s} - KH\right)\Sigma_a \tag{10.36}$$

and

$$m_p = m_a + K(y - Hm_a) \tag{10.37}$$

with $K = \Sigma_a H^H \left(H\Sigma_a H^H + \Sigma\right)^{-1}$. Observe the resemblance between m_p and the MMSE estimator from Section 3.2.2.1. From $p(A = a|Y = y, \mathcal{M})$, it is our goal to compute the message $\mu_{eq \to A_k}(A_k)$.

10.3.7.3   Step 2: determine $p(A_k = a_k | \mathbf{Y} = \mathbf{y}, \mathcal{M})$

The kth element of $\mathbf{m}_p$ is given by

$$m_{p,k} = \mathbf{e}_k^{\mathrm{T}} \mathbf{m}_p, \tag{10.38}$$

where $\mathbf{e}_k$ is an $N_s \times 1$ vector of all zeros, but with a 1 on the kth position. We find that

$$m_{p,k} = m_{a,k} + \sigma_{a,k}^2 \mathbf{h}_k^{\mathrm{H}} \left(\mathbf{H} \Sigma_a \mathbf{H}^{\mathrm{H}} + \Sigma \right)^{-1} (\mathbf{y} - \mathbf{H} \mathbf{m}_a), \tag{10.39}$$

where $\mathbf{h}_k$ is the kth column of $\mathbf{H}$ (or, $\mathbf{h}_k = \mathbf{H} \mathbf{e}_k$). Similarly, the kth diagonal element of Σ_p is given by

$$\sigma_{p,k}^2 = \mathbf{e}_k^{\mathrm{T}} \Sigma_p \mathbf{e}_k \tag{10.40}$$

$$= \sigma_{a,k}^2 - \mathbf{e}_k^{\mathrm{T}} \Sigma_a \mathbf{H}^{\mathrm{H}} \left(\mathbf{H} \Sigma_a \mathbf{H}^{\mathrm{H}} + \Sigma \right)^{-1} \mathbf{H} \Sigma_a \mathbf{e}_k \tag{10.41}$$

$$= \sigma_{a,k}^2 - \sigma_{a,k}^2 \mathbf{h}_k^{\mathrm{H}} \left(\mathbf{H} \Sigma_a \mathbf{H}^{\mathrm{H}} + \Sigma \right)^{-1} \mathbf{h}_k \sigma_{a,k}^2 \tag{10.42}$$

$$= \sigma_{a,k}^2 \left(1 - \mathbf{h}_k^{\mathrm{H}} \left(\mathbf{H} \Sigma_a \mathbf{H}^{\mathrm{H}} + \Sigma \right)^{-1} \mathbf{h}_k \sigma_{a,k}^2 \right). \tag{10.43}$$

10.3.7.4   Step 3: remove the dependence of $p(A_k = a_k | \mathbf{Y} = \mathbf{y}, \mathcal{M})$ on $p(A_k = a_k)$

We know that the extrinsic probability $\mu_{\mathrm{eq} \to A_k}(a_k)$ cannot depend on the a-priori probability $p(A_k = a_k)$. Taking into account that $1/|\Omega| \sum_{a \in \Omega} a = 0$ and $1/|\Omega| \sum_{a \in \Omega} |a|^2 = 1$, we replace $m_{a,k}$ by 0, and $\sigma_{a,k}^2$ by 1, resulting in

$$m_{e,k} = \underbrace{\mathbf{h}_k^{\mathrm{H}} \left(\mathbf{H} \Sigma_{\bar{k}} \mathbf{H}^{\mathrm{H}} + \Sigma \right)^{-1}}_{\mathbf{w}_k^{\mathrm{H}}} (\mathbf{y} - \mathbf{H} \mathbf{m}_{\bar{k}}), \tag{10.44}$$

$$\sigma_{e,k}^2 = 1 - \mathbf{w}_k^{\mathrm{H}} \mathbf{h}_k, \tag{10.45}$$

where $\mathbf{m}_{\bar{k}}$ is $\mathbf{m}_a$ with the kth entry replaced by 0, and $\Sigma_{\bar{k}}$ is Σ_a with the kth entry replaced by 1.

10.3.7.5   Step 4: convert the Gaussian distribution into a message over Ω

We consider two common techniques to convert the Gaussian distribution $\mathcal{N}_{a_k}^{\mathbb{C}}(m_{e,k}, \sigma_{e,k}^2)$ into the message $\mu_{\mathrm{eq} \to A_k}(a_k)$.

Probabilistic data association

We set [101]

$$\mu_{\mathrm{eq} \to A_k}(a_k) \propto \exp \left(-\frac{1}{\sigma_{e,k}^2} |a_k - m_{e,k}|^2 \right). \tag{10.46}$$

The normalization constant is found by setting $\sum_{a_k \in \Omega} \mu_{\mathrm{eq} \to A_k}(a_k) = 1$.

Equivalent Gaussian channel

The most common technique works under the assumption that $m_{e,k}$ is a noisy version of a_k [31,97]:

$$m_{e,k} = \mu_k a_k + \nu_k, \tag{10.47}$$

where $\nu_k \sim \mathcal{N}_{\nu_k}^{\mathbb{C}}(0, \sigma_{\text{ch}}^2)$. We find that

$$\mu_k = \mathbb{E}\left\{A_k M_{e,k}\right\} \tag{10.48}$$

$$= \mathbf{w}_k^{\mathsf{H}} \mathbb{E}\left\{A_k \left(\mathbf{Y} - \mathbf{H}\mathbf{m}_{\bar{k}}\right)\right\} \tag{10.49}$$

$$= \mathbf{w}_k^{\mathsf{H}} \mathbf{h}_k. \tag{10.50}$$

Note that $\mu_k \in \mathbb{R}$. We have

$$\sigma_{\text{ch}}^2 = \mathbb{E}\left\{|V_k|^2\right\} \tag{10.51}$$

$$= \mathbb{E}\left\{|M_{e,k} - \mu_k A_k|^2\right\} \tag{10.52}$$

$$= \mathbb{E}\left\{|M_{e,k}|^2\right\} - |\mu_k|^2 \tag{10.53}$$

$$= \mathbb{E}\left\{\left|\mathbf{w}_k^{\mathsf{H}}(\mathbf{Y} - \mathbf{H}\mathbf{m}_{\bar{k}})\right|^2\right\} - |\mu_k|^2 \tag{10.54}$$

$$= \mathbf{w}_k^{\mathsf{H}}\left(\mathbf{H}\Sigma_{\bar{k}}\mathbf{H}^{\mathsf{H}} + \Sigma\right)\mathbf{w}_k - |\mu_k|^2 \tag{10.55}$$

$$= \mathbf{w}_k^{\mathsf{H}} \mathbf{h}_k - |\mu_k|^2 \tag{10.56}$$

$$= \mu_k - \mu_k^2. \tag{10.57}$$

The equivalent Gaussian channel then tells us that

$$\mu_{\text{eq}\to A_k}(a_k) \propto \exp\left(-\frac{1}{\sigma_{\text{ch}}^2}|m_{e,k} - a_k\mu_k|^2\right), \tag{10.58}$$

where the normalization constant is found by setting $\sum_{a_k \in \Omega} \mu_{\text{eq}\to A_k}(a_k) = 1$.

10.3.7.6 Comments

In the above technique, we must invert the matrix $\left(\mathbf{H}\Sigma_{\bar{k}}\mathbf{H}^{\mathsf{H}} + \Sigma\right)$. This is an $N_{\text{o}} \times N_{\text{o}}$ matrix. The matrix $\mathbf{H}$, on the other hand, is an $N_{\text{o}} \times N_{\text{s}}$ matrix. When $N_{\text{o}} < N_{\text{s}}$, we say that the matrix is *fat*. When $N_{\text{o}} > N_{\text{s}}$, we say that the matrix is *tall*. In many inference problems, we like our matrices tall, since then we have a problem with more observations than unknowns. However, this also means that we have to invert large ($N_{\text{o}} \times N_{\text{o}}$) matrices. This can be avoided as follows. Using the matrix-inversion lemma,[3] we can write $\mathbf{K}$ in

---

[3] One form of the matrix-inversion lemma tells us that $(\mathbf{I} + \mathbf{AB})^{-1}\mathbf{A} = \mathbf{A}(\mathbf{I} + \mathbf{BA})^{-1}$.

(10.36) and (10.37) as

$$\mathbf{K} = \Sigma_a \mathbf{H}^H \left(\mathbf{H} \Sigma_a \mathbf{H}^H + \Sigma \right)^{-1} \tag{10.59}$$

$$= \left(\Sigma_a^{-1} + \mathbf{H}^H \Sigma^{-1} \mathbf{H} \right)^{-1} \mathbf{H}^H \Sigma^{-1}. \tag{10.60}$$

Assuming that Σ^{-1} can be computed efficiently (in many cases Σ^{-1} can be pre-computed), we need only invert an $N_s \times N_s$ matrix. Note that Σ_a is a non-singular diagonal matrix, so its inverse is easily found. This leads to an alternative formulation of $m_{e,k}$ and $\sigma_{e,k}^2$:

$$m_{e,k} = \mathbf{e}_k^T \left(\Sigma_{\bar{k}}^{-1} + \mathbf{H}^H \Sigma^{-1} \mathbf{H} \right)^{-1} \mathbf{H}^H \Sigma^{-1} (\mathbf{y} - \mathbf{H} \mathbf{m}_{\bar{k}}), \tag{10.61}$$

$$\sigma_{e,k}^2 = 1 - \mathbf{e}_k^T \left(\Sigma_{\bar{k}}^{-1} + \mathbf{H}^H \Sigma^{-1} \mathbf{H} \right)^{-1} \mathbf{H}^H \Sigma^{-1} \mathbf{h}_k. \tag{10.62}$$

We see that, depending on whether the matrix $\mathbf{H}$ is fat or tall, we can express $\mathbf{K}$ such that we need only invert matrices of size $\min\{N_s, N_o\} \times \min\{N_s, N_o\}$. Generally, the matrices to invert are highly structured, so the inverses can be computed efficiently [98].

10.4 Interaction with the demapping and the decoding node

As we have seen, the messages $\mu_{\text{eq} \to A_k}(A_k)$ from the equalizer to the demapper depend on the messages $\mu_{A_l \to \text{eq}}(A_l)$, $l \neq k$ from the demapper to the equalizer. This means that, even when the factor graph of $p(\mathbf{Y} = \mathbf{y}|\mathbf{A}, \mathcal{M})$ does not contain any cycles, the overall factor graph of $p(\mathbf{Y} = \mathbf{y}, \mathbf{B}|\mathcal{M})$ may contain cycles. This in turn implies that equalization can be performed in an iterative fashion: for the first iteration, the messages $\mu_{A_l \to \text{eq}}(A_l)$ are set to uniform distributions; at all subsequent iterations, these messages are obtained from the demapper. There are various ways to schedule the computation of messages, with different scheduling strategies leading to different performances.

10.5 Performance illustration

To illustrate the performance of these equalization strategies, let us consider an example. Consider an observation model of the form

$$y_k = \sqrt{E_s} h(a_k, a_{k-1}, a_{k-2}) + n_k,$$

where the noise samples are iid, $n_k \sim \mathcal{N}_n^{\mathbb{C}}(0, N_0)$ and $h(\cdot)$ is a third-order non-linear Volterra channel [15],

$$
\begin{aligned}
h(a_k, a_{k-1}, a_{k-2}) = {} & (0.780855 + \text{j}0.413469)a_k + (0.040323 - \text{j}0.000640)a_{k-1} \\
& + (0.015361 - \text{j}0.008961)a_{k-2} + (-0.04 - \text{j}0.009)a_k^2 a_k^* \\
& + (-0.035 + \text{j}0.035)a_k^2 a_{k-1}^* + (0.039 + \text{j}0.022)a_k^2 a_{k-2}^* \\
& + (-0.001 - \text{j}0.017)a_{k-1}^2 a_k^* + (0.018 - \text{j}0.018)a_{k-2}^2 a_k^*.
\end{aligned}
$$

We use QPSK signaling with Gray mapping, and employ a terminated recursive systematic convolutional code with $g_{\text{FF}}(D) = 1+D^4$ and $g_{\text{FB}}(D) = 1+D+D^2+D^3+D^4$. We set $N_s = 256$. The encoder and mapper are separated by a pseudo-random bit-interleaver. Three equalizers will be considered: the state-space-model (SSM) equalizer from Section 10.3.4, and two MC equalizers from Section 10.3.6:

- an MC equalizer using importance sampling, drawing $L = 10$ samples from $p(\mathbf{A}|\mathbf{Y} = \mathbf{y}, \mathcal{M})$ with a burn-in period of five samples; and
- an MC equalizer using distinct samples with a sliding window $\mathbf{y}_k = [y_k, y_{k+1}, y_{k+2}]^{\text{T}}$ and $\mathbf{a}_k = [a_{k-2}, a_{k-1}, a_k, a_{k+1}, a_{k+2}]^{\text{T}}$, drawing $L = 10$ samples from $p(\mathbf{A}_k|\mathbf{Y}_k = \mathbf{y}_k, \mathcal{M})$, with a burn-in period of five samples.

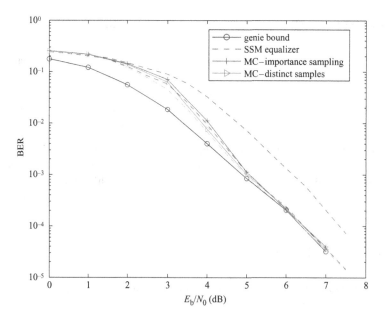

Figure 10.3. Equalization: BER versus SNR performance. The performance of the SSM equalizer is shown for three iterations, while the performance of two MC equalizers is shown for the third iteration only. The genie bound corresponds to perfect a-priori information from the demapper to the equalizer.

We evaluate the performance in terms of bit error rate (BER) versus signal-to-noise ratio (SNR). The SNR is expressed as E_b/N_0 (in decibels[4]), with $E_b = E_s/(R\log_2|\Omega|)$, where R denotes the code rate. The results are shown in Fig. 10.3. As a reference, we also include the genie bound, where the equalizer is provided with perfect a-priori information from the demapper. The SSM equalizer converges after three iterations, and, for BER below 10^{-3}, is quite close to the genie bound. Both MC equalizers exhibit a performance that almost coincides with the SSM equalizer.

10.6   Main points

The equalization process deals with computing messages $\mu_{eq \to A_k}(A_k)$ to the demapper, on the basis of the node $p(\mathbf{Y} = \mathbf{y}|\mathbf{A}, \mathcal{M})$ in the receivers factor graph, and the incoming messages $\mu_{A_l \to eq}(A_l)$. We have considered a model

$$\mathbf{y} = \mathbf{h}(\mathbf{a}, \tilde{\mathbf{a}}) + \mathbf{n}, \tag{10.63}$$

where $\mathbf{n}$ is a Gaussian noise-vector, $\tilde{\mathbf{a}}$ represents symbols that affect the observation $\mathbf{y}$, but are not part of $\mathbf{a}$, and $\mathbf{h}(\cdot)$ is a *known* transformation encapsulating the physical channels, as well as any processing at the transmitter and the receiver. We have described exact equalization methods and approximate equalization methods.

Exact methods:
- SPA equalizer,
- structured equalizer,
- SSM equalizer.

Approximate methods:
- sliding-window equalizer,
- MC equalizer (possibly combined with the structured equalizer or the sliding window equalizer),
- Gaussian/MMSE equalizer (usually combined with the structured equalizer or the sliding-window equalizer).

These general equalization techniques need to be combined and tailored to the specific structure of the problem. In the following chapters, we will revisit our digital communication schemes from Chapter 2 and derive suitable observation models of the form (10.63).

---

[4] We can express an SNR in decibels (dB) as $10\log_{10}$ SNR.

11   Equalization: single-user, single-antenna communication

11.1   Introduction

As depicted in Fig. 11.1, in single-user, single-antenna transmission, both the receiver and the transmitter are equipped with a single antenna. There are no other transmitters. This is the most conventional and well-understood way of communicating. Many receivers for such a set-up have been designed during the past few decades. These receivers usually consist of a number of stages. The first stage is a conversion from the continuous-time received waveform to a suitable observation (to allow digital signal processing), followed by equalization (to counteract inter-symbol interference), demapping (where decisions with respect to the coded bits are taken), and finally decoding (where we attempt to recover the original information sequence). This is a one-shot approach, whereby no information flows back from the decoder to the demapper or to the equalizer. Here the terms decoder, demapper, and equalizer pertain to the more conventional receiver tasks, not to nodes in any factor graph. In a conventional mind-set it is hard to come up with a non-ad-hoc way of exploiting information from the decoder during the equalization process. In the factor-graph framework, the flow of information between the various blocks appears naturally and explicitly. These two approaches to receiver design are depicted in Fig. 11.2.

In this chapter we will see how to convert the received waveform into a suitable observation $\mathbf{y}$. This conversion is exactly the same as in conventional receivers. From $\mathbf{y}$ we will then show how to perform equalization using factor graphs by computing the messages $\mu_{\mathrm{eq}\to A_k}(A_k)$ from the equalization node to the demapping node. The sequence of coded symbols of interest will generally consist of N_s consecutive symbols

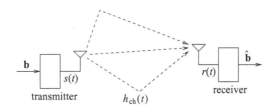

Figure 11.1. Single-user, single-antenna communication: the equivalent baseband transmitted signal $s(t)$ propagates through the equivalent baseband channel $h_{\mathrm{ch}}(t)$, and is corrupted by thermal noise at the receiver. The resulting equivalent baseband received signal is denoted $r(t)$.

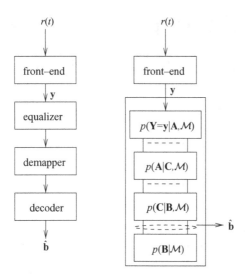

Figure 11.2. A conventional receiver (left) and a factor-graph receiver (right). The observation **y** is fed to the factor graph as a parameter. Implementing the SPA on the factor graph of $p(\mathbf{B}, \mathbf{Y} = \mathbf{y}|\mathcal{M})$ gives us the marginal a-posteriori distributions $p(B_k|\mathbf{Y} = \mathbf{y}, \mathcal{M})$ on the basis of which decisions on the information bits can be made.

$[a_{k_0}, a_{k_0+1}, \ldots, a_{k_0+N_s-1}]^{\mathrm{T}}$ for some $k_0 \in \mathbb{Z}$, though indexing will heavily depend on the transmission scheme under consideration.

This chapter is organized as follows.

- In **Section 11.2** we will deal with single-carrier modulation. We will describe three ways of converting the received waveform into an observation **y** and show how to compute the messages from the equalization node to the demapping node.
- Multi-carrier modulation will be the topic of **Section 11.3**, where we discuss the standard OFDM receiver.

Whenever applicable, equivalent receivers from the technical literature are mentioned.

11.2   Single-carrier modulation

11.2.1   The received waveform

Going back all the way to Section 2.3.1, we know that the baseband equivalent received signal is given by

$$r(t) = \sqrt{E_s} \sum_{k=-\infty}^{+\infty} a_k h(t - kT) + n(t), \tag{11.1}$$

where E_s is the energy per transmitted symbol, $h(t)$ is the convolution of the transmit pulse $p(t)$ and the baseband equivalent physical channel $h_{\mathrm{ch}}(t)$, and $n(t)$ is a complex white Gaussian noise process with power-spectral density $N_0/2$ for the independent real and imaginary components. We model the physical channel as a multi-path channel with L resolvable paths,

$$h_{\mathrm{ch}}(t) = \sum_{l=0}^{L-1} \alpha_l \delta(t - \tau_l), \tag{11.2}$$

where $\alpha_l \in \mathbb{C}$ and $\tau_l \in \mathbb{R}$ denote the complex gain and the propagation delay of the lth path, respectively. We will assume that the N_{s} data symbols of interest are $[a_0, a_1, \ldots, a_{N_{\mathrm{s}}-1}]^{\mathrm{T}}$. We will consider three types of receivers: a matched filter receiver, a whitened matched filter receiver, and an oversampling receiver.

On noise and sampling
Given a spectrally white Gaussian noise process $n(t)$ with independent real and imaginary components, each with power-spectral density $N_0/2$, and a filter $q(t)$, then the filtered noise is given by

$$w(t) = \int_{-\infty}^{+\infty} q(u) n(t - u) \mathrm{d}u. \tag{11.3}$$

Sampling $w(t)$ at time instants kT_{s} yields samples $w_k, k \in \mathbb{Z}$. These samples are Gaussian, and have zero mean, with

$$\mathbb{E}\{W_k W_{k'}^*\} = g((k - k')T_{\mathrm{s}})N_0, \tag{11.4}$$

where

$$g(t) = \int_{-\infty}^{+\infty} q^*(u) q(t + u) \mathrm{d}u. \tag{11.5}$$

In the particular case when $q(t)$ is a square-root Nyquist pulse for a rate $1/T_{\mathrm{s}}, g(kT_{\mathrm{s}}) = \delta_k$, so the noise samples $\{w_k\}$ are uncorrelated (and, since they are Gaussian, they are also independent).

11.2.2   Matched filter receiver

In a matched filter receiver, we filter the incoming signal $r(t)$ with a filter, $h^*(-t)$, that is matched to the equivalent channel, followed by sampling at the symbol rate [38, 102]. The signal at the output of the matched filter is

$$y_{\mathrm{MF}}(t) = \int_{-\infty}^{+\infty} h^*(u) r(t + u) \mathrm{d}u \tag{11.6}$$

$$= \sqrt{E_{\mathrm{s}}} \sum_{k=-\infty}^{+\infty} a_k g(t - kT) + n_{\mathrm{MF}}(t), \tag{11.7}$$

where $g(t) = \int h^*(u)h(t+u)\mathrm{d}u$ and $n_{\mathrm{MF}}(t)$ is the filtered noise. Sampling at time $k'T$ yields

$$y_{\mathrm{MF}}(k'T) = \sqrt{E_s} \sum_{k=-\infty}^{+\infty} a_k g_{k'-k} + n_{\mathrm{MF}}(k'T) \tag{11.8}$$

$$= \sqrt{E_s} \sum_{l=L_{\min}}^{L_{\max}} g_l a_{k'-l} + n_{\mathrm{MF}}(k'T), \tag{11.9}$$

where $g_l = g(lT)$ and $\mathbb{E}\{N_{\mathrm{MF}}(k'T)N_{\mathrm{MF}}^*(k''T)\} = g_{k'-k''}N_0$. We have assumed that $g(lT)$ takes on significant values only for $L_{\min} \le l \le L_{\max}$. Let us stack samples (for instance, for $k \ge L_{\min}$ until $k = N_s + L_{\max} - 1$) into a vector $\mathbf{y}_{\mathrm{MF}}$ and write

$$\mathbf{y}_{\mathrm{MF}} = \mathbf{H}_{\mathrm{MF}}\mathbf{a} + \mathbf{n}_{\mathrm{MF}}, \tag{11.10}$$

where $\mathbf{a}$ contains the coded data symbols of interest $[a_0, a_1, \ldots, a_{N_s-1}]^{\mathrm{T}}$, as well as several unknown data symbols $a_{k<0}$ and/or $a_{k>N_s-1}$. The matrix $\mathbf{H}_{\mathrm{MF}}$ is a Toeplitz matrix in which the elements on the diagonals are given by $\sqrt{E_s}g_k$, $k = L_{\min}, \ldots, L_{\max}$.

The equalizer

We discern two cases, depending on whether or not $g(t)$ is a scaled Nyquist pulse for rate $1/T$.

- **Nyquist pulse.** When $g(t)$ is a scaled Nyquist pulse for rate $1/T$, $g_l = 0$ for $l \ne 0$. This has two implications: on the one hand, that noise is white, and on the other hand, there is no inter-symbol interference:

$$y_{\mathrm{MF}}(kT) = \sqrt{E_s}g_0 a_k + n_{\mathrm{MF}}(kT). \tag{11.11}$$

This situation occurs when $p(t)$ is a square-root Nyquist pulse for rate $1/T$ and the physical channel is frequency-flat $h_{\mathrm{ch}}(t) = \alpha\delta(t-\tau)$ so that $h(t) = \alpha p(t-\tau)$. We find that, with $\sigma^2 = N_0 g_0/2$,

$$p(\mathbf{Y} = \mathbf{y}_{\mathrm{MF}}|\mathbf{A} = \mathbf{a}, \mathcal{M}) \propto \exp\left(-\frac{1}{2\sigma^2}\left\|\mathbf{y}_{\mathrm{MF}} - \sqrt{E_s}g_0\mathbf{a}\right\|^2\right) \tag{11.12}$$

$$= \prod_{k=0}^{N_s-1} \exp\left(-\frac{1}{2\sigma^2}\left|\sqrt{E_s}g_0 a_k - y_{\mathrm{MF}}(kT)\right|^2\right), \tag{11.13}$$

which immediately gives us a *structured equalizer*:

$$\mu_{\mathrm{eq}\to A_k}(a_k) \propto \exp\left(-\frac{1}{2\sigma^2}\left|\sqrt{E_s}g_0 a_k - y_{\mathrm{MF}}(kT)\right|^2\right). \tag{11.14}$$

- **Not a Nyquist pulse.** When $g(t)$ is not a scaled Nyquist pulse for rate $1/T$, the noise $\mathbf{n}_{\text{MF}}$ is not white (it is said to be colored) and inter-symbol interference occurs. We are constrained to use a *sliding-window MMSE equalizer*.

11.2.3 Whitened matched-filter receivers

A whitened matched filter processes the matched-filter samples $\{y_{\text{MF}}(kT)\}$ by filtering them with a suitable *whitening filter*[103]. It turns out that we then obtain samples of the form

$$y_{\text{WMF}}(kT) = \sum_{l=0}^{L_{\text{WMF}}-1} h_l a_{k-l} + n_{\text{WMF}}(kT), \tag{11.15}$$

where $\mathbb{E}\{N^*_{\text{WMF}}(kT)N_{\text{WMF}}(k'T)\} = N_0 \delta_{k-k'}$. Stacking samples (for instance for $k = 0$ until $k = N_s - 1 + L_{\text{WMF}}$) yields

$$\mathbf{y}_{\text{WMF}} = \mathbf{H}_{\text{WMF}}\mathbf{a} + \mathbf{n}_{\text{WMF}}, \tag{11.16}$$

where $\mathbf{a}$ contains the coded data symbols of interest $[a_0, a_1, \ldots, a_{N_s-1}]^T$, as well as several unknown data symbols $a_{k<0}$ and/or $a_{k>N_s-1}$. The matrix $\mathbf{H}_{\text{WMF}}$ is a Toeplitz matrix in which the elements on the diagonals are given by h_k, $k = 0, \ldots, L_{\text{WMF}} - 1$.

Example 11.1. *For $L_{\text{WMF}} = 2$, suppose that we take as observations $y_m = y_{\text{WMF}}(kT)$, for $k = -1, 0, 1, 2, 3$. Then $\mathbf{y}$ is given by*

$$
\begin{bmatrix} y_{-1} \\ y_0 \\ y_1 \\ y_2 \\ y_3 \end{bmatrix} =
\begin{bmatrix}
h_1 & h_0 & 0 & 0 & 0 & 0 \\
0 & h_1 & h_0 & 0 & 0 & 0 \\
0 & 0 & h_1 & h_0 & 0 & 0 \\
0 & 0 & 0 & h_1 & h_0 & 0 \\
0 & 0 & 0 & 0 & h_1 & h_0
\end{bmatrix}
\begin{bmatrix} a_{-2} \\ a_{-1} \\ a_0 \\ a_1 \\ a_2 \\ a_3 \end{bmatrix} +
\begin{bmatrix} n_{-1} \\ n_0 \\ n_1 \\ n_2 \\ n_3 \end{bmatrix}.
$$

The equalizer

Since the noise is white, the observation model allows for an *SSM equalizer* [96]. For long channels (large L_{WMF}) or large constellations (large $|\Omega|$), the SSM equalizer is computationally too demanding. We can then look to a *sliding-window MMSE equalizer* [99, 104] or a *(sliding-window) MC equalizer*. The Toeplitz structure of the matrix $\mathbf{H}_{\text{WMF}}$ can be exploited in the MMSE equalizer to reduce the computational complexity.

11.2.4 The oversampling receiver

The oversampling receiver forms an alternative to the matched-filter receivers. We now first filter the signal $r(t)$ with a square-root Nyquist pulse $q(t)$ for a rate N/T. The resulting signal can be written as

$$y_{OS}(t) = \int_{-\infty}^{+\infty} q(u)r(t-u)du \tag{11.17}$$

$$= \sum_{k=-\infty}^{+\infty} a_k h_{OS}(t-kT) + n_{OS}(t). \tag{11.18}$$

Sampling at time instants $k'T + mT/N$ (for $m = 0, \dots, N-1$ and $k' \in \mathbb{Z}$) yields $y_{OS}(k'T + mT/N)$, which we abbreviate by $y_{k'}^{(m)}$. We can express the observation as

$$y_{k'}^{(m)} = \sum_{k=-\infty}^{+\infty} a_k h_{k'-k}^{(m)} + n_{k'}^{(m)} \tag{11.19}$$

where $h_k^{(m)} = h_{OS}(kT + mT/N)$ and

$$\mathbb{E}\left\{ N_{k'}^{(m)} \left(N_{k''}^{(m')} \right)^* \right\} = N_0 \delta_{k'-k''} \delta_{m-m'}.$$

Since $h_{OS}(t)$ is usually (approximately) time-limited to $t \in [L_{\min}T, L_{\max}T]$, we can also write

$$y_k^{(m)} = \sum_{l=L_{\min}}^{L_{\max}} h_l^{(m)} a_{k-l} + n_k^{(m)}. \tag{11.20}$$

Stacking samples (first for fixed k, then for different successive k) yields

$$\mathbf{y}_{OS} = \mathbf{H}_{OS}\mathbf{a} + \mathbf{n}_{OS}. \tag{11.21}$$

Observe that $\mathbf{H}_{OS}$ is no longer a Toeplitz matrix, but is now a band matrix.

Example 11.2. *Suppose that* $N = 2$, $L_{\min} = 0$, *and* $L_{\max} = 1$, *and we observe* $y_{OS}(-T), \dots, y_{OS}(2T)$, *then*

$$
\begin{bmatrix} y_{-1}^{(0)} \\ y_{-1}^{(1)} \\ y_0^{(0)} \\ y_0^{(1)} \\ y_1^{(0)} \\ y_1^{(1)} \\ y_2^{(0)} \end{bmatrix}
=
\begin{bmatrix}
h_1^{(0)} & h_0^{(0)} & 0 & 0 & 0 \\
h_1^{(1)} & h_0^{(1)} & 0 & 0 & 0 \\
0 & h_1^{(0)} & h_0^{(0)} & 0 & 0 \\
0 & h_1^{(1)} & h_0^{(1)} & 0 & 0 \\
0 & 0 & h_1^{(0)} & h_0^{(0)} & 0 \\
0 & 0 & h_1^{(1)} & h_0^{(1)} & 0 \\
0 & 0 & 0 & h_1^{(0)} & h_0^{(0)}
\end{bmatrix}
\begin{bmatrix} a_{-2} \\ a_{-1} \\ a_0 \\ a_1 \\ a_2 \end{bmatrix}
+ \mathbf{n}_{OS}.
$$

Oversampling receivers are attractive from a practical point of view because the observation $\mathbf{y}_{OS}$ can be generated without explicit knowledge of the channel $h(t)$. The enables channel estimation in the digital domain. This is in contrast to the matched-filter receivers, which rely explicitly on $h(t)$ to obtain the observation $\mathbf{y}_{MF}$ or $\mathbf{y}_{WMF}$.

The equalizer

Since the noise is white, we could in principle use an SSM equalizer. However, the channel length $L_{OS} = L_{max} - L_{min} + 1$ is usually fairly high, so that in practice only *sliding window MMSE equalization* [105] or *(sliding window) MC equalization* can be applied.

11.3 Multi-carrier modulation

11.3.1 The received waveform

For OFDM, we know from Section 2.3.1 that the received waveform can be expressed as

$$r(t) = \sum_{k=-\infty}^{+\infty} \sum_{l=-N_{CP}}^{N_{FFT}-1} \breve{a}_{l,k} h(t - lT - kT_{OFDM}) + n(t), \tag{11.22}$$

where $\breve{a}_{l,k}$ is the lth time-domain value of the kth OFDM symbol, $h(t)$ is the equivalent channel (including the transmit energy factor $\sqrt{E_s N_{FFT}/(N_{FFT} + N_{CP})}$), $n(t)$ is a complex white Gaussian noise process with power-spectral density $N_0/2$ for the real and imaginary components, and T_{OFDM} is the OFDM symbol duration. Suppose that our vector $\mathbf{a}$ corresponds to N OFDM symbols, corresponding to $k = 0, \ldots, N-1$ in (11.22). In other words, $N_s = N N_{FFT}$ (assuming that all the subcarriers are used for data transmission).

11.3.2 The OFDM receiver

We first filter $r(t)$ with a unit-energy square-root Nyquist filter $q(t)$ for a rate $1/T$, resulting in

$$\breve{y}(t) = \int_{-\infty}^{+\infty} q(u) r(t - u) \mathrm{d}u \tag{11.23}$$

$$= \sum_{k=-\infty}^{+\infty} \sum_{l=-N_{CP}}^{N_{FFT}-1} \breve{a}_{l,k} \breve{h}(t - lT - kT_{OFDM}) + \breve{n}(t). \tag{11.24}$$

Let us introduce the notion of the *support* of $\breve{h}(t)$: let t_{min} be the largest time instant for which $\breve{h}(t) = 0$, for all $t < t_{min}$. Similarly, let t_{max} be the smallest time instant for which $\breve{h}(t) = 0$, for all $t > t_{max}$. The support of $\breve{h}(t)$ is then given by the interval $[t_{min}, t_{max}]$. One of the key properties of an OFDM systems is that the delay spread of the channel

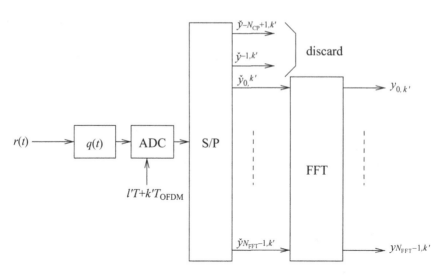

Figure 11.3. OFDM receiver with filter, analog-to-digital conversion (ADC) and serial-to-parallel (S/P) conversion. The samples corresponding to the cyclic prefix are discarded. The remaining N_{FFT} samples are converted to the frequency domain and provided to the equalization node.

$\check{h}(t)$ may not exceed the duration of the cyclic prefix $N_{\text{CP}}T$. To be more exact:

$$\frac{t_{\max} - t_{\min}}{T} < N_{\text{CP}} + 1. \tag{11.25}$$

For the sake of convenience, let us assume that the support of $\check{h}(t)$ is $[0, N_{\text{DS}}T)$, with $N_{\text{DS}} \leq N_{\text{CP}} + 1$. Sampling $\check{y}(t)$ at time instants $l'T + k'T_{\text{OFDM}}$, $l' = 0, \ldots, N_{\text{FFT}} - 1$, $k' = 0, \ldots, N - 1$, yields (see Fig. 11.3)

$$\check{y}_{l',k'} = \sum_{k=-\infty}^{+\infty} \sum_{l=-N_{\text{CP}}}^{N_{\text{FFT}}-1} \check{a}_{l,k}\check{h}\big(l'T + k'T_{\text{OFDM}} - lT - kT_{\text{OFDM}}\big) + \check{n}_{l',k'} \tag{11.26}$$

with $\mathbb{E}\{\check{N}_{l',k'}\check{N}^*_{l'',k''}\} = \delta_{l'-l''}\delta_{k'-k''}N_0$. Note that the N_{CP} samples corresponding to the cyclic prefix are discarded (this correspond to $l' = -N_{\text{CP}}, \ldots, -1$). The finite support of $\check{h}(t)$ allows us to write $\check{y}_{l',k'}$ in a more compact form,

$$\check{y}_{l',k'} = \sum_{l=-N_{\text{CP}}}^{N_{\text{FFT}}-1} \check{a}_{l,k'}\check{h}_{l'-l} + \check{n}_{l',k'}, \tag{11.27}$$

where $\check{h}_l = \check{h}(lT)$. Observe that, for a fixed k', the samples $\check{y}_{l',k'}$, for $l = 0, \ldots, N_{\text{FFT}} - 1$, do not suffer from interference of the time-domain values $\check{a}_{l,k}$ from other OFDM symbols (i.e., $k \neq k'$). Note also that $\check{h}_l$ is zero for $l < 0$ and $l \geq N_{\text{DS}}$. For a fixed k', stacking the N_{FFT} samples yields

$$\check{\mathbf{y}}_{k'} = \check{\mathbf{H}}\check{\mathbf{a}}_{k'} + \check{\mathbf{n}}_{k'}, \tag{11.28}$$

where $\mathbb{E}\{\check{\mathbf{N}}_{k'}\check{\mathbf{N}}_{k'}^H\} = \mathbf{I}_{N_{FFT}}N_0$, $\check{\mathbf{a}}_{k'} = [\check{a}_{0,k'},\ldots,\check{a}_{N_{FFT}-1,k'}]^T$ and $\check{\mathbf{H}}$ is an $N_{FFT} \times N_{FFT}$ circulant matrix. For $N_{FFT} = 8$ and $N_{DS} = 3$, the circulant matrix looks as follows:

$$\check{\mathbf{H}} = \begin{bmatrix} \check{h}_0 & 0 & 0 & 0 & 0 & 0 & \check{h}_2 & \check{h}_1 \\ \check{h}_1 & \check{h}_0 & 0 & 0 & 0 & 0 & 0 & \check{h}_2 \\ \check{h}_2 & \check{h}_1 & \check{h}_0 & 0 & 0 & 0 & 0 & 0 \\ 0 & \check{h}_2 & \check{h}_1 & \check{h}_0 & 0 & 0 & 0 & 0 \\ 0 & 0 & 0 & \check{h}_1 & \check{h}_0 & 0 & 0 & 0 \\ 0 & 0 & 0 & \check{h}_2 & \check{h}_1 & \check{h}_0 & 0 & 0 \\ 0 & 0 & 0 & 0 & \check{h}_2 & \check{h}_1 & \check{h}_0 & 0 \\ 0 & 0 & 0 & 0 & 0 & \check{h}_2 & \check{h}_1 & \check{h}_0 \end{bmatrix}. \tag{11.29}$$

Any circulant matrix $\check{\mathbf{H}}$ has the property that $\mathbf{F}^H\check{\mathbf{H}}\mathbf{F}$ is a diagonal matrix, so

$$\mathbf{y}_{k'} = \mathbf{F}^H\check{\mathbf{y}}_{k'} \tag{11.30}$$

$$= \mathbf{F}^H\check{\mathbf{H}}\mathbf{F}\mathbf{a}_{k'} + \mathbf{F}^H\check{\mathbf{n}}_{k'} \tag{11.31}$$

$$= \mathbf{H}\mathbf{a}_{k'} + \mathbf{n}_{k'}, \tag{11.32}$$

where $\mathbb{E}\{\mathbf{N}_{k'}\mathbf{N}_{k'}^H\} = \mathbf{I}_{N_{FFT}}N_0$ and $\mathbf{H} = \mathrm{diag}(H_0, H_1, \ldots, H_{N_{FFT}-1})$ with

$$H_q = \frac{1}{N_{FFT}} \sum_{l=0}^{N_{FFT}-1} \check{h}_l e^{j 2\pi q l/N_{FFT}}. \tag{11.33}$$

In other words,

$$y_{k',q} = H_q a_{k',q} + n_{k',q} \tag{11.34}$$

for $k' = 0,\ldots,N-1$ and $q = 0,\ldots,N_{FFT}-1$. The complete observation $\mathbf{y}$ is obtained by stacking $\mathbf{y}_{k'}$ for $k' = 0,\ldots,N-1$.

The equalizer

Our observation model allows us to derive a *structured equalizer* [7] by factorization of $p(\mathbf{Y} = \mathbf{y}|\mathbf{A},\mathcal{M})$. Since the noise samples are independent across time instants (k) and subcarriers (q),

$$p(\mathbf{Y} = \mathbf{y}|\mathbf{A} = \mathbf{a},\mathcal{M}) \propto \prod_{k=0}^{N-1} \exp\left(-\frac{1}{N_0}\|\mathbf{y}_k - \mathbf{H}\mathbf{a}_k\|^2\right) \tag{11.35}$$

$$= \prod_{k=0}^{N-1} \prod_{q=0}^{N_{FFT}-1} \exp\left(-\frac{1}{N_0}|y_{k,q} - H_q a_{k,q}|^2\right). \tag{11.36}$$

Hence, the factor graph of $p(\mathbf{Y} = \mathbf{y}|\mathbf{A},\mathcal{M})$ consists of $N \times N_{FFT}$ disjoint graphs. It then follows immediately that $\mu_{eq \to A_{k,q}}(a) \propto \exp(-|y_{k,q} - H_q a|^2/N_0)$. Note that the messages from the equalizer to the demapper do not require the messages $\mu_{A_{k',q'} \to \mathrm{eq}}(a)$ from the demapper to the equalizer. Hence, equalization for OFDM needs to be performed once only; no iterations are necessary.

In practical OFDM systems, different subcarriers may use different constellations (for instance, depending on the channel quality $|H_q|$ on subcarrier q). The extension to such a scheme is straightforward.

11.4   Main points

In this chapter, we have shown how to obtain suitable observation models $\mathbf{y}$ from the received signal $r(t)$, for transmission schemes with a single user, where both the transmitter and the receiver are equipped with one antenna. We have considered both single-carrier and multi-carrier modulation. In contrast to the conventional one-shot approach in receiver design, factor graphs allow us to derive near-optimal iterative receivers in an elegant way.

For single-carrier transmission, we have described the matched-filter receiver, the whitened matched-filter receiver, and the oversampling receiver. In most cases a sliding-window equalizer combined with MMSE or MC equalization needs to be used. For multi-carrier modulation, the standard OFDM receiver was derived. The observation leads to the well-known structured (non-iterative) equalizer for OFDM.

12 Equalization: multi-antenna communication

12.1 Introduction

From Chapter 2 we know that in multi-antenna (or MIMO) communication the transmitter and/or the receiver are equipped with multiple antennas (see Fig. 12.1). Multiple receive antennas allow the reception of multiple independent copies of the transmitted signal, translating into more reliable communications through diversity: when one receive antenna is in a deep fade, another antenna may experience a better channel. Multiple transmit antennas can be used either to increase throughput (since independent data streams can be transmitted on the different antennas) or to ensure more reliable communication (through diversity). The receiver has the task of combining the information on its antennas and of possibly separating the signals coming from different transmit antennas. In this chapter, we will apply the factor-graph framework and derive receivers for single- and multi-carrier modulation. In particular, receivers for space–time coding, spatial multiplexing, and MIMO-OFDM will be detailed.

This chapter is organized as follows.

- In **Section 12.2** we focus on single-carrier modulation and describe how a suitable observation $\mathbf{y}$ can be obtained, and how messages from the equalizer node to the

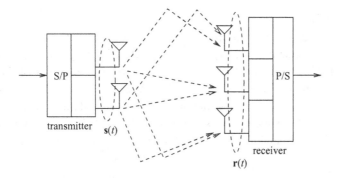

Figure 12.1. Single-user, multi-antenna communication with $N_T = 2$ transmit and $N_R = 3$ receive antennas: the N_T-dimensional equivalent baseband transmitted signal $\mathbf{s}(t)$ propagates through the equivalent baseband channel, and is corrupted by thermal noise at the receiver. The resulting equivalent baseband received signal is denoted by an N_R-dimensional vector $\mathbf{r}(t)$.

demapper node can be computed. Both spatial multiplexing and space–time coding will be covered.

- **Section 12.3** deals with multi-carrier modulation in the form of MIMO-OFDM.

Whenever applicable, equivalent receivers from the technical literature are cited.

12.2   Single-carrier modulation

12.2.1   The received waveform

As we have seen in Section 2.4.1, we can write the transmitted signal at a particular time t as an $N_T \times 1$ vector $\mathbf{s}(t)$,

$$\mathbf{s}(t) = \sqrt{E_s} \sum_{k=-\infty}^{+\infty} \mathbf{a}_k p(t - kT), \tag{12.1}$$

where $\mathbf{a}_k$ is the vector of transmitted symbols for the kth symbol period. At the receiver side, we can stack the received signals at time t in a vector of length N_R:

$$\mathbf{r}(t) = \sum_{k=-\infty}^{+\infty} \sum_{n=1}^{N_T} a_k^{(n)} \mathbf{h}^{(n)}(t - kT) + \mathbf{n}(t), \tag{12.2}$$

where $\mathbf{h}^{(n)}(t)$ represents the equivalent channels between the nth transmit antenna and the various receive antennas. The symbol $a_k^{(n)}$ is the symbol transmitted over the nth transmit antenna during the kth symbol period. The Gaussian noise is spectrally white per antenna, and independent from antenna to antenna (we say that the noise is spatially white). We distinguish between two transmission schemes: space–time coding and spatial multiplexing.

Space–time coding
Let us revisit the Alamouti scheme from Chapter 2. Our goal is to transmit the sequence $a_0, a_1, \ldots, a_{N_s-1}$ over an $N_T = 2$, $N_R = 1$ MIMO channel using the Alamouti space–time block code [22]. Assuming that N_s is even, we group the data symbols in couples so that the kth group is given by (a_{2k}, a_{2k+1}), $k = 0, \ldots, N_s/2 - 1$. For the transmission of each couple we use two symbol durations. During symbol duration $2k$, we transmit $\mathbf{a}_{2k} = [a_{2k} \quad a_{2k+1}]^T$, while during symbol duration $2k + 1$, we transmit $\mathbf{a}_{2k+1} = [-a_{2k+1}^* \quad a_{2k}^*]^T$. The Alamouti code is a special case of an orthogonal space–time block code [23, 24].

Spatial multiplexing
In spatial multiplexing, the symbols are sent over the MIMO channel directly: a sequence of N_s data symbols (say $a_0, \ldots, a_{N_s-1}$) can be sent in N_s/N_T symbol durations. At time

$k \in \{0, \ldots, N_s/N_T - 1\}$ we transmit

$$\mathbf{a}_k = \left[a_k^{(1)}, a_k^{(2)}, \ldots, a_k^{(N_T)} \right]^T \tag{12.3}$$

$$= [a_{kN_T}, a_{kN_T+1}, \ldots, a_{(k+1)N_T-1}]^T. \tag{12.4}$$

12.2.2 The receiver for a frequency-flat channel

In a frequency-flat channel, we can express the channel between the nth transmit antenna and the mth receive antenna as

$$h_m^{(n)}(t) = \alpha_m^{(n)} p(t - \tau), \tag{12.5}$$

where $p(t)$ is the square-root Nyquist transmit pulse for rate $1/T$, τ is the propagation delay between transmitter and receiver (which is equal for all transmit and receive antennas), and $\alpha_m^{(n)}$ represents the complex channel gain (including transmit energy) between the nth transmit antenna and the mth receive antenna. In other words,

$$\mathbf{h}^{(n)}(t) = \begin{bmatrix} \alpha_1^{(n)} \\ \vdots \\ \alpha_{N_R}^{(n)} \end{bmatrix} p(t - \tau). \tag{12.6}$$

On the mth receive antenna, a filter $p^*(-t - \tau)$ is applied. Sampling the output of the filter at time $k'T$, $k' \in \mathbb{Z}$, yields the observation

$$y_{m,k'} = \sum_{k=-\infty}^{+\infty} \int_{-\infty}^{+\infty} p^*(-u - \tau) r(-u + k'T) du \tag{12.7}$$

$$= \sum_{n=1}^{N_T} \alpha_m^{(n)} a_{k'}^{(n)} + n_{m,k'}. \tag{12.8}$$

The noise samples are both spatially uncorrelated and temporally uncorrelated: $\mathbb{E}\{N_{m,k} N_{m',k'}^*\} = N_0 \delta_{k-k'} \delta_{m-m'}$. Stacking for fixed k' gives us an $N_R \times 1$ vector

$$\mathbf{y}_{k'} = \mathbf{H} \mathbf{a}_{k'} + \mathbf{n}_{k'}, \tag{12.9}$$

where $\mathbf{H}$ is the $N_R \times N_T$ channel matrix, with $[\mathbf{H}]_{m,n} = \alpha_m^{(n)}$.

Space–time coding
The channel matrix in (12.9) is now a 1×2 vector. We can write the observation at time $2k$ as

$$y_{2k} = [\alpha^{(1)} \quad \alpha^{(2)}] \begin{bmatrix} a_{2k} \\ a_{2k+1} \end{bmatrix} + n_{2k} \tag{12.10}$$

and that at time $2k + 1$ as

$$y_{2k+1} = [\alpha^{(1)} \quad \alpha^{(2)}] \begin{bmatrix} -a_{2k+1}^* \\ a_{2k}^* \end{bmatrix} + n_{2k+1}. \tag{12.11}$$

Taking the complex conjugate of y_{2k+1} and stacking yields

$$\begin{bmatrix} y_{2k} \\ y_{2k+1}^* \end{bmatrix} = \begin{bmatrix} \alpha^{(1)} & \alpha^{(2)} \\ (\alpha^{(2)})^* & -(\alpha^{(1)})^* \end{bmatrix} \begin{bmatrix} a_{2k} \\ a_{2k+1} \end{bmatrix} + \begin{bmatrix} n_{2k} \\ n_{2k+1}^* \end{bmatrix} \tag{12.12}$$

$$= \tilde{\mathbf{H}} \begin{bmatrix} a_{2k} \\ a_{2k+1} \end{bmatrix} + \begin{bmatrix} n_{2k} \\ n_{2k+1}^* \end{bmatrix}. \tag{12.13}$$

By virtue of the design of the Alamouti scheme, the matrix $\tilde{\mathbf{H}}$ is an orthogonal matrix, so

$$\tilde{\mathbf{H}}^H \tilde{\mathbf{H}} = \begin{bmatrix} E_h & 0 \\ 0 & E_h \end{bmatrix}, \tag{12.14}$$

where $E_h = |\alpha^{(1)}|^2 + |\alpha^{(2)}|^2$. Hence

$$\tilde{\mathbf{y}}_k = \frac{1}{E_h} \tilde{\mathbf{H}}^H \begin{bmatrix} y_{2k} \\ y_{2k+1}^* \end{bmatrix} \tag{12.15}$$

$$= \begin{bmatrix} a_{2k} \\ a_{2k+1} \end{bmatrix} + \tilde{\mathbf{n}}_k \tag{12.16}$$

$$= \begin{bmatrix} \tilde{y}_{2k} \\ \tilde{y}_{2k+1} \end{bmatrix}, \tag{12.17}$$

where $\mathbb{E}\{\tilde{\mathbf{N}}_k \tilde{\mathbf{N}}_{k'}^H\} = \delta_{k-k'} \mathbf{I}_2 N_0 / E_h$. Let us stack $\tilde{\mathbf{y}}_k$ for $k = 0, \ldots, N_s/2 - 1$ into $\tilde{\mathbf{y}}$, then

$$\tilde{\mathbf{y}} = \mathbf{a} + \mathbf{n}. \tag{12.18}$$

This results in the following likelihood function:

$$p(\mathbf{Y} = \tilde{\mathbf{y}} | \mathbf{A} = \mathbf{a}, \mathcal{M}) \propto \exp\left(-\frac{E_h}{N_0} \|\tilde{\mathbf{y}} - \mathbf{a}\|^2\right) \tag{12.19}$$

$$= \prod_{k=0}^{N_s-1} \exp\left(-\frac{E_h}{N_0} |\tilde{y}_k - a_k|^2\right). \tag{12.20}$$

Thus the factor graph of $p(\mathbf{Y} = \mathbf{y} | \mathbf{A}, \mathcal{M})$ consists of N_s disjoint graphs. Applying the SPA to these disjoint graphs results in a *structured equalizer* [22] with

$$\mu_{\text{eq} \to A_k}(a_k) \propto \exp\left(-\frac{E_h}{N_0} |\tilde{y}_k - a_k|^2\right).$$

Note that the messages from the equalizer to the demapper do not require the messages from the demapper to the equalizer $\mu_{A_l \to eq}(a_l)$. Hence, equalization for space–time block codes needs to be performed once only; no iterations are necessary. The Alamouti scheme can also be used when the channel varies slowly: the channel needs to remain constant for only two successive symbol durations.

Spatial multiplexing

We can use a structured approach based on the observation (12.9). We stack the observations $y_0, \ldots, y_{N_s/N_T-1}$ to obtain a long vector $\mathbf{y}$. Since the noise samples for different time instants k are independent,

$$p(\mathbf{Y} = \mathbf{y}|\mathbf{A} = \mathbf{a}, \mathcal{M}) = \prod_{k=0}^{N_s/N_T-1} p(\mathbf{Y}_k = \mathbf{y}_k|\mathbf{A}_k = \mathbf{a}_k, \mathcal{M}) \qquad (12.21)$$

$$\propto \prod_{k=0}^{N_s/N_T-1} \exp\left(-\frac{1}{N_0}\|\mathbf{y}_k - \mathbf{Ha}_k\|^2\right) \qquad (12.22)$$

with the factor graph shown in Fig. 12.2. This leads to a *structured equalizer* [106–108]:

$$\mu_{eq \to A_k^{(n)}}(a) \propto \sum_{\mathbf{a}_k \in \Omega^{N_T}: a_k^{(n)} = a} \exp\left(-\frac{1}{N_0}\|\mathbf{y}_k - \mathbf{Ha}_k\|^2\right) \prod_{n' \neq n} \mu_{A_k^{(n')} \to eq}\left(a_k^{(n')}\right).$$

$$(12.23)$$

The overall complexity of the equalizer scales as $\mathcal{O}(N_s|\Omega|^{N_T})$. The complexity can be further reduced by using an *MMSE equalizer* [109] or an *MC equalizer* [100] on the observation $\mathbf{y}_k = \mathbf{Ha}_k + \mathbf{n}_k$. Note that both these approximate techniques operate only on a single node in Fig 12.2. For $N_T > 1$, the equalizers are all iterative, since messages from equalizer to demapper depend on messages from demapper to equalizer.

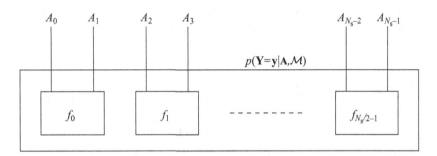

Figure 12.2. Factor graph of $p(\mathbf{Y} = \mathbf{y}|\mathbf{A}, \mathcal{M})$ for a frequency-flat MIMO system with spatial multiplexing. In this example, $N_T = 2$ with $f_m(a_{2m}, a_{2m+1}) = \exp(-\|\mathbf{y}_m - \mathbf{Ha}_m\|^2/N_0)$, where $\mathbf{a}_m = [a_{2m} \quad a_{2m+1}]^T$.

12.2.3   The receiver for a frequency-selective channel

A possible receiver operates by oversampling the signals on the various receive antennas as follows. We filter the received waveforms on the receive antennas by a square-root Nyquist pulse for a rate N/T. We then sample the filtered signals at time instants $kT + mT/N$, for $m = 0, \ldots, N - 1$ and $k \in \mathbb{Z}$. Stacking the N_R observations at time instant $kT + mT/N$ yields (see also Section 11.2.4)

$$\mathbf{y}_k^{(m)} = \sum_{n=1}^{N_T} \sum_{l=L_{min}}^{L_{max}} \mathbf{h}_l^{(n,m)} a_{k-l}^{(n)} + \mathbf{n}_k^{(m)}, \tag{12.24}$$

where $\mathbf{h}_l^{(n,m)}$ is an $N_R \times 1$ vector containing the values of the channel impulse responses at time $lT + mT/N$ (with $L_{min} \le l \le L_{max}$ and $0 \le m \le N - 1$) between the nth transmit antenna and the N_R receive antennas and $\mathbf{n}_k^{(m)}$ is the noise component at time $kT + mT/N$ on the N_R receive antennas. The noise samples are both spatially uncorrelated and temporally uncorrelated:

$$\mathbb{E}\{\mathbf{N}_k^{(m)}(\mathbf{N}_{k'}^{(m')})^H\} = \delta_{k-k'}\delta_{m-m'}\mathbf{I}_{N_R}N_0. \tag{12.25}$$

We introduce $\mathbf{a}_k = [a_k^{(1)}, \ldots, a_k^{(N_T)}]^T$ and the $N_R \times N_T$ matrix $\mathbf{H}_l^{(m)} = [\mathbf{h}_l^{(1,m)}, \ldots, \mathbf{h}_l^{(N_T,m)}]$ and write

$$\mathbf{y}_k^{(m)} = \sum_{l=L_{min}}^{L_{min}} \mathbf{H}_l^{(m)} \mathbf{a}_{k-l} + \mathbf{n}_k^{(m)}. \tag{12.26}$$

Stacking the samples (first for fixed k and then for all k) again leads to

$$\mathbf{y} = \mathbf{H}\mathbf{a} + \mathbf{n}. \tag{12.27}$$

Example 12.1. *For $L_{min} = 0$, $L_{max} = 1$, $N = 1$, $N_T = 2$, and $N_R = 2$, since $N = 1$, only $m = 0$ can occur, and we can safely drop the superscript m. Suppose that we take as observations $\mathbf{y}_k$, for $k = -1, 0, 1, 2, 3$. Then*

$$\mathbf{y} = [\mathbf{y}_{-1}^T \mathbf{y}_0^T \mathbf{y}_1^T \mathbf{y}_2^T \mathbf{y}_3^T]^T$$

$$= \begin{bmatrix} \mathbf{H}_1 & \mathbf{H}_0 & 0 & 0 & 0 & 0 \\ 0 & \mathbf{H}_1 & \mathbf{H}_0 & 0 & 0 & 0 \\ 0 & 0 & \mathbf{H}_1 & \mathbf{H}_0 & 0 & 0 \\ 0 & 0 & 0 & \mathbf{H}_1 & \mathbf{H}_0 & 0 \\ 0 & 0 & 0 & 0 & \mathbf{H}_1 & \mathbf{H}_0 \end{bmatrix} \begin{bmatrix} \mathbf{a}_{-2} \\ \mathbf{a}_{-1} \\ \mathbf{a}_0 \\ \mathbf{a}_1 \\ \mathbf{a}_2 \\ \mathbf{a}_3 \end{bmatrix} + \begin{bmatrix} \mathbf{n}_{-1} \\ \mathbf{n}_0 \\ \mathbf{n}_1 \\ \mathbf{n}_2 \\ \mathbf{n}_3 \end{bmatrix}.$$

The equalizer

We consider only spatial multiplexing: an *MMSE equalizer with windowing* [110–112] or possibly *MC equalizers (with windowing)* can be applied. Observe that MIMO equalizers for spatial multiplexing always require the messages from the demapper. Hence, equalization can be performed in an iterative fashion.

12.3 Multi-carrier modulation

12.3.1 The received waveform

We know from Section 2.4.2 that the transmitted signal at the nth transmit antenna can be written as

$$s^{(n)}(t) = \sqrt{\frac{E_s N_{FFT}}{N_{FFT} + N_{CP}}} \sum_{k=-\infty}^{+\infty} \sum_{l=-N_{CP}}^{N_{FFT}-1} \check{a}_{l,k}^{(n)} p^{(n)}(t - lT - kT_{OFDM}), \qquad (12.28)$$

where $\check{a}_{l,k}^{(n)}$ is the lth time-domain value in the kth OFDM symbol, transmitted on the nth transmit antenna. At the receiver, equipped with N_R antennas, the received signal at time t can be expressed as an $N_R \times 1$ vector

$$\mathbf{r}(t) = \sum_{k=-\infty}^{+\infty} \sum_{n=1}^{N_T} \sum_{l=-N_{CP}}^{N_{FFT}-1} \check{a}_{l,k}^{(n)} \mathbf{h}^{(n)}(t - lT - kT_{OFDM}) + \mathbf{n}(t), \qquad (12.29)$$

where $\mathbf{h}^{(n)}(t)$ is an $N_R \times 1$ vector representing the equivalent channel between the nth transmit antenna and the various receive antennas (including the factor $\sqrt{E_s N_{FFT}/(N_{FFT} + N_{CP})}$). The sequence of N_s coded data symbols is sent during $N = N_s/(N_{FFT}N_T)$ consecutive OFDM symbols. We will denote by $a_{q,k}^{(n)}$ the symbol over the qth subcarrier during the kth OFDM symbol on the nth transmit antenna.

12.3.2 MIMO-OFDM receivers

At every receive antenna we have a basic OFDM receiver (see Fig. 12.3): we filter the signal with a square-root Nyquist pulse for a rate $1/T$, resulting in an $N_R \times 1$ observation at time t:

$$\check{\mathbf{y}}(t) = \sum_{k=-\infty}^{+\infty} \sum_{n=1}^{N_T} \sum_{l=-N_{CP}}^{N_{FFT}-1} \check{a}_{l,k}^{(n)} \check{\mathbf{h}}^{(n)}(t - lT - kT_{OFDM}) + \check{\mathbf{n}}(t). \qquad (12.30)$$

The support of the channel $\check{\mathbf{h}}^{(n)}(t)$ is defined in a similar way to that for the single-antenna case: the channel $\check{\mathbf{h}}^{(n)}(t) = \mathbf{0}$ for all t outside the support $[t_{min}, t_{max}]$, as depicted in Fig. 12.4. For notational convenience, we assume the support of $\check{\mathbf{h}}^{(n)}(t)$ to be $[0, N_{DS}T)$, for all $n \in \{1, \ldots, N_T\}$, with $N_{DS} \leq N_{CP} + 1$. Sampling at time instants $l'T + k'T_{OFDM}$,

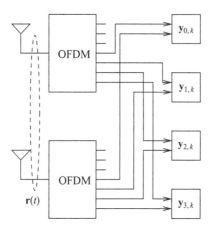

Figure 12.3. A MIMO-OFDM receiver for $N_R = 2$, with $N_{FFT} = 4$ subcarriers and a cyclic prefix of length 3.

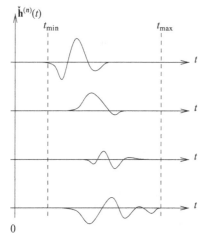

Figure 12.4. Support for a MIMO-OFDM channel $\check{\mathbf{h}}^{(n)}(t)$ for $N_R = 4$.

$l' = 0, \ldots, N_{FFT} - 1$, $k' = 0, \ldots, N - 1$, discarding the samples corresponding to the cyclic prefix, and applying the DFT matrix $\mathbf{F}^H$ yields the following observation for OFDM symbol k, receive antenna m, and subcarrier q (see also Section 11.3.2):

$$y_{q,k,m} = \sum_{n=1}^{N_T} H_{q,m}^{(n)} a_{q,k}^{(n)} + n_{q,k,m}. \tag{12.31}$$

If we focus on a single subcarrier and a single OFDM symbol, we can stack the observations from the various receive antennas into an $N_R \times 1$ vector, and write

$$\mathbf{y}_{q,k} = \mathbf{H}_q \mathbf{a}_{q,k} + \mathbf{n}_{q,k}, \tag{12.32}$$

where $\mathbb{E}\{\mathbf{N}_{q,k}\mathbf{N}_{q',k'}^{\mathrm{H}}\} = \delta_{k-k'}\delta_{q-q'}N_0\mathbf{I}_{N_\mathrm{R}}$, $\mathbf{a}_{q,k} = [a_{q,k}^{(1)}, \ldots, a_{q,k}^{(N_\mathrm{T})}]^\mathrm{T}$, and $\mathbf{H}_l$ is an $N_\mathrm{R} \times N_\mathrm{T}$ matrix containing the channel gains on the lth subcarrier between the various transmit and receive antennas. The MIMO-OFDM receiver boils down to a bank of standard MIMO receivers for frequency-flat channels, one for each subcarrier. The final observation $\mathbf{y}$ is obtained by stacking $\mathbf{y}_{q,k}$ for all q and k.

The equalizer

Since noise samples at different time instants (k) and for different subcarriers (q) are independent, we can express the likelihood function as

$$p(\mathbf{Y} = \mathbf{y}|\mathbf{A} = \mathbf{a}, \mathcal{M}) = \prod_{q=0}^{N_{\mathrm{FFT}}-1}\prod_{k=0}^{N-1} p(\mathbf{Y}_{q,k} = \mathbf{y}_{q,k}|\mathbf{A}_{q,k} = \mathbf{a}_{q,k}, \mathcal{M}) \qquad (12.33)$$

$$\propto \prod_{q=0}^{N_{\mathrm{FFT}}-1}\prod_{k=0}^{N-1} \exp\left(-\frac{1}{N_0}\|\mathbf{y}_{q,k} - \mathbf{H}_q\mathbf{a}_{q,k}\|^2\right), \qquad (12.34)$$

leading to a *structured equalizer* [113, 114]:

$$\mu_{\mathrm{eq}\to A_{q,k}^{(n)}}(a) \propto \sum_{\mathbf{a}_{q,k}\in\Omega^{N_\mathrm{T}}:a_{q,k}^{(n)}=a} \exp\left(-\frac{1}{N_0}\|\mathbf{y}_{q,k} - \mathbf{H}_q\mathbf{a}_{q,k}\|^2\right)\prod_{n'\neq n}\mu_{A_{q,k}^{(n')}\to\mathrm{eq}}\left(a_{q,k}^{(n')}\right).$$

$$(12.35)$$

The overall complexity of the equalizer scales as $\mathcal{O}(N_\mathrm{s}|\Omega|^{N_\mathrm{T}})$. Alternatively, we can apply an *MMSE equalizer* [113] or an *MC equalizer* to reduce the overall complexity. Note that for $N_\mathrm{T} > 1$ the equalizers are iterative.

12.4   Main points

The MIMO systems theoretically exhibit significant capacity gains compared with single-antenna transmission. In order to exploit these potential gains fully in a practical setting, state-of-the-art iterative receivers must be considered. We have derived such receivers both for single- and for multi-carrier modulation. With the exception of orthogonal space–time codes over frequency-flat channels, all receivers require iterating between equalization and demapping/decoding.

13 Equalization: multi-user communication

13.1 Introduction

In multi-user communications, users transmit signals to a single receiver over the same channel (see Fig. 13.1). For the receiver to recover the information from the various users successfully, the channel must be shared between the users.

Multiple-access techniques such as time-division multiple access (TDMA) and frequency-division multiple access (FDMA) make users completely orthogonal to each other, resulting in simple receivers, but at the same time giving rise to significant losses in terms of bandwidth efficiency. To overcome this, techniques such as direct-sequence code-division multiple access (DS-CDMA) were developed, whereby all users transmit simultaneously over the same frequency band. A conventional DS-CDMA receiver would consist of a bank of single-user detectors (so-called Rake receivers [115]), one for each user. Multiple-access interference (MAI) is ignored at the output of the Rake receiver.

Thanks to the pioneering work of Verdú [31], it has become clear that detecting the information from the various users *jointly* can lead to significant performance gains. This

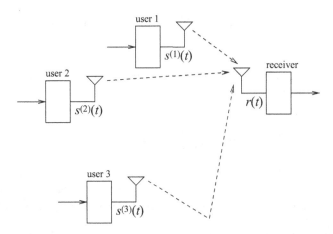

Figure 13.1. Multi-user, single-antenna communication with $N_u = 3$ active users transmitting to a single receiver. The equivalent baseband transmitted signal from user k is $s^{(k)}(t)$. After propagation through the channel, the receiver obtains the superposition of N_u signals, corrupted by thermal noise.

is known as multi-user detection (MUD). Since optimal MUD is usually intractable, iterative MUD has received a great deal of attention from the research community [97, 116]. The standard reference work on iterative MUD is by Wang and Poor [98]. In this chapter, we will see how such iterative MUD schemes for DS-CDMA fit into the general factor-graph framework.

Another popular multiple-access technique is orthogonal frequency-division multiple access (OFDMA), whereby users are assigned different subcarriers. As we will see, provided that some synchronization constraints are met, users remain orthogonal at the receiver, resulting in simple non-iterative equalization.

Although many other multiple-access schemes exist, the equalizers we will derive here will give the reader a flavor of the general techniques. The reader should also bear in mind that multi-user communication is very similar to multi-antenna communication: a multi-user system can be considered as a MIMO system in which the transmit antennas are separated in space (since they belong to different users).

This chapter is organized as follows.

- We will start with DS-CDMA in **Section 13.2**, considering both synchronous and asynchronous transmission.
- We then move on to OFDMA in **Section 13.3**.

13.2   Direct-sequence code-division multiple access

13.2.1   The received waveform

The transmitted signal of the nth user can be written as (see Section 2.5)

$$s^{(n)}(t) = \sqrt{E_s^{(n)}} \sum_{k=-\infty}^{+\infty} a_k^{(n)} p^{(n)}(t - kT), \tag{13.1}$$

where $a_k^{(n)}$ is the kth symbol sent by the nth user and $p^{(n)}(t)$ is the nth user's transmit pulse. This pulse is given by

$$p^{(n)}(t) = \frac{1}{\sqrt{N_{SG}}} \sum_{i=0}^{N_{SG}-1} d_i^{(n)} p_S\left(t - i\frac{T}{N_{SG}}\right), \tag{13.2}$$

where $\mathbf{d}^{(n)} = [d_0^{(n)}, \ldots, d_{N_{SG}-1}^{(n)}]^T$ is the kth user's spreading sequence and $p_S(t)$ is a square-root Nyquist pulse for a rate N_{SG}/T. For a total of N_u users, the received signal can be written as

$$r(t) = \sum_{n=1}^{N_u} \sum_{k=-\infty}^{+\infty} a_k^{(n)} h^{(n)}(t - kT) + n(t), \tag{13.3}$$

where $h^{(n)}(t)$ is the equivalent channel for the nth user:

$$h^{(n)}(t) = \sqrt{E_s^{(n)}} \int_{-\infty}^{+\infty} h_{ch}^{(n)}(u)p^{(n)}(t-u)du. \tag{13.4}$$

13.2.2 The receiver for synchronous transmission

In the conventional synchronous transmission scheme, the transmitters are synchronized in such a way that the signals arrive at the same time at the receiver. Additionally, a frequency-flat channel is assumed, so that

$$h_{ch}^{(n)}(t) = \alpha^{(n)}\delta(t-\tau) \tag{13.5}$$

and thus

$$h^{(n)}(t) = \sqrt{E_s^{(n)}}\alpha^{(n)}p^{(n)}(t-\tau). \tag{13.6}$$

We filter the received signal by a bank of N_u matched filters, one for every user: $(h^{(n')}(-t))^*$, $n' = 1, \ldots, N_u$. Sampling the output of the n'th filter at time $k'T$ yields

$$y_{k'}^{(n')} = \sum_{n=1}^{N_u} \sum_{k=-\infty}^{+\infty} a_k^{(n)} g_{k'-k}^{(n',n)} + n_{k'}^{(n')} \tag{13.7}$$

with

$$g_{k'}^{(n',n)} = \int_{-\infty}^{+\infty} (h^{(n')}(u))^* h^{(n)}(k'T+u)du \tag{13.8}$$

$$= \frac{A^{(n',n)}\delta_{k'}}{N_{SG}}(d^{(n')})^H d^{(n)} \tag{13.9}$$

where we have introduced $A^{(n',n)} = \sqrt{E_s^{(n)}E_s^{(n')}}(\alpha^{(n')})^*\alpha^{(n)}$. Stacking $y_{k'}^{(n')}$ for all n' yields an $N_u \times 1$ observation vector at time $k'T$:

$$\mathbf{y}_{k'} = \mathbf{H}\mathbf{a}_{k'} + \mathbf{n}_{k'}, \tag{13.10}$$

where $\mathbf{H}$ is an $N_u \times N_u$ matrix such that

$$[\mathbf{H}]_{n',n} = \frac{A^{(n',n)}}{N_{SG}}(d^{(n')})^H d^{(n)} \tag{13.11}$$

and

$$\mathbb{E}\{\mathbf{N}_{k'}\mathbf{N}_{k''}^H\} = \delta_{k'-k''}N_0\mathbf{H}. \tag{13.12}$$

In some cases, the spreading sequences of the users are designed so that they are orthogonal. In those cases, (13.8) simplifies to

$$\frac{A^{(n',n)}}{N_{SG}}(\mathbf{d}^{(n')})^{H}\mathbf{d}^{(n)} = A^{(n,n)}\delta_{n-n'} \tag{13.13}$$

so that $\mathbf{H}$ becomes a diagonal matrix. The final observation $\mathbf{y}$ is obtained by stacking $\mathbf{y}_{k'}$ for all k'.

The equalizer

Since the noise vectors $\mathbf{n}_k$ at different time instants k are independent,

$$p(\mathbf{Y} = \mathbf{y}|\mathbf{A} = \mathbf{a}, \mathcal{M}) = \prod_{k=0}^{N_s-1} p(\mathbf{Y}_k = \mathbf{y}_k|\mathbf{A}_k = \mathbf{a}_k, \mathcal{M}) \tag{13.14}$$

$$\propto \prod_{k=0}^{N_s-1} \exp\left(-\frac{1}{N_0}(\mathbf{y}_k - \mathbf{Ha}_k)^{H}\mathbf{H}^{-1}(\mathbf{y}_k - \mathbf{Ha}_k)\right) \tag{13.15}$$

with a factor-graph shown in Fig. 13.2. This leads to a *structured equalizer*[116]:

$$\mu_{eq \to A_k^{(n)}}(a) \propto \sum_{\mathbf{a}_k \in \Omega^{N_u}:a_k^{(n)}=a} \exp\left(-\frac{1}{N_0}(\mathbf{y}_k - \mathbf{Ha}_k)^{H}\mathbf{H}^{-1}(\mathbf{y}_k - \mathbf{Ha}_k)\right)$$

$$\times \prod_{n' \neq n} \mu_{A_k^{(n')} \to eq}\left(a_k^{(n')}\right). \tag{13.16}$$

The complexity scales as $\mathcal{O}(N_s|\Omega|^{N_u})$. This equalizer can be simplified by using an *MMSE equalizer* [98] or an *MC equalizer* [100]. Observe that, when $N_u > 1$, the equalizers will be iterative.

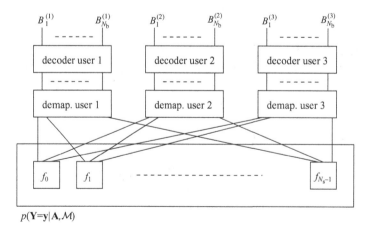

$p(\mathbf{Y}=\mathbf{y}|\mathbf{A},\mathcal{M})$

Figure 13.2. Factor graph for DS-CDMA with synchronous transmission for $N_u = 3$ users. The node f_k represents the function $p(\mathbf{Y}_k = \mathbf{y}_k|\mathbf{A}_k = \mathbf{a}_k, \mathcal{M})$, for $k = 0, \ldots, N_s - 1$.

In the particular case when users have orthogonal spreading codes, $p(\mathbf{Y} = \mathbf{y}|\mathbf{A}, \mathcal{M})$ can be factorized as

$$p(\mathbf{Y} = \mathbf{y}|\mathbf{A} = \mathbf{a}, \mathcal{M}) \propto \prod_{k=0}^{N_s-1} \exp\left(-\frac{1}{N_0}(\mathbf{y}_k - \mathbf{H}\mathbf{a}_k)^H \mathbf{H}^{-1}(\mathbf{y}_k - \mathbf{H}\mathbf{a}_k)\right) \qquad (13.17)$$

$$\propto \prod_{k=0}^{N_s-1} \prod_{n=1}^{N_u} \exp\left(-\frac{1}{N_0 A^{(n,n)}}|y_k^{(n)} - A^{(n,n)} a_k^{(n)}|^2\right), \qquad (13.18)$$

so that

$$\mu_{\text{eq} \to A_k^{(n)}}(a) \propto \exp\left(-\frac{1}{N_0 A^{(n,n)}}|y_k^{(n)} - A^{(n,n)} a|^2\right), \qquad (13.19)$$

leading to a non-iterative *structured equalizer*.

13.2.3 Receivers for asynchronous transmission

In asynchronous transmission, users are no longer synchronized and the channels may be frequency-selective. This means that the nth user's channel, as seen by the receiver is given by

$$h_{\text{ch}}^{(n)}(t) = \sum_{l=0}^{L-1} \alpha_l^{(n)} \delta(t - \tau_l^{(n)}). \qquad (13.20)$$

Two types of receivers will be considered: the matched-filter receiver and the oversampling receiver.

13.2.3.1 The matched-filter receiver

The matched-filter receiver operates as in the synchronous case (see Fig. 13.3): we filter the received signal by a bank of matched filters, one for every user: $(h^{(n')}(-t))^*$ for

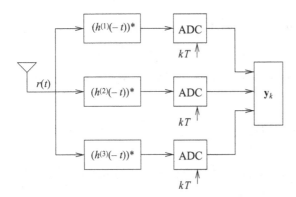

Figure 13.3. Matched-filter receiver for asynchronous DS-CDMA with $N_u = 3$ users.

$n' = 1, \ldots, N_u$. Sampling the output of the n'th filter at time $k'T$ yields

$$y_{k'}^{(n')} = \sum_{n=1}^{N_u} \sum_{k=-\infty}^{+\infty} a_k^{(n)} g_{k'-k}^{(n',n)} + n_{k'}^{(n')} \tag{13.21}$$

$$= \sum_{n=1}^{N_u} \left\{ \sum_{l=L_{\min}^{(n)}}^{L_{\max}^{(n)}} g_l^{(n',n)} a_{k'-l}^{(n)} \right\} + n_{k'}^{(n')} \tag{13.22}$$

for some $L_{\min}^{(n)} \in \mathbb{Z}$ and $L_{\max}^{(n)} \in \mathbb{Z}$, where $g_l^{(n',n)}$ was defined in (13.8). The noise is now correlated with

$$\mathbb{E}\left\{ N_{k'}^{(n')} \left(N_{k''}^{(n'')} \right)^* \right\} = N_0 g_{k'-k''}^{(n',n'')}. \tag{13.23}$$

Introducing $L_{\min} = \min_n L_{\min}^{(n)}$ and $L_{\max} = \max_n L_{\max}^{(n)}$, we can write

$$y_{k'}^{(n')} = \sum_{n=1}^{N_u} \sum_{l=L_{\min}}^{L_{\max}} g_l^{(n',n)} a_{k'-l}^{(n)} + n_{k'}^{(n')}. \tag{13.24}$$

Stacking for a fixed k' gives us

$$\mathbf{y}_{k'} = \tilde{\mathbf{H}} \tilde{\mathbf{a}}_{k'} + \mathbf{n}_{k'} \tag{13.25}$$

where $\tilde{\mathbf{a}}_{k'}$ is an $N_u(L_{\max} - L_{\min} + 1) \times 1$ vector: $\tilde{\mathbf{a}}_{k'} = \left[\mathbf{a}_{k'-L_{\max}}^{\mathrm{T}}, \ldots, \mathbf{a}_{k'-L_{\min}}^{\mathrm{T}} \right]^{\mathrm{T}}$ with $\mathbf{a}_{k'} = \left[a_{k'}^{(1)}, \ldots, a_{k'}^{(N_u)} \right]$. $\tilde{\mathbf{H}}$ is an $N_u \times N_u(L_{\max} - L_{\min} + 1)$ matrix. Stacking for different k' gives rise to

$$\mathbf{y} = \mathbf{H}\mathbf{a} + \mathbf{n}, \tag{13.26}$$

where $\mathbf{H}$ is a band matrix. Note that the noise generally is not white.

Example 13.1. *Consider a system with $N_u = 2$ users, $N_{SG} = 2$, $L_{\min} = 1$ and $L_{\max} = 3$. Then*

$$\mathbf{y}_{k'} = \left[y_{k'}^{(1)} \; y_{k'}^{(2)} \right]^{\mathrm{T}},$$

$$\mathbf{a}_{k'} = \left[a_{k'}^{(1)} \; a_{k'}^{(2)} \right]^{\mathrm{T}},$$

$$\tilde{\mathbf{a}}_{k'} = \left[a_{k'-3}^{(1)} \; a_{k'-3}^{(2)} \; a_{k'-2}^{(1)} \; a_{k'-2}^{(2)} \; a_{k'-1}^{(1)} \; a_{k'-1}^{(2)} \right]^{\mathrm{T}},$$

and

$$\tilde{\mathbf{H}} = \begin{bmatrix} g_3^{(1,1)} & g_3^{(1,2)} & g_2^{(1,1)} & g_2^{(1,2)} & g_1^{(1,1)} & g_1^{(1,2)} \\ g_3^{(2,1)} & g_3^{(2,2)} & g_2^{(2,1)} & g_2^{(2,2)} & g_1^{(2,1)} & g_1^{(2,2)} \end{bmatrix}.$$

Stacking for k' and $k'+1$ then yields

$$\begin{bmatrix} y_{k'}^{(1)} \\ y_{k'}^{(2)} \\ y_{k'+1}^{(1)} \\ y_{k'+1}^{(2)} \end{bmatrix} = \begin{bmatrix} g_3^{(1,1)} & g_3^{(1,2)} & g_2^{(1,1)} & g_2^{(1,2)} & g_1^{(1,1)} & g_1^{(1,2)} & 0 & 0 \\ g_3^{(2,1)} & g_3^{(2,2)} & g_2^{(2,1)} & g_2^{(2,2)} & g_1^{(2,1)} & g_1^{(2,2)} & 0 & 0 \\ 0 & 0 & g_3^{(1,1)} & g_3^{(1,2)} & g_2^{(1,1)} & g_2^{(1,2)} & g_1^{(1,1)} & g_1^{(1,2)} \\ 0 & 0 & g_3^{(2,1)} & g_3^{(2,2)} & g_2^{(2,1)} & g_2^{(2,2)} & g_1^{(2,1)} & g_1^{(2,2)} \end{bmatrix} \begin{bmatrix} a_{k'-3}^{(1)} \\ a_{k'-3}^{(2)} \\ a_{k'-2}^{(1)} \\ a_{k'-2}^{(2)} \\ a_{k'-1}^{(1)} \\ a_{k'-1}^{(2)} \\ a_{k'}^{(1)} \\ a_{k'}^{(2)} \end{bmatrix}$$

$$+ \begin{bmatrix} n_{k'}^{(1)} \\ n_{k'}^{(2)} \\ n_{k'+1}^{(1)} \\ n_{k'+1}^{(2)} \end{bmatrix}.$$

13.2.3.2 The oversampling receiver

We filter $r(t)$ by a square-root Nyquist filter for a rate T/N and obtain

$$y_{OS}(t) = \sum_{n=1}^{N_u} \sum_{k=-\infty}^{+\infty} a_k^{(n)} h_{OS}^{(n)}(t - kT) + n_{OS}(t). \tag{13.27}$$

Sampling at time $k'T + mT/N$ ($m = 0, \ldots, N-1$ and $k' \in \mathbb{Z}$) yields

$$y_{k'}^{(m)} = \sum_{n=1}^{N_u} \sum_{k=-\infty}^{+\infty} a_k^{(n)} \underbrace{h_{OS}^{(n)}\left(k'T + m\frac{T}{N} - kT\right)}_{h_{k'-k}^{(n,m)}} + n_{k'}^{(m)} \tag{13.28}$$

$$= \sum_{n=1}^{N_u} \sum_{k=L_{\min}}^{L_{\max}} h_k^{(n,m)} a_{k'-k}^{(n)} + n_{k'}^{(m)}, \tag{13.29}$$

where now the noise samples are uncorrelated: $\mathbb{E}\left\{N_k^{(m)}\left(N_{k'}^{(m')}\right)^*\right\} = N_0 \delta_{m-m'} \delta_{k-k'}$. This is exactly the same model as for the oversampling MIMO receiver with $N_R = 1$. Stacking the observations (first for fixed k' and then for all k') leads again to the model

$$\mathbf{y} = \mathbf{Ha} + \mathbf{n}. \tag{13.30}$$

Equalizers

Both the matched-filter receiver and the oversampling receiver require either a *sliding-window MMSE equalizer [97,98]* or an *MC equalizer*, possibly combined with a *sliding window*.

13.3   Orthogonal frequency-division multiple access

We introduced OFDMA in Section 2.5. In OFDMA, there are N_u users with a single antenna. Every user applies standard OFDM for data transmission. From the point of view of the receiver, this makes OFDMA very similar to MIMO-OFDM. It should come as no great surprise that we can essentially re-use the techniques from MIMO-OFDM (see Section 12.3).

13.3.1   The received waveform

At the receiver, equipped with N_R antennas, the received signal at time t can be expressed as an $N_R \times 1$ vector

$$\mathbf{r}(t) = \sum_{k=-\infty}^{+\infty} \sum_{n=1}^{N_u} \sum_{l=-N_{CP}}^{N_{FFT}-1} \check{a}_{l,k}^{(n)} \mathbf{h}^{(n)}(t - lT - kT_{OFDM}) + \mathbf{n}(t), \qquad (13.31)$$

where $\mathbf{h}^{(n)}(t)$ is an $N_R \times 1$ vector representing the equivalent channel between the nth user and the various receive antennas.

In OFDMA, different users (n) are assigned different subcarriers (q): user n is assigned set $\mathcal{S}_n \subset \{0, \ldots, N_{FFT}-1\}$, with $\mathcal{S}_n \cap \mathcal{S}_{n'} = \phi$, for all $n' \neq n$. Users are allowed only to transmit on the assigned subcarriers. For the sake of simplicity, let us assume that $|\mathcal{S}_n| = N_{FFT}/N_u$, for all n. Hence, the sequence of N_s coded data symbols of user n is sent using $N = N_s/|\mathcal{S}_n| = N_s N_u/N_{FFT}$ consecutive OFDM symbols. We will denote by $a_{q,k}^{(n)}$ the symbol sent over the qth subcarrier of the nth user during the kth OFDM symbol, with $q \in \mathcal{S}_n$. When $q \notin \mathcal{S}_n$, $a_{q,k}^{(n)} = 0$.

13.3.2   The OFDMA receiver

At every receive antenna we have a basic OFDM receiver: we filter the signal with a square-root Nyquist pulse for a rate $1/T$, resulting in the following $N_R \times 1$ observation at time t:

$$\check{\mathbf{y}}(t) = \sum_{k=-\infty}^{+\infty} \sum_{n=1}^{N_u} \sum_{l=-N_{CP}}^{N_{FFT}-1} \check{a}_{l,k}^{(n)} \check{\mathbf{h}}^{(n)}(t - lT - kT_{OFDM}) + \check{\mathbf{n}}(t). \qquad (13.32)$$

The support of the channel $\check{\mathbf{h}}^{(n)}(t)$ is defined in a similar way to that for the MIMO-OFDM case: $\check{\mathbf{h}}^{(n)}(t) = \mathbf{0}$ for all t outside the support $[t_{min}, t_{max}]$. For notational convenience, we assume the support of $\check{\mathbf{h}}^{(n)}(t)$ to be $[0, N_{DS}T)$ for all $n \in \{1, \ldots, N_u\}$, with $N_{DS} \leq N_{CP} + 1$. Note that this constraint on the support requires the users to be synchronized to a certain extent. This synchronization constraint can be loosened by increasing N_{CP}: the supports $\check{\mathbf{h}}^{(n)}(t)$ of the different users must overlap in a time window not exceeding $(N_{CP} + 1)T$. Sampling at time instants $l'T + k'T_{OFDM}$, $l' = 0, \ldots, N_{FFT} - 1$, $k' = 0, \ldots, N - 1$, discarding the samples corresponding to the cyclic prefix, and applying the FFT matrix $\mathbf{F}^H$ yields the following observation of OFDM symbol k, receive antenna m, and subcarrier q (see also Section 12.3):

$$y_{q,k,m} = \sum_{n:q\in\mathcal{S}_n} H^{(n)}_{q,m} a^{(n)}_{q,k} + n_{q,k,m} \tag{13.33}$$

with

$$\mathbb{E}\{N_{q,k,m} N^*_{q',k',m'}\} = \delta_{k-k'} \delta_{m-m'} \delta_{q-q'} N_0. \tag{13.34}$$

Since $S_n \cap S_{n'} = \phi$, for all $n' \neq n$, (13.33) simplifies to

$$y_{q,k,m} = H^{(i_q)}_{q,m} a^{(i_q)}_{q,k} + n_{q,k,m}, \tag{13.35}$$

where i_q is the index of the (unique) user assigned to subcarrier q. If we focus on a single subcarrier and stack the observations from the different receive antennas into an $N_R \times 1$ vector, we can write

$$\mathbf{y}_{q,k} = \mathbf{H}^{(i_q)}_q a^{(i_q)}_{q,k} + \mathbf{n}_{q,k}, \tag{13.36}$$

where $\mathbf{H}^{(i_q)}_q$ is an $N_R \times 1$ vector $\mathbf{H}^{(i_q)}_q = [H^{(i_q)}_{q,1}, \ldots, H^{(i_q)}_{q,N_R}]^T$. Note that

$$\mathbb{E}\{\mathbf{N}_{q,k} \mathbf{N}^H_{q',k'}\} = N_0 \mathbf{I}_{N_R} \delta_{k-k'} \delta_{q-q'}. \tag{13.37}$$

The final observation $\mathbf{y}$ is obtained by stacking $\mathbf{y}_{q,k}$ for all q and k.

The equalizer

Because users are assigned non-overlapping sets of subcarriers, and thanks to the noise being independent across time and subcarriers, the likelihood function can now be expressed as

$$p(\mathbf{Y} = \mathbf{y} | \mathbf{A} = \mathbf{a}, \mathcal{M}) = \prod_{q=0}^{N_{FFT}-1} \prod_{k=0}^{N-1} p\left(\mathbf{Y}_{q,k} = \mathbf{y}_{q,k} \Big| A^{(i_q)}_{q,k} = a^{(i_q)}_{q,k}, \mathcal{M}\right) \tag{13.38}$$

$$\propto \prod_{q=0}^{N_{FFT}-1} \prod_{k=0}^{N-1} \exp\left(-\frac{1}{N_0} \left\| \mathbf{y}_{q,k} - \mathbf{H}^{(i_q)}_q a^{(i_q)}_{q,k} \right\|^2\right), \tag{13.39}$$

leading to a non-iterative *structured equalizer*:

$$\mu_{\text{eq} \to A_{q,k}^{(i_q)}}(a) \propto \exp\left(-\frac{1}{N_0} \left\| \mathbf{y}_{q,k} - \mathbf{H}_q^{(i_q)} a_{q,k}^{(i_q)} \right\|^2 \right). \tag{13.40}$$

The overall complexity scales as $\mathcal{O}(N_{\text{s}} N_{\text{u}} |\Omega|)$. The receivers can easily be extended to a scenario in which different users use different constellations.

13.4   Main points

Receivers for multi-user communication closely resemble those for multi-antenna communication. In this chapter we have considered two multiple-access schemes: DS-CDMA and OFDMA. For DS-CDMA both the synchronous and the asynchronous scenario were investigated, leading to iterative equalizers. In OFDMA where users are assigned non-overlapping subcarriers, the factor-graph framework leads to a non-iterative equalizer.

14 Synchronization and channel estimation

14.1 Introduction

In the previous chapters, we invariably ended up with a model $y = h(a) + n$, where $h(\cdot)$ was a *known* function (often in the form of a matrix $\mathbf{H}$). The function $h(\cdot)$ depends on the physical channel ($h_{ch}(t)$) as well as on any processing done in the receiver. Since the physical channel may vary in time, the receiver needs to estimate the channel. Furthermore, transmission may be of a bursty nature, so that, for every incoming burst, the receiver has to lock on to the signal. This requires synchronization in terms of timing, carrier phase, and carrier frequency.

During the past few decades a wide variety of channel estimation and synchronization algorithms has been developed for just about any digital transmission scheme imaginable. They usually exploit statistical properties of the received signal, or sequences of known symbols in the data stream (training symbols). The resulting algorithms are known as non-data-aided (NDA) and data-aided (DA), respectively. Standard works on channel estimation and synchronization are [117, 118]. Iterative channel-estimation algorithms, which iterate between decoding/demapping/equalization and estimation, have recently become more popular. The resulting algorithms are known as code-aided (CA) [119, 120]. These algorithms generally make use of the expectation–maximization (EM) algorithm [121], or variations thereof.

Since factor graphs are well suited to solving inference problems, it makes sense to try to apply them to the channel estimation and synchronization. This idea was originally proposed in [122] and later successfully applied to phase-noise tracking in [123, 124].

This chapter is organized as follows.

- In **Section 14.2** we describe how to perform channel estimation using factor graphs and illustrate how this approach differs from conventional channel-estimation algorithms.
- An example is provided in **Section 14.3** in the context of a tracking a frequency-flat time-varying channel.

14.2 Channel estimation, synchronization, and factor graphs

14.2.1 The conventional approach

Our goal has been to create a factor graph of the distribution $p(\mathbf{B}, \mathbf{Y} = \mathbf{y}|\mathcal{M})$. The model $\mathcal{M}$ included the channel function $h(\cdot)$. In most cases, this function can be fully described by a small set of parameters, say $\mathbf{d}$. For instance, in a multi-path channel, $\mathbf{d}$ consists of the L channel gains and L propagation delays: $\mathbf{d} = [(\tau_0, \alpha_0), \ldots, (\tau_{L-1}, \alpha_{L-1})]$. Prior to data detection, a conventional receiver can determine an estimate $\hat{\mathbf{d}}$ of $\mathbf{d}$. How exactly $\hat{\mathbf{d}}$ is determined depends on the specific scenario. Once $\mathbf{d}$ has been estimated, we can create a factor graph of $p(\mathbf{B}, \mathbf{Y} = \mathbf{y} | \mathbf{D} = \hat{\mathbf{d}}, \mathcal{M})$. Now the model $\mathcal{M}$ no longer includes the channel function. Implementing the SPA on this factor graph yields (approximations of) the marginal distributions $p(B_k | \mathbf{Y} = \mathbf{y}, \mathbf{D} = \hat{\mathbf{d}}, \mathcal{M})$. Final decisions with respect to the information bits can then be made. Throughout this book, we have implicitly assumed this approach.

14.2.2   The factor-graph approach

The conventional approach is somewhat ad-hoc. A more systematic way of dealing with $\mathbf{D}$ is as follows. As always, we create a factor graph of $p(\mathbf{B}, \mathbf{Y} = \mathbf{y} | \mathcal{M})$:

$$p(\mathbf{B}, \mathbf{Y} = \mathbf{y} | \mathcal{M}) = p(\mathbf{Y} = \mathbf{y} | \mathbf{B}, \mathcal{M}) p(\mathbf{B} | \mathcal{M}) \tag{14.1}$$

Opening up the node $p(\mathbf{Y} = \mathbf{y} | \mathbf{B}, \mathcal{M})$ to include the coded bits and the coded symbols gives us

$$p(\mathbf{Y} = \mathbf{y}, \mathbf{A}, \mathbf{C} | \mathbf{B}, \mathcal{M}) = p(\mathbf{Y} = \mathbf{y} | \mathbf{A}, \mathcal{M}) p(\mathbf{C} | \mathbf{A}, \mathcal{M}) p(\mathbf{C} | \mathbf{B}, \mathcal{M}). \tag{14.2}$$

We can now open up the node $p(\mathbf{Y} = \mathbf{y} | \mathbf{A}, \mathcal{M})$ to include the channel parameters and obtain

$$p(\mathbf{Y} = \mathbf{y}, \mathbf{D} | \mathbf{A}, \mathcal{M}) = p(\mathbf{Y} = \mathbf{y} | \mathbf{A}, \mathbf{D}, \mathcal{M}) p(\mathbf{D} | \mathcal{M}), \tag{14.3}$$

assuming the channel parameters to be independent from the data symbols (this is generally true). The resulting factor graph is shown in Fig. 14.1.

Applying the SPA to this graph results in $p(B_k | \mathbf{Y} = \mathbf{y}, \mathcal{M})$, on which final decisions with respect to the information bits can be made. Observe the difference from the previous section. No explicit estimate regarding the channel parameters $\mathbf{d}$ needs to be made. Hence, the factor-graph receiver can be interpreted as being non-coherent. The reader can easily verify that the conventional approach can be interpreted as the node $p(\mathbf{D} | \mathcal{M})$ sending Dirac distributions over the D_k edges:

$$\mu_{D_k \to \mathrm{eq}}(d_k) = \delta(d_k - \hat{d}_k). \tag{14.4}$$

Comments

There are some important differences between the parameter $\mathbf{D}$ and the other parameters (information bits $\mathbf{B}$, coded bits $\mathbf{C}$, and coded symbols $\mathbf{A}$).

- Usually $\mathbf{D}$ belongs to a continuous (instead of a discrete) domain. This means that messages will be probability density functions (instead of probability mass functions). Consequently, we must represent messages as described in Chapter 5, Section 5.3.4.

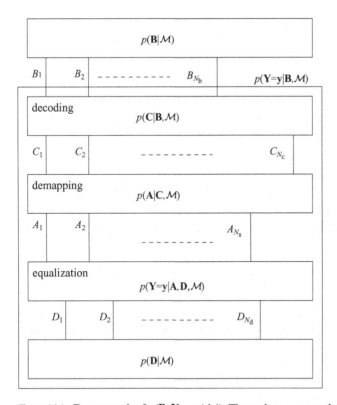

Figure 14.1. Factor graph of $p(\mathbf{B}, \mathbf{Y} = \mathbf{y}|\mathcal{M})$. The nodes are opened to reveal its structure. Apart from the standard tasks of decoding, demapping, and equalization, there is now a new node $p(\mathbf{D}|\mathcal{M})$ related to channel estimation/synchronization.

- The conversion from the received waveform $r(t)$ to the observation $\mathbf{y}$ may depend on the value of $\mathbf{D}$. For instance, a matched-filter receiver relies on knowledge of the equivalent channel $h(t)$. In such cases, it is not meaningful to create a factor graph of $p(\mathbf{B}, \mathbf{Y} = \mathbf{y}|\mathcal{M})$. Instead, we must create a factor graph of $p(\mathbf{B}, \mathbf{R} = \mathbf{r}|\mathcal{M})$, where $\mathbf{r}$ is a vector representation of $r(t)$.
- The parameter $\mathbf{D}$ is related to the model of the physical channel. This model may but need not reflect the true behavior of the physical channel. The corresponding a-priori distribution $p(\mathbf{D}|\mathcal{M})$ is also based on a model.

14.3   An example

14.3.1   Problem description

To see how to perform channel estimation using factor graphs, let us consider an example (inspired by [123, 124]): channel estimation for frequency-flat time-varying channels in a single-carrier, single-user, single-antenna setting. Suppose after suitable matched filtering and sampling, that we can express the observation at time

k $(k = 0, \ldots, N_s - 1)$ as

$$y_k = d_k a_k + n_k, \tag{14.5}$$

where $d_k \in \mathbb{C}$ is the (unknown) channel gain at time k and $\mathbb{E}\{N_k N_{k'}^*\} = \delta_{k-k'} N_0/E_s$. Stacking the observations gives us

$$\mathbf{y} = \mathbf{H}\mathbf{a} + \mathbf{n}, \tag{14.6}$$

where $\mathbf{H} = \text{diag}\{\mathbf{d}\}$ with $\mathbf{d} = [d_0 \quad d_1 \quad \cdots \quad d_{N_s-1}]^{\mathrm{T}}$. In order to apply the technique from the previous section, we require the a-priori distribution $p(\mathbf{D})$. Let us assume that the channel varies according to a first-order Markov model:

$$p(\mathbf{D}) = p(D_0) \prod_{k=1}^{N_s-1} p(D_k | D_{k-1}), \tag{14.7}$$

where both the initial distribution $p(D_0)$ and the transition distribution $p(D_k | D_{k-1})$ are known to the receiver.

14.3.2   The factor graph

The factor graph of $p(\mathbf{B}, \mathbf{Y} = \mathbf{y})$ can be opened up to (using a notational shorthand)

$$p(\mathbf{b}, \mathbf{a}, \mathbf{c}, \mathbf{d}, \mathbf{y}) \propto p(\mathbf{y}|\mathbf{a}, \mathbf{d}) p(\mathbf{a}|\mathbf{c}) p(\mathbf{c}|\mathbf{b}) p(\mathbf{b}) p(\mathbf{d}) \tag{14.8}$$

$$= p(d_0) p(y_0 | a_0, d_0) \prod_{k=1}^{N_s-1} p(y_k | a_k, d_k) p(d_k | d_{k-1}) p(\mathbf{a}|\mathbf{c}) p(\mathbf{c}|\mathbf{b}) p(\mathbf{b}). \tag{14.9}$$

The corresponding factor graph is shown in Fig. 14.2.

14.3.3   The sum-product algorithm

We will denote the message as shown in Fig. 14.3. Since there are cycles in the graph, there are many ways to schedule the messages. Let us consider the following scheduling.

Initialization
- We initialize $\mu_{f \to D_k}(D_k)$ to uniform distributions.
- We set $\mu_{g \to D_0}(d_0) = p(D_0 = d_0)$.

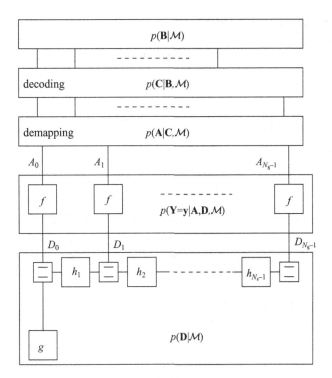

Figure 14.2. Factor graph of $p(\mathbf{B}, \mathbf{Y} = \mathbf{y} | \mathcal{M})$. The nodes are opened to reveal their structure. The node $f(A_k, D_k)$ represents the function $p(y_k | A_k, D_k)$, $g(D_0)$ represents $p(D_0)$ and $h_k(D_k, D_{k-1})$ represents $p(D_k | D_{k-1})$.

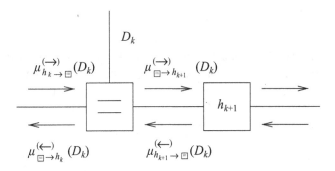

Figure 14.3. A detailed view of messages

Messages update

(1) Compute forward messages from left to right:

$$\mu^{(\rightarrow)}_{\boxminus \to h_1}(d_0) \propto \mu_{f \to D_0}(d_0)\mu_{g \to D_0}(d_0),$$

$$\mu^{(\rightarrow)}_{h_k \to \boxminus}(d_k) \propto \int p(D_k = d_k | D_{k-1} = d_{k-1})\mu^{(\rightarrow)}_{\boxminus \to h_k}(d_{k-1})\mathrm{d}d_{k-1},$$

$$\mu^{(\rightarrow)}_{\boxminus \to h_{k+1}}(d_k) \propto \mu_{f \to D_k}(d_k)\mu^{(\rightarrow)}_{h_k \to \boxminus}(d_k).$$

(2) Compute messages from right to left:

$$\mu^{(\leftarrow)}_{\boxminus \to h_{N_s-1}}(d_{N_s-1}) = \mu_{f \to D_{N_s-1}}(d_{N_s-1}),$$

$$\mu^{(\leftarrow)}_{h_k \to \boxminus}(d_{k-1}) \propto \int p(D_k = d_k | D_{k-1} = d_{k-1}) \mu^{(\leftarrow)}_{\boxminus \to h_k}(d_k) dd_k,$$

$$\mu^{(\leftarrow)}_{\boxminus \to h_k}(d_k) \propto \mu^{(\leftarrow)}_{h_{k+1} \to \boxminus}(d_k) \mu_{f \to D_k}(d_k).$$

(3) Compute upward messages:

$$\mu_{D_k \to f}(d_k) \propto \mu^{(\leftarrow)}_{h_{k+1} \to \boxminus}(d_k) \mu^{(\to)}_{h_k \to \boxminus}(d_k),$$

$$\mu_{A_k \to \mathrm{dem}}(a_k) \propto \int p(y_k | a_k, d_k) \mu_{D_k \to f}(d_k) dd_k.$$

(4) Perform demapping and decoding. This results in downward messages $\mu_{\mathrm{dem} \to A_k}(A_k)$.
(5) Compute downward messages:

$$\mu_{f \to D_k}(D_k) \propto \sum_{a_k \in \Omega} p(y_k | a_k, d_k) \mu_{\mathrm{dem} \to A_k}(a_k).$$

(6) Go back to step 1.

Termination
- After a number of iterations, we compute the approximate marginals $p(B_k | \mathbf{Y} = \mathbf{y}, \mathcal{M})$ and make final decisions:

$$\hat{b}_k = \arg \max_{b \in \mathbb{B}} p(B_k = b | \mathbf{Y} = \mathbf{y}, \mathcal{M}).$$

Since the messages over the D_k-edges have no convenient closed form, we must resort to approximate techniques such as quantization or particle representations to represent and compute messages, as discussed in Section 5.3.4. Furthermore, the temporal model (14.7) we assume may be incorrect, leading to an estimator that is not robust. These practical aspects are treated in detail in [123, 124].

14.4   Main points

Factor graphs can be applied to scenarios in which the channel function is unknown by including channel and synchronization parameters in the factor graph of $p(\mathbf{B}, \mathbf{Y} = \mathbf{y} | \mathcal{M})$, albeit with the introduction of additional cycles into the graph. The receiver requires explicit knowledge of the a-priori distribution of the channel parameters. This a-priori distribution may but might not be accurate (or even available). The channel function is usually characterized by a set of continuous variables, so messages over the corresponding edges must be represented in an efficient way. For this, we must look to quantization, parametric representations, and non-parametric representations.

15 Appendices

15.1 Useful matrix types

- A *Toeplitz matrix* $\mathbf{A}$ from a sequence $\mathbf{a} = [a_{-N+1}, \ldots, a_{M-1}]$ of length $M + N - 1$ is an $M \times N$ matrix where $A_{i,j} = a_{i-j}$, $i = 0, \ldots, M - 1, j = 0, \ldots, N - 1$, in other words, a matrix where every diagonal from left to right is constant. For instance a 4×5 Toeplitz matrix is given by

$$\mathbf{A} = \begin{bmatrix} a_0 & a_{-1} & a_{-2} & a_{-3} & a_{-4} \\ a_1 & a_0 & a_{-1} & a_{-2} & a_{-3} \\ a_2 & a_1 & a_0 & a_{-1} & a_{-2} \\ a_3 & a_2 & a_1 & a_0 & a_{-1} \end{bmatrix}.$$

- An $M \times N$ *circulant matrix* $\mathbf{A}$ from a sequence $\mathbf{a} = [a_0, \ldots, a_{M-1}]$ is a Toeplitz matrix where $a_i = a_{i+N}$, so that

$$\mathbf{A} = \begin{bmatrix} a_0 & a_3 & a_2 & a_1 & a_0 \\ a_1 & a_0 & a_3 & a_2 & a_1 \\ a_2 & a_1 & a_0 & a_3 & a_3 \\ a_3 & a_2 & a_1 & a_0 & a_3 \end{bmatrix}.$$

- A *band matrix* (or banded matrix) $\mathbf{A}$ is a matrix for which $A_{i,j} = 0$ for $j - i < -K_1$ or $j - i > K_2$ for some $K_1, K_2 > 0$.
- The concept of Toeplitz, circulant, and band matrices can be extended to block matrices, by replacing scalars in the above definitions by matrices.

15.2 Random variables and distributions

We will write random variables using capital letters, and realizations/instances of random variables by small letters. So Z is a random variable and z is a realization belonging to a set $\mathcal{Z}$. This has as a disadvantage that matrices and vector random variables use the same notation: $\mathbf{Z}$ can be a vector random variable or a matrix. Random variables can be either discrete or continuous. Continuous random variables are defined over a closed subset of $\mathbb{R}^N$ or $\mathbb{C}^N$ for some $N \geq 1$. Discrete random variables can take on values in either a finite or an infinite set. The distribution (the probability mass function or the probability density function) of Z, evaluated in $z \in \mathcal{Z}$, is written as $p_Z(z)$, $p(Z = z)$, or $p(z)$. The distribution itself (the function from $\mathcal{Z}$ to $\mathbb{R}^+$) will be denoted by $p_Z(\cdot)$ or[1] $p(Z)$. We will use the notation $z \sim p_Z(z)$ or $z \sim p_Z(\cdot)$ when z is drawn from the distribution $p_Z(z)$. The *expectation* of a function $f(z)$ with regard to a random variable Z is denoted as

$$\mathbb{E}_Z\{f(Z)\} = \int_{\mathcal{Z}} f(z)p_Z(z)\mathrm{d}z.$$

For discrete random variables, integrations should be replaced by summations.

The Dirac distribution

The *Dirac distribution* $\delta(z)$ is defined as

$$\int_{-\infty}^{+\infty} h(z)\delta(z - z_0)\mathrm{d}z = h(z_0)$$

for any $z_0 \in \mathbb{R}$ and any integrable function $h(z)$. The *discrete Dirac distribution* δ_k $(k \in \mathbb{Z})$ is defined as $\delta_k = 1$ when $k = 0$ and $\delta_k = 0$ otherwise.

The Gaussian distribution

Arguably the most important distribution is the *Gaussian (or normal) distribution*. Any Gaussian random variable $\mathbf{Z}$ is fully determined by its mean

$$\mathbb{E}\{\mathbf{Z}\} = \mathbf{m}$$

and its covariance matrix

$$\mathbb{E}\{(\mathbf{Z} - \mathbf{m})(\mathbf{Z} - \mathbf{m})^{\mathrm{H}}\} = \mathbf{\Sigma},$$

where $\mathbf{A}^{\mathrm{H}}(\mathbf{A}^{\mathrm{T}})$ stands for the conjugate transpose (transpose) of the matrix $\mathbf{A}$. When $\mathbf{Z}$ is real, we write $p_{\mathbf{Z}}(\mathbf{z}) = \mathcal{N}_{\mathbf{z}}(\mathbf{m}, \mathbf{\Sigma})$, while for complex $\mathbf{Z}$ with independent real and imaginary parts, each with the same covariance matrix $\frac{1}{2}\mathbf{\Sigma}$, we write $\mathcal{N}_{\mathbf{x}}^{\mathbb{C}}(\mathbf{m}, \mathbf{\Sigma})$. When $\mathbf{\Sigma}$ is non-singular, the probability distribution of $\mathbf{Z}$ can be expressed in closed form:

$$\mathcal{N}_{\mathbf{z}}(\mathbf{m}, \mathbf{\Sigma}) = \frac{1}{\sqrt{(2\pi)^N \det \mathbf{\Sigma}}} \exp\left(-\frac{1}{2}(\mathbf{z} - \mathbf{m})^{\mathrm{T}}\mathbf{\Sigma}^{-1}(\mathbf{z} - \mathbf{m})\right)$$

and

$$\mathcal{N}_{\mathbf{z}}^{\mathbb{C}}(\mathbf{m}, \mathbf{\Sigma}) = \frac{1}{\pi^N \det \mathbf{\Sigma}} \exp\left(-(\mathbf{z} - \mathbf{m})^{\mathrm{H}}\mathbf{\Sigma}^{-1}(\mathbf{z} - \mathbf{m})\right).$$

When $\mathbf{\Sigma}$ is singular, Gaussian distributions cannot be described through their probability distributions. Other techniques[2] exist but will not be pursued in this book: all covariance

---

[1] This is a somewhat unorthodox notation, but it will serve us well later.
[2] Such as the Cramer–Wold argument.

matrices are assumed to be well-behaved. Gaussian distributions have some very useful properties. We will focus on real distributions. For complex distributions transposition should be replaced by conjugate transposition.

- When $\mathbf{Z}_1$ and $\mathbf{Z}_2$ are independent Gaussian random variables with $p_{\mathbf{Z}_1}(\mathbf{z}_1) = \mathcal{N}_{\mathbf{z}_1}(\mathbf{m}_1, \Sigma_1)$ and $p_{\mathbf{Z}_2}(\mathbf{z}_2) = \mathcal{N}_{\mathbf{z}_2}(\mathbf{m}_2, \Sigma_2)$, then $\mathbf{Z}_3 = \mathbf{AZ}_1 + \mathbf{BZ}_2$, for matrices $\mathbf{A}$ and $\mathbf{B}$ (of the proper dimensions), is also a Gaussian random variable with $p_{\mathbf{Z}_3}(\mathbf{z}_3) = \mathcal{N}_{\mathbf{z}_3}(\mathbf{m}_3, \Sigma_3)$, where

$$\mathbf{m}_3 = \mathbf{Am}_1 + \mathbf{Bm}_2$$

and

$$\Sigma_3 = \mathbf{A}\Sigma_1\mathbf{A}^\mathsf{T} + \mathbf{B}\Sigma_2\mathbf{B}^\mathsf{T}.$$

- Multiplying two Gaussian distributions $p_{\mathbf{Z}}(\mathbf{z}) = \mathcal{N}_{\mathbf{z}}(\mathbf{m}_1, \Sigma_1)$ and $p_{\mathbf{Z}}(\mathbf{z}) = \mathcal{N}_{\mathbf{z}}(\mathbf{m}_2, \Sigma_2)$ gives a Gaussian distribution, up to a multiplicative constant,

$$\mathcal{N}_{\mathbf{z}}(\mathbf{m}_1, \Sigma_1)\mathcal{N}_{\mathbf{z}}(\mathbf{m}_2, \Sigma_2) = C \times \mathcal{N}_{\mathbf{z}}(\mathbf{m}_3, \Sigma_3),$$

where C does not depend on $\mathbf{z}$, and

$$\Sigma_3^{-1} = \Sigma_1^{-1} + \Sigma_2^{-1},$$
$$\Sigma_3^{-1}\mathbf{m}_3 = \Sigma_1^{-1}\mathbf{m}_1 + \Sigma_2^{-1}\mathbf{m}_2.$$

- When $\mathbf{Z}$ is Gaussian with $\mathbf{z} = [\mathbf{z}_1\ \mathbf{z}_2]^\mathsf{T} \sim \mathcal{N}_{\mathbf{z}}(\mathbf{m}, \Sigma)$, where

$$\mathbf{m} = \begin{bmatrix} \mathbf{m}_1 \\ \mathbf{m}_2 \end{bmatrix},$$

$$\Sigma = \begin{bmatrix} \Sigma_{11} & \Sigma_{12}^\mathsf{T} \\ \Sigma_{12} & \Sigma_{22} \end{bmatrix},$$

then the marginal of $\mathbf{Z}_1$ is given by $p_{\mathbf{Z}_1}(\mathbf{z}_1) = \mathcal{N}_{\mathbf{z}_1}(\mathbf{m}_1, \Sigma_{11})$ and the conditional distribution by $p_{\mathbf{Z}_1|\mathbf{Z}_2}(\mathbf{z}_1|\mathbf{z}_2) = \mathcal{N}_{\mathbf{z}_1}(\mathbf{m}_{1|2}(\mathbf{z}_2), \Sigma_{1|2})$, where

$$\mathbf{m}_{1|2}(\mathbf{z}_2) = \mathbf{m}_1 + \Sigma_{12}\Sigma_{22}^{-1}(\mathbf{z}_2 - \mathbf{m}_2),$$
$$\Sigma_{1|2} = \Sigma_1 - \Sigma_{12}\Sigma_{22}^{-1}\Sigma_{12}^\mathsf{T}.$$

Bayes' rule

This entire book builds on the work of the Reverend Thomas Bayes (1702–1761), who introduced the concept of *inverse probability*. While the term "Bayesian" is now much broader than inverse probability, it is still useful to recall the basic idea. We wish to estimate a parameter $\mathbf{x}$ from an observation $\mathbf{y}$. The direct probability is the function $p_{\mathbf{Y}|\mathbf{X}}(\mathbf{y}|\mathbf{x})$ (which now goes under the name "likelihood function").

The inverse probability is the a-posteriori distribution $p_{X|Y}(x|y.)$. One form of Bayes' rule gives the following relations for a joint distribution, the marginals, and the conditional distributions:

$$p_{Z_1,Z_2}(z_1, z_2) = p_{Z_1|Z_2}(z_1|z_2)p_{Z_2}(z_2)$$
$$= p_{Z_2|Z_1}(z_2|z_1)p_{Z_1}(z_1).$$

Assuming that z_2 is a discrete random variable, $p_{Z_1}(z_1) = \sum_{z_2} p_{Z_1,Z_2}(z_1, z_2)$, so we can also write

$$p_{Z_2|Z_1}(z_2|z_1.) = \frac{p_{Z_1|Z_2}(z_1|z_2)p_{Z_2}(z_2)}{\sum_z p_{Z_1|Z_2}(z_1|z)p_{Z_2}(z)}.$$

This equation nicely encapsulates the various operations we use throughout this book: we multiply probabilities, add probabilities, and normalize functions so that they become distributions.

15.3   Signal representations

In general, a vector representation $\mathbf{r}$ of a signal $r(t)$ is obtained by expanding $r(t)$ onto a set of orthonormal basis functions: $\{\phi_0(t), \phi_1(t), \ldots, \phi_{N-1}(t)\}$ as follows:

$$r_k = \int_{-\infty}^{+\infty} r(t)\phi_k(t)\mathrm{d}t$$

so that $\mathbf{r} = [r_0, r_1, \ldots, r_{N-1}]^T$. Since the basis functions are orthonormal, we have

$$r(t) = \sum_{k=0}^{N-1} r_k\phi_k(t).$$

Note that sampling a bandlimited signal at a sufficiently high rate corresponds to a particular set of basis functions (i.e., delay-shifted sinc pulses). In many cases $N = +\infty$. When the signal is a random process, the values $\{r_k\}$ will be random variables.

References

[1] C. E. Shannon. "A mathematical theory of communication." *Bell System Technical Journal*, 27:379–423, July 1948.

[2] G. D. Forney and G. Ungerboeck. "Modulation and coding for linear Gaussian channels." *IEEE Transactions of Information Theory*, 44(6):2384–2415, October 1998.

[3] C. Berrou, A. Glavieux, and P. Thitimajshima. "Near Shannon limit error-correcting coding and decoding: turbo codes." In *Proc. IEEE International Conference on Communications (ICC)*, pages 1064–1070, Geneva, Switzerland, May 1993.

[4] S. Aji and R. McEliece. "The generalized distributive law." *IEEE Transactions on Information Theory*, 46:325–353, March 2000.

[5] B. J. Frey. *Graphical Models for Machine Learning and Digital Communications*. MIT Press, 1998.

[6] J. M. Wozencraft and I. M. Jacobs. *Principles of Communications Engineering*. Wiley, 1967.

[7] J. G. Proakis. *Digital Communications*. McGraw-Hill, 4th edition, 2001.

[8] S. Haykin. *Communication Systems*. Wiley, 2000.

[9] T. S. Rappaport. *Wireless Communications: Principles and Practice*. Prentice-Hall, 2001.

[10] B. Sklar. *Digital Communications: Fundamentals and Applications*. Prentice-Hall, 2001.

[11] J. R. Barry, E. A. Lee, and D. G. Messerschmitt. *Digital Communication*. Springer, 2003.

[12] D. Tse and P. Viswanath. *Fundamentals of Wireless Communication*. Cambridge University Press, 2005.

[13] A. Goldsmith. *Wireless Communications*. Cambridge University Press, 2005.

[14] A. Molisch. *Wireless Communications*. Wiley–IEEE Press, 2005.

[15] S. Benedetto and E. Biglieri. *Principles of Digital Transmission with Wireless Applications*. Springer, 1999.

[16] J. D. Parsons. *The Mobile Radio Propagation Channel*. Wiley, 2000.

[17] J. A. C. Bingham. "Multicarrier modulation for data transmission: an idea whose time has come." *IEEE Communications Magazine*, 28(5):5–14, May 1990.

[18] S. B. Weinstein and P. M. Ebert. "Data transmission by frequency-division multiplexing using the discrete Fourier transform." *IEEE Transactions on Communication Technology*, 19(5):628–634, 1971.

[19] J. H. Winters, J. Salz, and R. D. Gitlin. "The impact of antenna diversity on the capacity of wireless communication systems." *IEEE Transactions on Communications*, COM-42:1740–1751, 1994.

[20] G. J. Foshini and M. J. Gans. "On limits of wireless communication in a fading environment when using multiple antennas." *Wireless Personal Communications*, 6(3):311–335, March 1998.

[21] I. Telatar. "Capacity of multi-antenna Gaussian channels." Technical report, AT&T Bell Labs internal Technical Memorandum, 1995.

[22] S. M. Alamouti. "A simple transmit diversity technique for wireless communications." *IEEE Journal on Selected Areas in Communications*, 16(8):1451–1458, October 1998.

[23] V. Tarokh, N. Seshadri, and A. R. Calderbank. "Space–time codes for high data rate wireless communications: performance criterion and code construction." *IEEE Transactions on Communications*, 44(2):744–765, March 1998.

[24] V. Tarokh, H. Jafarkhani, and A. R. Calderbank. "Space–time block codes from orthogonal design." *IEEE Transactions on Information Theory*, 45(5):1456–1467, July 1999.

[25] D. Gesbert, M. Shafi, D. Shiu, P. J. Smith, and A. Naguib. "From theory to practice: an overview of MIMO space–time coded wireless systems." *IEEE Journal on Selected Areas in Communications*, 21(3):281–302, April 2003.

[26] A. Paulraj, R. Nabar, and Dhananjay Gore. *Introduction to Space–Time Wireless Communications*. Cambridge University Press, 2003.

[27] L. Zheng and D. Tse. "Diversity and multiplexing: a fundamental tradeoff in multiple antenna channels." *IEEE Transactions on Information Theory*, 49:1073–1096, May 2003.

[28] G. D. Golden, G. J. Foschini, R. A. Valenzuela, and P. W. Wolniansky. "Detection algorithm and initial laboratory results using the V-BLAST space–time communication architecture." *Electronics Letters*, 35(1):14–15, January 1999.

[29] G. G. Raleigh, and J. M. Cioffi. "Spatio-temporal coding for wireless communication." *IEEE Transactions on Communications*, 46(3):57–366, March 1998.

[30] H. Bölcskei, D. Gesbert, and A. J. Paulraj. "On the capacity of OFDM-based spatial multiplexing systems." *IEEE Transactions on Communications*, 50(2):225–234, February 2002.

[31] S. Verdú. *Multiuser Detection*. Cambridge University Press, New York, 1998.

[32] A. Jamalipour, T. Wada, and T. Yamazato. "A tutorial on multiple access technologies for beyond 3G mobile networks." *IEEE Communications Magazine*, 43(2):110–117, February 2005.

[33] R. Pickholtz, D. Schilling, and L. Milstein. "Theory of spread-spectrum communications – a tutorial." *IEEE Transactions on Communications*, 30(5):855–884, May 1982.

[34] R. Scholtz. "The spread spectrum concept." *IEEE Transactions on Communications*, 25(8):748–755, August 1977.

[35] R. L. Peterson, R. E. Ziemer, and D. E. Borth. *Introduction to Spread Spectrum Communications*. Prentice-Hall, 1995.

[36] H. Sari and G. Karam. "Orthogonal frequency division multiple access and its applications to CATV networks." *European Transactions on Telecommunications*, 9(6):507–516, November 1998.

[37] Cheong Yui Wong, R. S. Cheng, K. B. Lataief, and R. D. Murch. "Multiuser OFDM with adaptive subcarrier, bit, and power allocation." *IEEE Journal on Selected Areas in Communications*, 17(10):1747–1758, October 1999.

[38] H. L. Van Trees. *Detection, Estimation, and Modulation Theory, Part I*. Wiley, October 2001.

[39] S. M. Kay. *Fundamentals of Statistical Signal Processing, Volume I: Estimation Theory*. Prentice-Hall, 1993.

[40] E. L. Lehmann and G. Casella. *Theory of Point Estimation*. Springer, 2003.

[41] V. Poor. *An Introduction to Signal Detection and Estimation*. Springer, 1998.

[42] C. P. Robert and G. Casella. *Monte Carlo Statistical Methods*. Springer, 2005.

[43] W. R. Gilks, S. Richardson, and D. J. Speigelhalter. *Markov Chain Monte Carlo in Practice*. Chapman & Hall, 1995.

[44] G. Fishman. *Monte Carlo: Concepts, Algorithms and Applications*. Springer, 2003.

[45] B. W. Silverman. *Density Estimation for Statistics and Data Analysis*. Chapman & Hall, 1986.

[46] D. J. C. MacKay. *Information Theory, Inference and Learning Algorithms*. Cambridge University Press, 2003.

[47] R. M. Neal. Probabilistic inference using Markov chain Monte Carlo methods. Technical Report CRG-TR-93-1, Dept. of Computer Science, University of Toronto, 1993.

[48] A. Doucet and X. Wang. "Monte Carlo methods for signal processing: a review in the statistical signal processing context." *IEEE Signal Processing Magazine*, 22(6):152–170, November 2005.

[49] S. Geman and D. Geman. "Stochastic relaxation, Gibbs distributions, and the Bayesian restoration of images." *IEEE Transactions on Pattern Analysis and Machine Intelligence*, 6:721–741, 1984.

[50] R. G. Gallager. *Low Density Parity Check Codes*. MIT Press, 1963.

[51] F. Spitzer. "Random fields and interacting particle systems." In *M.A.A. Summer Seminar Notes*, Mathematical Association of America, 1971.

[52] G. D. Forney, Jr. "The Viterbi algorithm." *Proceedings of the IEEE*, 61:286–278, March 1973.

[53] R. Kindermann and J. Snell. *Markov Random Fields and their Applications*. American Mathematical Society, 1980.

[54] R. M. Tanner. "A recursive approach to low complexity codes." *IEEE Transactions on Information Theory*, IT-27(5):533–547, September 1981.

[55] S. J. Lauritzen and D. J. Spiegelhalter. "Local computations with probabilities on graphical structures and their application to expert systems." *Journal of the Royal Statistical Society B*, 50:157–224, 1988.

[56] J. Pearl. *Probabilistic Reasoning in Intelligent Systems: Networks of Plausible Inference*. Morgan Kaufmann, San Mateo, 1988.

[57] N. Wiberg. Codes and decoding on general graphs. PhD thesis, Linköping University, Sweden, 1996.

[58] F. Kschischang, B. Frey, and H.-A. Loeliger. "Factor graphs and the sum-product algorithm." *IEEE Transactions on Information Theory*, 47(2):498–519, February 2001.

[59] G. D. Forney. "Codes on graphs: normal realizations." *IEEE Transactions on Information Theory*, 47(2):520–545, February 2001.

[60] H.-A. Loeliger. "An introduction to factor graphs." *IEEE Signal Processing Magazine*, 21(1):28–41, January 2004.

[61] J. S. Yedidia, W. T. Freeman, and Y. Weiss. "Constructing free energy approximations and generalized belief propagation algorithms." *IEEE Transactions on Information Theory*, 51(7):2282–2312, July 2005.

[62] J. Dauwels. On graphical models for communications and machine learning: algorithms, bounds, and analog implementation. PhD thesis, ETH Zürich, December 2005.

[63] M. J. Wainwright and M. I. Jordan. "Graphical models, exponential families, and variational inference." Technical Report 649, UC Berkeley, Dept. of Statistics, September 2003.

[64] P. Robertson, P. Hoeher, and E. Villebrun. "Optimal and sub-optimal maximum a posteriori algorithms suitable for turbo decoding." *European Transactions on Telecommunications (ETT)*, 8(2):119–125, March 1997.

[65] M. S. Arulampalam, S. Maskell, N. Gordon, and T. Clapp. "A tutorial on particle filters for online nonlinear/non-Gaussian Bayesian tracking." *IEEE Transactions on Signal Processing*, 50(2):174–188, February 2002.

[66] E. Sudderth, A. Ihler, W. Freeman, and A. Willsky. "Nonparametric belief propagation." Technical Report 551, MIT, Laboratory for Information and Decision Systems, October 2002.

[67] Y. Weiss. "Correctness of local probability propagation in graphical models with loops." *Neural Computation*, 12:1–41, 2000.

[68] S. Ikeda, T. Tanaka, and S. Amari. "Information geometry of turbo and low-density parity-check codes." *IEEE Transactions on Information Theory*, 50(6):1097–1114, June 2004.

[69] M. J. Wainwright, T. S. Jaakkola, and A. S. Willsky. "A new class of upper bounds on the log partition function." *IEEE Transactions on Information Theory*, 51(7):2313–2335, July 2005.

[70] B. D. O. Anderson and J. B. Moore. *Optimal Filtering*. Prentice-Hall, 1979.

[71] A. C. Harvey. *Forecasting, Structural Time Series Models and the Kalman Filter.* Cambridge University Press, 1991.

[72] J. Durbin and S. J. Koopman. *Time Series Analysis by State Space Methods*. Oxford University Press, 2001.

[73] A. Doucet, N. de Freitas, and N. Gordon. *Sequential Monte Carlo Methods in Practice*. Springer, 2001.

[74] L. R. Rabiner. "A tutorial on hidden Markov models and selected applications in speech recognition." *Proceedings of the IEEE*, 77(2):257–286, 1989.

[75] D. Simon. *Optimal State Estimation: Kalman, H Infinity, and Nonlinear Approaches*. Wiley, 2006.

[76] S. Korl. A factor graph approach to signal modelling, system identification, and filtering. Dissertation 16 170, ETH Zürich, July 2005.

[77] H.-A. Loeliger. "Least squares and Kalman filtering on Forney graphs." In *Festschrift in Honour of David Forney on the Occasion of His 60th Birthday*, pages 113–135. Kluwer, 2002.

[78] A. Doucet, S. Godsill, and C. Andrieu. "On sequential Monte Carlo sampling methods for Bayesian filtering." *Statistics and Computing*, 10(3):197–208, 2000.

[79] P. M. Djuric, J. H. Kotecha, Jianqui Zhang, Yufei Huang, T. Ghirmai, M. F. Bugallo, and J. Miguez. "Particle filtering." *IEEE Signal Processing Magazine*, 20(5):19–38, 2003.

[80] N. J. Gordon, D. J. Salmond, and A. F. M. Smith. "Novel approach to nonlinear/non-Gaussian Bayesian state estimation." *IEE Proceedings – F, Radar and Signal Processing*, 140(2):107–113, 1993.

[81] D. Divsalar, H. Jin, and R. J. McEliece. "Coding theorems for 'turbo-like codes'." In *Proc. 36th Allerton Conf. on Communications, Control and Computing*, pages 201–210, September 1998.

[82] D. J. C. MacKay. "Good error-correcting codes based on very sparse matrices." *IEEE Transactions on Information Theory*, 45(2):399–431, March 1999.

[83] P. Elias. "Coding for noisy channels." *IRE Convention Record*, 3, pt. 4:37–46, 1955.

[84] Shu Lin and D. J. Costello. *Error Control Coding*. Prentice-Hall, 2004.

[85] E. Biglieri. *Coding for Wireless Channels*. Springer, 2005.

[86] S. Benedetto and G. Montorsi. "Design of parallel concatenated convolutional codes." *IEEE Transactions on Communications*, 42(5):409–429, May 1996.

[87] D. J. C. MacKay. "Online database of low-density parity check codes." http://www.inference.phy.cam.ac.uk/mackay/codes/data.html.

[88] L. R. Bahl, J. Cocke, F. Jelinek, and J. Raviv. "Optimal decoding of linear codes for minimising symbol error rate." *IEEE Transactions on Information Theory*, 20:284–287, March 1974.

[89] S. Benedetto, D. Divsalar, G. Montorsi, and F. Pollara. "Serial concatenation of interleaved codes: performance analysis, design, and iterative decoding." *IEEE Transactions on Information Theory*, 44(3):909–926, May 1998.

[90] E. Zehavi. "8-PSK trellis codes for Rayleigh fading channels." *IEEE Transactions on Communications*, 41:873–883, May 1992.

[91] G. Caire, G. Taricco, and E. Biglieri. "Bit-interleaved coded modulation." *IEEE Transactions on Information Theory*, 44:927–946, May 1998.

[92] S. ten Brink, J. Speidel, and J. C. Yan. "Iterative demapping and decoding for multilevel modulation." In *IEEE GLOBECOM'98*, volume 1, pages 579–584. IEEE, 1998.

[93] X. Li and J. A. Ritcey. "Trellis-coded modulation with bit interleaving and iterative decoding." *IEEE Journal on Selected Areas in Communications*, 17(4):715–724, April 1999.

[94] X. Li, A. Chindapol, and J. A. Ritcey. "Bit-interleaved coded modulation with iterative decoding and 8PSK signaling." *IEEE Transactions on Communications*, 50(8):1250–1257, August 2002.

[95] G. Ungerboeck. "Channel coding with multilevel/phase signal." *IEEE Transactions on Information Theory*, 28:55–66, January 1982.

[96] A. Glavieux, C. Laot, and J. Labat. "Turbo equalization over a frequency selective channel." In *Proceedings of the International Symposium on Turbo Codes*, pages 96–102. Brest, France, September 1997.

[97] X. Wang and H. V. Poor. "Iterative (turbo) soft interference cancellation and decoding for coded CDMA." *IEEE Transactions on Communications*, 47(7):1046–1061, July 1999.

[98] X. Wang and H. V. Poor. *Wireless Communication Systems: Advanced Techniques for Signal Reception*. Prentice-Hall, 2003.

[99] R. Koetter, A. C. Singer, and M. Tüchler. "Turbo equalization." *Signal Processing Magazine*, 21(1):67–80, January 2004.

[100] B. Farhang-Boroujeny, H. Zhu, and S. Shi. "Markov chain Monte Carlo algorithms for CDMA and MIMO communication systems." *IEEE Transactions on Signal Processing*, 54(5):1896–1909, May 2006.

[101] J. Luo, K. R. Pattipati, P. K. Willett, and F. Hasegawa. "Near-optimal multiuser detection in synchronous CDMA using probabilistic data association." *IEEE Communications Letters*, 5(9):361–363, September 2001.

[102] D. O. North. "An analysis of the factors which determine signal/noise discrimination in pulsed carrier systems." Technical Report PTR-6C, RCA Labs, Princeton, NJ, 1943.

[103] G. D. Forney. "Maximum-likelihood sequence estimation of digital sequences in the presence of intersymbol interference." *IEEE Transactions on Information Theory*, 18(3):363–378, May 1972.

[104] M. Tüchler, R. Koetter, and A. C. Singer. "Turbo-equalization: principles and new results." *IEEE Transactions on Communications*, 50(5):754–767, May 2002.

[105] D. Reynolds and X. Wang. "Low-complexity turbo-equalization for diversity channels." *Signal Processing*, 81(5):989–995, 2000.

[106] B. M. Hochwald and S. ten Brink. "Achieving near-capacity on a multiple-antenna channel." *IEEE Transactions on Communications*, 51(3):389–399, March 2003.

[107] S. Haykin, M. Sellathurai, Y. de Jong, and T. Willink. "Turbo-MIMO for wireless communications." *IEEE Communications Magazine*, 42(10):48–53, October 2004.

[108] R. Visoz and A. O. Berthet. "Iterative decoding and channel estimation for space–time BICM over MIMO block fading multipath AWGN channel." *IEEE Transactions on Communications*, 51(8):1358–1367, August 2003.

[109] M. Witzke, S. Baro, F. Schreckenbach, and J. Hagenauer. "Iterative detection of MIMO signals with linear detectors." In *Conference Record of the Thirty-Sixth Asilomar Conference*, volume 1, pages 289–293, 2002.

[110] G. Bauch and N. Al-Dhahir. "Reduced-complexity space–time turbo-equalization for frequency-selective MIMO channels." *IEEE Transactions on Wireless Communications*, 1(4):819–828, October 2002.

[111] Shoumin Liu and Zhi Tian. "Near-optimum soft decision equalization for frequency selective MIMO channels." *IEEE Transactions on Signal Processing*, 52(3):721–733, May 2004.

[112] X. Wautelet, A. Dejonghe, and L. Vandendorpe. "MMSE-based fractional turbo receiver for space–time BICM over frequency selective MIMO fading channels." *IEEE Transactions on Signal Processing*, 52(6):1804–1809, June 2004.

[113] B. Lu, Guosen Yue, and X. Wang. "Performance analysis and design optimization of LDPC-coded MIMO OFDM systems." *IEEE Transactions on Signal Processing*, 52(2):348–361, February 2004.

[114] R. Piechocki. Space–time techniques for W-CDMA and OFDM. PhD thesis, University of Bristol, 2002.

[115] R. Price and P. E. Green. "A communication technique for multi-path channels." *Proceedings of the IRE*, pages 555–570, 1958.

[116] J. Boutros and G. Caire. "Iterative multiuser joint decoding: unified framework and asymptotic analysis." *IEEE Transactions on Information Theory*, 48(7):1772–1793, July 2002.

[117] H. Meyr, M. Moeneclaey, and S. A. Fechtel. *Synchronization, Channel Estimation, and Signal Processing*, volume 2 of *Digital Communication Receivers*. Wiley, 1997.

[118] U. Mengali and A. N. D'Andrea. *Synchronization Techniques for Digital Receivers*. Plenum Press, 1997.

[119] C. N. Georghiades and J. C. Han. "Sequence estimation in the presence of random parameters via the EM algorithm." *IEEE Transactions on Communications*, 45(3):300–308, March 1997.

[120] N. Noels, V. Lottici, A. Dejonghe, H. Steendam, M. Moeneclaey, M. Luise, and L. Vandendorpe. "A theoretical framework for soft information based synchronization in iterative (turbo) receivers." *EURASIP Journal on Wireless Communications and Networking JWCN, Special issue on Advanced Signal Processing Algorithms for Wireless Communications*, 2005(2):117–129, April 2005.

[121] A. P. Dempster, N. M. Laird, and D. B. Rubin. "Maximum likelihood from incomplete data via the EM algorithm." *Journal of the Royal Statistical Society B*, 39(1):1–38, 1977.

[122] A. P. Worthen and W. E. Stark. "Unified design of iterative receivers using factor graphs." *IEEE Transactions on Information Theory*, 47(2):843–849, February 2001.

[123] J. Dauwels and H.-A. Loeliger. "Phase estimation by message passing." In *Proc. IEEE International Conference on Communications (ICC)*, pages 523–527, Paris, France, June 2004.

[124] G. Colavolpe, A. Barbieri, and G. Caire. "Algorithms for iterative decoding in the presence of strong phase noise." *IEEE Journal on Selected Areas in Communications*, 23(9):1748–1757, September 2005.

Index

a posteriori distribution, 21, 78
a priori distribution, 20, 79, 89, 137
adjacency, 42
alamouti space–time block code, 218
Alamouti space–time block code, 13, 219

Band matrix, 243
Baseband signal, 7
Bayes' Rule, 245
BCJR algorithm, 169
Belief, 100, 137, 191
Belief propagation, 89
Bit-interleaved coded modulation, 178
Bootstrap filter, 128
Burn-in time, 32

CDMA, 16, 228
Channel estimation, 237
Channel model, 9
Circulant matrix, 243
Coding, 6, 143
Commutative semi-ring, 71
Continuous variables, 69, 95, 238
Convergence, 100
Convolution, 8
Convolutional code, 163
Cost function, 20
Cycles, 44, 69, 100
Cyclic prefix, 11

Decibel, 175
Decoding, 143
Degree, 43
Demapping, 177
Dirac distribution, 244
Discrete variables, 90

Edge, 42
Equality function, 27
Equalization, 187

Error probability, 24, 136
Error-floor region, 175
Expectation operator, 244

Factor Graph
    definition, 37, 47
    Normal graphs, 62
Factorization, 44
    Acyclic factorization, 46
    Connected factorization, 46
    Cyclic factorization, 69
Filtering, 112
Forest, 44
Forward-backward algorithm, 113, 169, 193
Frequency-flat channel, 9, 210, 219, 229
Frequency-selective channel, 9, 208, 213, 222, 223, 231, 234

Gaussian distribution, 244
Gaussian equalizer, 200
Generator matrix, 146
Graph, 42
    Bipartite graph, 43
    Connected graph, 44

Importance sampling distribution, 29
Incidence, 43
Indicator function, 27, 63
Interleaver, 145, 149, 170, 178
Invariant distribution, 32
Iterated expectation, 125

Jacobian logarithm, 91

Kalman filter, 122
Kalman smoother, 124

LDPC code, 154
Leaf, 44

Likelihood, 78
Likelihood function, 21, 78, 79, 137
Limiting distribution, 32
Log-likelihood, 90
Log-likelihood function, 80
Log-likelihood ratio, 94
Loopy inference, 100

MAP estimator, 23, 78, 136
Mapping, 6, 177
Marginal, 27, 45
Markov chain, 31, 106, 240
Matched filter receiver, 209, 229, 231
Matrix inversion lemma, 121, 202
Max-sum algorithm, 72, 80
Messages, 57, 86
Mixture sampling, 98, 129
MMSE equalizer, 200
MMSE estimator, 21, 200
Monte Carlo equalizer, 196
Monte Carlo techniques, 25
Multi-antenna communication, 12, 217, 234
Multi-carrier transmission
    Multi-user, 16, 234
    Single user, multi-antenna, 14, 223
    Single user, single antenna, 10, 213
Multi-path channel, 9
Multi-user communication, 15, 227

Non-parametric representations, 96
Normalization, 80, 88
Nyquist pulse, 10
    square root Nyquist pulse, 9

OFDM, 10, 14, 213, 223
OFDMA, 16, 234
Opening nodes, 66, 78
Oversampling receiver, 211, 222, 233

Parametric representation, 96, 122
Parity check matrix, 146
Particle methods
    Message representation, 96
    Particle filter, 127
    Particle representation, 25
    Particle smoother, 131
Path, 43
Pinch-off point, 175
Prediction, 112

Pulse-shaping, 7
Puncturing, 149

Quantization, 96

RA code, 149
Regularization, 27, 99
Resampling, 27
RF signal, 7

Sampling methods
    Gibbs sampling, 32, 199
    Importance sampling, 28, 30, 97, 128, 198
Sequential processing, 112
Signaling constellation, 6
Single-carrier transmission
    Multi-user, 16, 228
    Single user, multi-antenna, 12, 218
    Single user, single antenna, 9, 208
Sliding window equalizer, 194
Smoothing, 112, 124
Space-time code, 13, 218
Spatial multiplexing, 13, 218
State, 31, 106, 165
State-Space model equalizer, 193
State-space models, 105
Statistical inference, 77
Subcarrier, 12
Sufficient statistic, 136
Sum-product algorithm, 58
Support, 213
Synchronization, 237

Target distribution, 29
Termination, 147
Toeplitz matrix, 243
Transition kernel, 31
Tree, 44
Trellis coded modulation, 182
Turbo code, 170

Vertex, 42
Viterbi algorithm, 116, 169, 183

Waterfall region, 175
White Gaussian noise, 7

Printed in the United States
By Bookmasters